MODERN SCIENCE SERIES

Biology for the Modern Mind

Hans Joachim Bogen

with a Foreword by Professor Adolf Butenandt

translated from the German
by Harold Oldroyd

THE MACMILLAN COMPANY
NEW YORK

This book was originally published by Droemersche Verlagsanstalt Th. Knaur Nachf., Münich/Zürich under the title EXAKTE GEHEIMNISSE: KNAURS BUCH DER MODERNEN BIOLOGIE.

Authorized translation of the work by Hans Joachim Bogen, EXAKTE GEHEIMNISSE: KNAURS BUCH DER MODERNED BIOLOGIE.

Library of Congress Catalog Card Number 69–10608
The Macmillan Company New York
Collier-Macmillan Canada Ltd., Toronto, Ontario
Printed in Germany.

CONTENTS

FOREWORD

by Nobel prize-winner Professor Adolf Butenandt

It happened by chance that I finished reading this fascinating book about modern biology on the very same day that I completed exactly forty years of studying chemistry and biology since my graduation from the University of Göttingen. This prompted me to ask myself what a book about 'modern biology' would have looked like at that time, and how it would have differed from a book with a similar title today. The answer to this question showed what a revolutionary development had taken place in the last few decades, and what a deep insight we had gained into the laws that govern the phenomena of life.

Since I was convinced from the beginning that many of the basic aspects of life could be elucidated by systematically thinking about them in chemical and physical terms, I tried to direct my studies towards a synthesis of chemistry, physics, and biology. At that time this was not an easy thing to do. Among the first lecture notes that I took as a young student at Marburg in the summer term of 1921, there can still clearly be read the dogma that natural science can be divided into two fields, which differ fundamentally in the way that they set out their problems and in their methods of solving them, physics and chemistry being 'exact' sciences, and biology and geology, in contrast, being 'descriptive.' During the course of my studies I realized more clearly each year that such a division was a dubious one. I have lived through the period during which an experimental biology has developed out of zoology and botany.

During my early years as a student at Marburg, the study of biology meant almost exclusively learning about the multifariousness of the world of living organisms – that is, of life as it manifests itself at the present time. In 1924 I went to Göttingen, to study chemistry under Adolf Windaus. On taking leave of one of my teachers at Marburg, I was given the warning: 'It is a good thing that here in Marburg you have learned about the animals as they really exist, because in Göttingen you will now be learning about organisms as they never were!'

This warning was meant as a joke, but it contained a grain of truth. In my second study period with Alfred Kühn in Göttingen I learned to look for the universal features in living organisms, and to pick out those aspects of appearance and behaviour that are common to and characteristic of all the organisms of a particular group, or even of all living things. This was the first hint of what we understand by modern biology today. Whereas classical biology studies each individual object separately and tries to bring the immense variety of living things into a systematic arrangement, modern biology selects out of these a few that seem to be specially suitable material for tackling particular problems, for studying some of the basic phenomena of life, and for analyzing and interpreting phenomena that are common to all living things, and characteristic of life.

Although such a development was already recognizable by 1927, a book about our state of knowledge at that time would have contained little of the 'exact science' that Hans Joachim Bogen is able to give the readers of his *Modern Biology* today. Between then and now we have learned how to take advantage of the theory and practice of chemistry and physics. Today the question is, How far can the basic problems of biology be solved by a systematic use of chemistry and physics?

The method that was adopted turned out to be unusually successful and made possible a deep insight into the mysteries of living matter. It achieved a revolution in biology, the results of which encourage us to predict that the second half of this century will be not the age of physics, but the age of modern biology. Very many biological mechanisms can be traced back to molecular phenomena – that is, back to the structure and characteristics of chemical molecules, and how these react in living cells. Hence one branch of modern biology leads to 'molecular biology.' This expression is more than just a catchword of the moment; it also implies a programme. Molecular biology is a scientific discipline directed towards both investigation and interpretation, and its aim is to use atomic and molecular science to clarify all the basic phenomena of life, such as inheritance, growth, development, differentiation, irritability, movement, and memory.

This programme of molecular biology gives support to the classical disciplines of biology, and its basic plan shows its origin in natural philosophy. In contrast to any form of vitalism, modern biology makes use only of arguments that are admissible in natural science. We shall have to wait and see how far modern biology succeeds in its aims; the problems it has to solve are great and difficult to compass. At the present time we are seeing such tempestuous progress that we are led to expect that the same reasoning, applied to an ever widening sphere of knowledge, will give us ever new insight, and enable us to go on interpreting facts that had previously seemed unrelated. This striking progress has been achieved by a trick of experimental limitation. Modern biology experiments with a few objects, which are systematically studied by every suitable method of natural science; we use biophysics as well as biochemistry, X-ray crystallography as well as genetics. So far as its methods are concerned, this is a derivative science, which acquires its individual character from the kind of interpretive way in which it frames its questions – a way that is not natural to traditional biophysics and biochemistry – and by trying to deduce the biological function of a molecule from its structure and its structural changes.

This book leads us into this new world in a fascinating and easily comprehensible way, and by convincing examples, shows how biological phenomena can be explained. Models are used to explain things that previously could only be described in words. The thoughtful reader, however, is also made conscious of the limits of the comparative method. Even though modern biology provides us with a new outlook, and a deeper insight into the nature of many living phenomena, it still does not answer the question of what life itself is.

In the words of our author: 'It is in no sense the duty or the aim of natural science to prove that . . . life can be reduced to laws that are exclusively those of physics and chemistry. It is the duty of natural science to investigate how far physical and chemical laws operate in living systems, and if these laws are possibly sufficient to leave nothing that is unexplained. . . . This does not exclude the possibility that other methods of investigation exist.'

Everything that we learn from modern biology can be eminently practical when applied to human societies, and the reader will be glad to find this aspect touched upon in the last chapter. Since a working knowledge of modern biology can only increase one's respect for the wonders of this world, we may hope that people will become aware of the responsibility that they bear towards all the living things on earth, towards plants, towards animals, and towards man himself.

BIOLOGY IN THE DAILY PRESS THE 'PUNCHED CARDS OF LIFE'

The 'general reader' to whom many explanatory books are addressed is a statistical abstraction; the present book has been written not for an abstraction, but for a particular kind of person. He is someone who learned biology at school, and still remembers some of it. He reads the more serious daily and weekly periodicals and tries to follow the progress of modern biology from them. Here are the sort of headlines that he might read:

Modern Biology Solves Riddle of Inheritance: Nucleic Acids the Life Code

Synthetic Virus – RNA: Life from the Test Tube?

Molecular Memory?

Cancer Cure by Tissue Grafting?

Such headlines make him think that the modern biologist is very clever, and he is curious to know more about new biological discoveries, but when he reads the articles that follow he is rarely much wiser. The authors go too far, and too fast, and take too much for granted. They bandy about such terms as *DNA*, *messenger-RNA*, *immunity tolerance*, and *plasma coding*, as if they were expressions in everyday use. Perhaps they are to the writers, but they must know that the terms are strange and mystifying to everyone else.

The reader of journalistic articles about modern biology often gets an uneasy feeling that their writers are trying to impress rather than to instruct. Sometimes the reader's misgivings go deeper. His own knowledge is insufficient to tell him whether the new theories are likely to be correct, whether they are soundly based on previous work, and whether the deductions made from them are justified. He may begin to wonder whether this new knowledge is being exploited for purposes of which he does not approve, and used to justify social and political policies with which he does not agree.

That is why this book is being written. It attempts to fill in a little of the gap that exists between traditional biology and the biology of today. Without assuming too much previous knowledge, it will try to explain some of the progress made by biologists in the last thirty years. Perhaps the reader will then be able to form his own ideas of what the biologists of today are up to, what they can do – and what they still cannot.

Look at genetics, or the science of heredity.

Secret of Heredity Unveiled: DNA Code Is the Material of Heredity. From this the reader cannot really get very far, even if he has heard that DNA is only the universally accepted abbreviation for deoxyribonucleic acid. He is accustomed to explaining heredity in his own way: it is more or less obvious to him that children resemble their parents, or differ from them only slightly. He must accept that deoxyribonucleic acid plays an important part in this, but this tells him nothing. If he remembers the Mendelian laws, through which the mechanism of heredity may be understood, it is not any easier for him to see a connection between DNA and the law of segregation. He is all the more likely to ask whether the new theory has anything in common with his present ideas of heredity. He is afraid that 'molecular genetics' is a completely new science, which with new methods, new ideas, and new ways of thinking, has arrived at a revolutionary and incomprehensible result.

Fortunately it is not as bad as that. True, it is a big step from the well-known chromosomes, visible under the microscope, to molecules of nucleic acid, at least in terms of size. True, the biologists of today work with new methods and new ideas; true, their results are revolutionary. But hardly any of the simple classical explanations need to be discarded. In this way the continuity of science is preserved; the objectives of the research have not changed, and so the journey continues as before, even though the coachman and the horses have been changed en route. We still lack a complete understanding of heredity, however, and cannot achieve one with our present resources.

It is obvious that Mendel's Laws cannot explain the entire mechanism of inheritance. Their shortcomings were realized long ago, and have been a handicap. Mendelian genetics is actually a genetics of 'crossing.' Two organisms are crossed with one another, and their offspring then segregate themselves into one, or several, forms – snapdragons with white blossoms, for instance, and those with red–which nevertheless belong to the same species. Now biologists can trace out what the offspring of the first generation look like and which colours emerge if individual flowers of that generation are crossed, either among themselves or with their parents (this sort of thing can be done without hesitation to *flowers*).

What can be discovered and controlled in this way, is, of course, only the *mechanism* by which the hereditary factors are distributed, mixed, and recombined. It is important to realize that in fact *patterns* and not complete attributes are produced. However, the most important part of inheritance is not the mechanism by which characters are handed on, but the nature and the potentialities of such patterns, a point that is often overlooked. Many people – specialists among them – are content to label a hereditary factor a 'gene' or an 'allele.' If they are then asked what is the significance of a gene, and how a gene for red colour or blue colour operates, the question is passed on to the related discipline of developmental physiology. Evasion like this makes one's interest quickly evaporate.

So when a new branch of genetic science, *mutation genetics*, comes into being, there is not much co-operation from the older branches. Mutations are discontinuous changes in hereditary textures, and are themselves inherited. Individual mutations appear 'spontaneously,' but they can be induced in quantity artificially by using X-rays, ultra-violet rays, and various chemical substances.

Of course, this has been known for a long time, and the prospects of getting an insight into the nature and structure of the gene by these means appears small. But the mutation geneticists have co-operated with the biochemists and by patient work have shown that the gene must have something to do with the production of an enzyme (or ferment). Because of the war, these discoveries, which were made in the United States, were not immediately known throughout the scientific world; when, afterwards, scientific links were again joined, 'biochemical genetics' had developed so far that the multiplicity of new facts and new ideas confused everyone, so that no conclusions could be drawn from it.

What is a gene? And how does it cause a particular characteristic to develop? These are the two precise questions to which modern molecular geneticists want answers, and which today they are able to answer in principle.

Their path through the individual stages of discovery was never as straight as the above would suggest, and it has not yet reached its goal; but the progress to date makes it seem likely that the end will be reached. Travelling with them along this road were physicists, chemists, and mathematicians. They would not have come so far, or so quickly, if they had not been able to rely on the assistance of engineers, especially of communications engineers. To the unbiased – and molecular geneticists were this – this fact is neither strange nor surprising. Every higher organism, whether plant, animal, or human being, starts off as a single cell, the fertilized ovum. Single-celled organisms multiply by splitting and giving rise to daughter cells, which are essentially the same as their parent. How, then, does an ovum know what it has to develop into? How does it happen that a single ovum in the ovary of a poppy always gives rise to another poppy, and never to a lime tree, or a hedgehog? Obviously this 'know-how' is transmitted in the form of a complete package of 'information,' and the theory of inheritance can be treated as information theory. The question now arises, How and where is the information concealed, and how does it become 'revealed,' i.e., realized?

Elements of information are no more than signals that stand out above the 'noise level' of the unorganized background. Such elements of information are not scattered at random throughout the cell, but must be located on some carrier, where they are 'written down.' There they can be assembled, as letters are into words and sentences. Assembly of information can also be achieved with the help of dots and dashes, as in the Morse code, or on a punched card. By means of punched cards, for example, the career – the work done and the value to a business or government organization of each individual employee – is often registered. The individual cannot read his punched card; as far as he is concerned, the information on it is 'secret,' *coded.* But an accounting machine, the computer, can decipher the information, and gives it out to authorized users in legible characters, i.e., letters.

At first sight this seems strange and unbiological when applied to genetics, since it involves only technical ideas; yet modern techniques of taxation and administration, cybernetics and information theory have – independently, and on a broader basis – discovered and used what living organisms have been practicing for millions of years with complete success.

In fact, all living cells carry the complete information for their particular species recorded inside them, filed away on a sort of punched card. Not quite literally, of course, though the principle of coding is broadly similar, and a gene is in fact something like a punched card.

This provides a life code of really magnificent simplicity. With four different symbols, which are composed of four slightly different arrangements of the basic material, the complete information is 'written down,' just as it is in a fertilized poppy head, where two green sepals, four red petals, many black stamens, and a capsule with many blue seeds are all 'decreed' in the ovum. The cell also contains a set of complete operating instructions, described later on, with four 'letters.' The cell 'reads these over,' and now 'knows' what it must do, and when it must be done.

It is not to be wondered at that the process of encoding and releasing the orders implicit in the 'information' by such an ingeniously simple key is itself somewhat complex; though this, in turn, is comparatively simple, by comparison with many phenomena of atomic physics. It is not too complicated for research to be undertaken, leading to an understanding of coding of biological information, transfer of information, and the utilization of data.

By this means, the contact between genetics and the old science of heredity, lost at one time, can be regained. Yet genetics cannot be studied for its 'own sake,' any more than we can study the phenomena of heredity if they are isolated from all aspects of life. We are already trespassing on the ground of metabolism, with its 'regulation,' since this is the source of the essential substances. In addition, attention must be given to other, more divergent fields, as, for example, the biology of immunity, or the basis of memory – and then these also will be mastered, or at least controlled, by the 'punched cards of life.'

And finally, the reader will no doubt ask the question, Where do these punched cards actually come from? So far we have said that they were provided for each species, as if they were arranged in advance. But undoubtedly they, too, must be *understood*. To ask *how* and *when* they came into being is to ask how life itself originated on the earth. Naturally, no precise answer can be given, but at most the first instalments, fragments of information, and hypotheses. Nevertheless these are vital to the understanding of this world, of which an authoritative picture will be built up by physics, chemistry, and biology.

1. LIFE IN THE CELL

1.01 Nucleic Acids Control the Proteins

Can and Should Life Be Defined?

If we want to talk about anything, it is important as a first step to mark the limits of the subject and give it a basic definition. Sometimes this definition is good and indeed essential, but often it may be tedious and platitudinous, and very often it may be of little real value. It is completely impossible to think of defining life, the subject matter of biology. On the contrary it must be said the life is characterized by the very fact that it defies any comprehensive definition.

This is a remarkable fact. The problem of life interests everyone, and everyone thinks about it at one time or another. Many men, especially artists, philosophers, and teachers, have given the world their thoughts, ideas, and speculations about life, in an abundance that cannot be ignored. Yet no one has been able to tell us exactly what life is.

What science has to say on this subject is less verbose and more precise. It is more modest, because no scientist is rash enough to attempt to comprehend life as a whole; instead, the scientist contents himself with observing living creatures and their manifestations of life, describing, analyzing, and elucidating these. The result, however, is something more than a few vague ideas and the knowledge thus obtained can be passed on to other people, can be amplified by further experiment, and thus, for scientists, becomes evidence.

Life has been studied since the time of Aristotle; but in the last few decades our knowledge has increased at such a rate that it has become an avalanche.

Enough is known already to throw light on many very important precursors of life, to track these down to molecular level, and to separate them into passive and active components.

The latter are the precursors that have a decisive effect on the future life of the organism. From these we can already 'make a picture.' And in the present book it is hoped to make it possible for the reader also to form a picture, a picture of life as biologists of today present it, of what is going on today in biology, what can be done and what can not. So we must give up the attempt to define life itself, and turn our attention back to living beings. Life is not an object that occupies space and has a material existence; life is always tied to an organism, in its simplest form to a living cell. The fact that the cell is alive is evident from its signs of life: from the metabolism of respiration or digestion, from growth, from development, from ageing, and even more, from death. These are thus certain functions which we associate with the phenomenon of life.

But these functions do not exist in their own right; they are always tied to an organism, to cells, or more precisely, *to the structure of the cell. Structure and function* – in other words, the signs of life – are the twin aspects of life, and they mutually influence each other.

They must and should not develop along different lines, but convergently, unless there is to be a state of chaos. In this respect it is all the same which of the two aspects we study first.

The most obvious aspect to begin with is that of cell structure. The reader will remember this from his school days, all about cell walls, protoplasm, nuclei, grains of chlorophyll (in green plants), cytoplasm, and so on. But we live in a 'functional'age, and it is no accident that prominence is given to the analysis of functions. Above all, it is a mistake to think that to be clear is to be simple. Even if the field of metabolic function is much clearer than it was, the whole subject has become much more extensive. And now, since we must make a beginning somewhere, we will start with the scope of metabolic function.

1.02 Life as a Chemical Reaction

The Transformation of Foodstuffs

Metabolism is the transformation of substances. In the process of digestion, for example, we take in nutritive substances and pass out waste products, which no longer resemble the substances taken in. An organism is truly alive if it consumes food and yet maintains the same weight. If an organism overeats, it grows in bulk, and lays down fat in its bodily tissues, even if only a little actual fat is eaten. If, instead of fat, chocolates, say, or beer is consumed, a transformation of carbohydrates (sugar) into fat takes place.

The breaking down of food – not the mechanical breaking of food by chewing, but the splitting up by chemical means of the bigger molecules of foodstuffs into smaller and smaller fragments – is a part of metabolism. (Here we speak of *molecules* of foodstuffs, using this term in the same way as the modern biologists do.) It is just the same whether we take as our food materials carbohydrates with sugar and starch, or proteins, or fat: the original substances have molecular weights of the order of 1,000 to 100,000; the first fragments, of about one hundred; and the end-products, of between twenty and forty. For comparison, water, with the well-known formula H_2O – two atoms of hydrogen, each with atomic weight 1, and one atom of oxygen, atomic weight 16 – has a molecular weight of $2 + 16 = 18$. During the chemical breakdown of food molecules, *energy*, which was present as chemical energy in the original food, becomes freely available to the cell in which the breakdown is taking place – freely available, first, for the maintenance of the activities of the living cell, which are very extravagant of energy, and second, for the synthesis of new substances from which new cells are constructed, as replacement for worn-out substances, or for special demands that may have arisen.

The *balance* between breakdown and synthesis, between processes that release energy and those which absorb it deserves a special chapter; at this point we can concern ourselves in passing only with the mutual *exchanges*. These can take place without any participation by the living cell, but so infinitely slowly that, in many cases, during the lifetime of a human being the total effect cannot be measured. This is because the products of the breakdown of organic substances are carbon compounds which are very sluggish in their reactions, and very different in this respect from inorganic ions. The latter react quickly with each other: if we make a solution of common salt and add a drop of silver nitrate solution, an insoluble precipitate of silver chloride appears immediately. If we mix glycerine with fatty acids (the ingredients of fat), we cannot detect any fat even after several days.

What do carbon compounds look like?

We shall have much to do with carbon compounds and their molecular structure; we cannot avoid having to repeat such structures in formula, and so we must say a little about the most practical way of representing them:

A carbon atom (C) readily combines with atoms of hydrogen (H), oxygen (O), and nitrogen (N) – to mention only the most important examples. Take as an example ethyl alcohol. One molecule of this compound comprises two carbon atoms, six hydrogen atoms, and one oxygen atom. Its general formula is thus C_2H_6O. This tells us little, since there are carbon compounds that have the same general formula as each other but are quite differently put together.

To avoid confusion, let us employ a compromise between the general formula and a complete structural arrangement of the atoms. For ethyl alcohol we can use either

(1a) CH_3CH_2OH

or

(1b)
$$\begin{array}{l} CH_3 \\ | \\ CH_2OH \end{array}$$

The chemist (or the molecular biologist) knows clearly what are the relationships between atoms set out in this way, but the reader of this book may not be so experienced in this respect; perhaps he can understand the relationships better by means of the structural formula

(2)
$$\begin{array}{ccccccc} & & H & & & & \\ & & | & & & & \\ H & - & C & - & H & & \\ & & | & & & & \\ H & - & C & - & O & - & H \\ & & | & & & & \\ & & H & & & & \end{array}$$

It is even clearer if we use different type:

(3)
$$\begin{array}{ccccccc} & & H & & & & \\ & & | & & & & \\ H & - & \mathbf{C} & - & H & & \\ & & | & & & & \\ H & - & \mathbf{C} & - & \mathit{O} & - & H \\ & & | & & & & \\ & & H & & & & \end{array}$$

Or finally the atoms can be represented by little coloured balls fitted together.

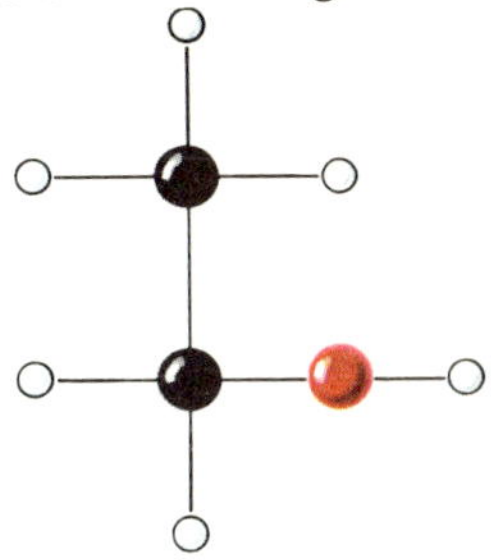

However artistic this is, it involves additional, irrelevant effort on the part of the reader, who must now first translate: big black balls = C; big red balls = O; little blue balls = H. And if more kinds of atoms are involved, and the molecule becomes still bigger, the reader has too many colours and sizes to distinguish. Furthermore, there are two gross errors associated with this kind of arrangement with balls. Firstly, the distance

1. A three-dimensional model of the molecule of ethyl alcohol, showing the relative size, distance apart, and spatial arrangement of the component atoms.

apart of the atoms in actual fact is much less than in the formulae shown above; so the balls are packed tightly together and make the model incomprehensible. Secondly, the bonds within a molecule do not lie in one plane, i.e. the plane of the diagram, but extend in all three dimensions of space. Three-dimensional models are even more confusing than flat ones, especially when huge molecules with hundreds of atoms are concerned. So we shall consequently use only formulae of the type of 1b; these have the fewest drawbacks, while the gain in clarity is substantial.

1.03 Enzymes Bring About Transformations

Let us now move on from the structural formulae to the organic carbon compounds themselves. These react only very slowly with each other, and naturally no living system could be controlled through such sluggish reactions: the living cell – which we shall take as representative of all multicellular organisms – has provided itself with a means of speeding up these reactions.

This instrument is known as a ferment, or *enzyme*.

In the past a distinction was drawn between ferments and enzymes. Ferments (from the Latin *fermentum,* yeast) were non-specific substances, which could be extracted from the cell and used elsewhere, as, for example, in the intestine; enzymes (probably from the Greek *zymoun,* causing fermentation, perhaps also from the Greek *kytos*, a container, a hollow space), on the contrary, were specific substances, effective only within the cell. However, since the German chemist Edward Buchner showed (1897) that the yeast from alcoholic fermentation, which had long been considered to be the origin of the enzyme, can be extracted from the yeast cell, and that it will work unchanged in a test tube, *in vitro,* the distinction appears to be pedantic; today the two terms are interchangeable.

One single enzyme splits the molecule of cane sugar (sucrose) into one molecule of grape sugar (glucose) and one molecule of fruit sugar (fructose). While the substrate – the name for the substance upon which the enzyme works, so in our case the sucrose – is emphatically changed by the process, the enzyme emerges unchanged, and so can immediately begin to work on the next molecule of sucrose. This means that one molecule of the enzyme can act upon a mass of substrate that is incomparably greater than the cells originally involved. Figure 2 symbolizes the activity of two enzymes: enzyme 1 splits up a tree by cutting it into disks; enzyme 2 splits these disks into halves. The daily output of an enzyme is considerable. It is known that a single specimen (for our purposes a 'molecule', though a very bulky one) in less than one minute can split off up to five million molecules of the substrate, and that at 0°C!

Enzymes Are Specialists

The figure illustrates, though not very explicitly, a more general property of enzymes: that is, that they are not all-round instruments, but specialized tools. Thus, just as a screwdriver will loosen a screw but will not saw a piece of wood – and for a saw the opposite is true – enzymes have their own particular jobs. Each enzyme can perform only its own – special – functions. Moreover, this specialization is a double one: first, the enzyme is specialized in relation to the particular substrate it operates upon, and second it is specialized in the kind of results it produces.

Let us look again at the enzyme that acts on sucrose. One molecule of cane sugar (sucrose)

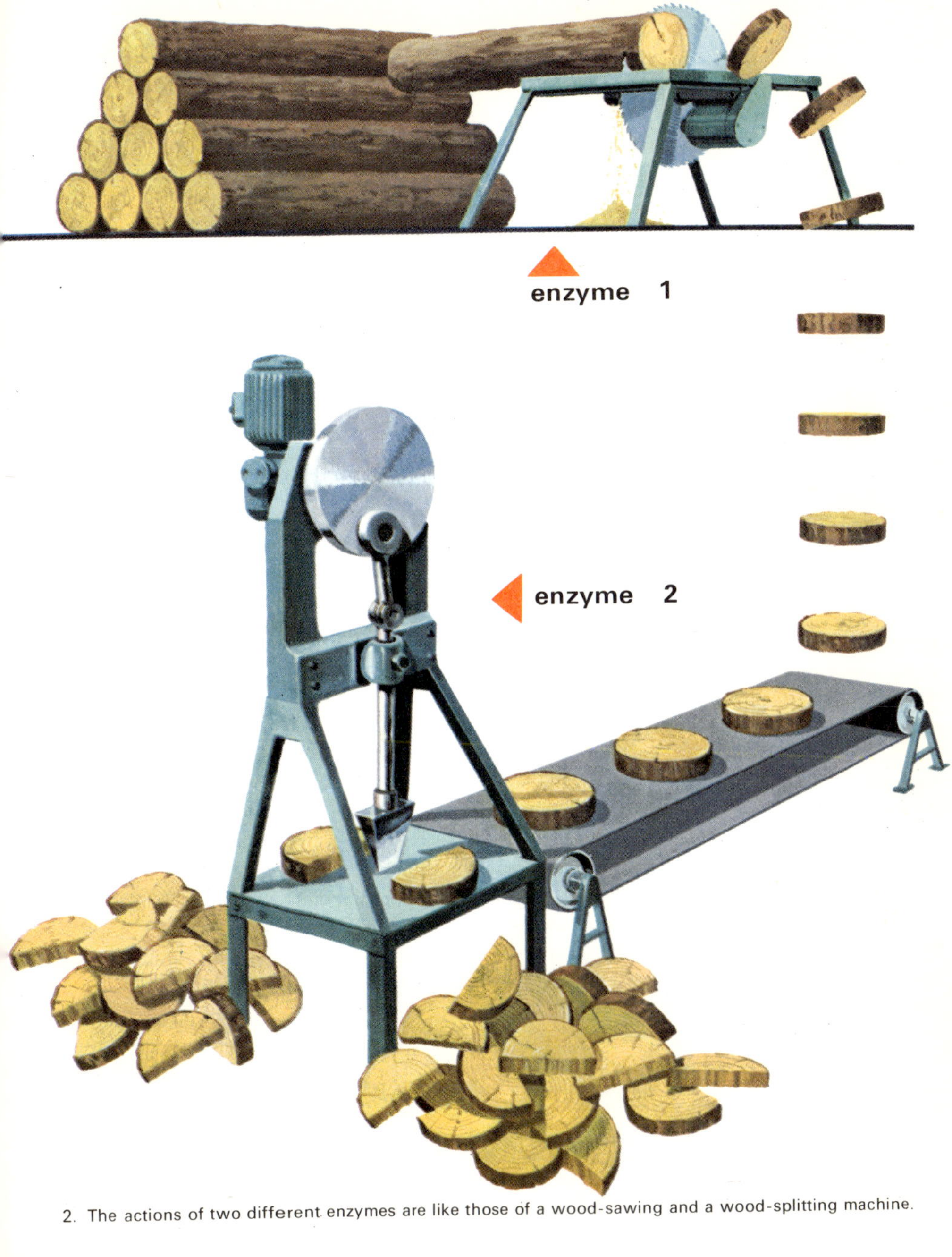

2. The actions of two different enzymes are like those of a wood-sawing and a wood-splitting machine.

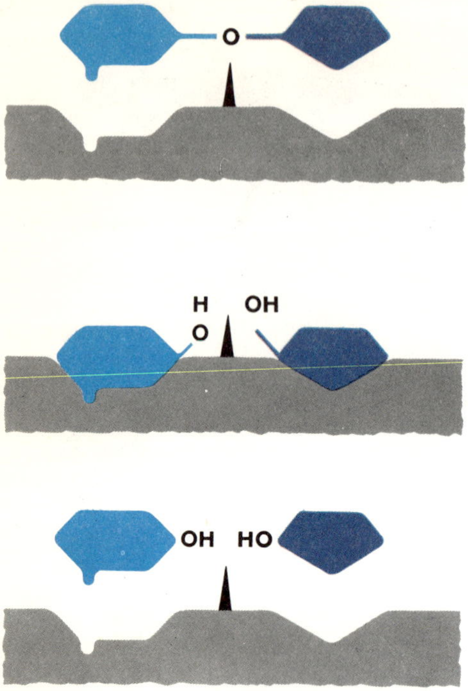

3. A molecule of cane sugar (blue), on impact with the surface of the enzyme (grey), is split into two halves, which then combine with water (H_2O) to form one molecule of glucose and one of fructose.

consists of one molecule of glucose and one molecule of fructose. The enzyme does no more than split the molecule of sucrose into its two components, and that with the help of water. The enzyme works on sucrose, and only on sucrose, and can do only one thing with it: in the presence of water the enzyme can break the link that binds the two components – can *hydrolyze* the sucrose, as the chemists say. Its operation is *substrate specific* (for sucrose only) and *effect specific* (it only hydrolyzes). The fact that it is an enzyme is denoted by giving it the ending '-ase,' which follows the stem of the name of the substrate (sugar = saccharum): hence *saccharase*. To be more precise, and include in the name also the nature of the operation, the enzyme should be called sucrose-hydrolase.

A different group of enzymes splits off hydrogen from the substrate; these enzymes are called dehydrogenases. All of them have effect specificity in common – they remove hydrogen; but they differ in substrate specificity, i.e., in which substrate they extract hydrogen from. There is a lactic acid dehydrogenase, a succinic acid dehydrogenase, etc. Many other enzymes are less selective. The lipases – in general, that is enzymes that split fats (lipoids), and which are especially produced in the pancreas – can separate the fatty acids from the glycerine in all fats.

These are matched by another group of enzymes, which have even specialized themselves upon one or two 'stereo-isomers,' and of which the following may be said: many organic compounds can exist in two or more forms, which differ from each other only in the position of particular groups. Thus, for example, the lactic acids:

$$\begin{array}{ccc} \mathrm{C-OOH} & & \mathrm{C-OOH} \\ | & & | \\ \mathrm{H-C-OH}\leftarrow & & \rightarrow\mathrm{HO-C-H} \\ | & & | \\ \mathrm{CH_3} & & \mathrm{CH_3} \end{array}$$

dextro-lactic acid | laevo-lactic acid

Generally they are completely alike in their chemical properties; physically, on the other hand, they differ in rotating the plane of polarized light in different directions – and so are known as 'optical isomers' – though this does not concern us here. If a mixture of the two isomers is treated with a specialized enzyme, it is fascinating to watch how the enzyme operates exclusively upon one form, while the other is left completely unaffected.

Some enzyme molecules, as we have seen, can split off five million molecules of the substrate in a single minute. This suggests that the enzyme must operate on 'its' substrate only from a distance; otherwise, how could it achieve so much? Actually, the opposite is true: the enzyme has close contact with its substrate; we even speak of an *enzyme-substrate bond.* When the substrate is bonded to the enzyme it is chemically changed (e.g., split into two components); then the products of the change are released, and the enzyme is ready for a new bond.

A Marriage of Unequal Partners

It follows from this (as do many other consequences which are of less significance for the following short account) that a twofold specificity corresponds with a twofold action: first, 'recognition' and binding of the substrate, and second, the chemical transforma-

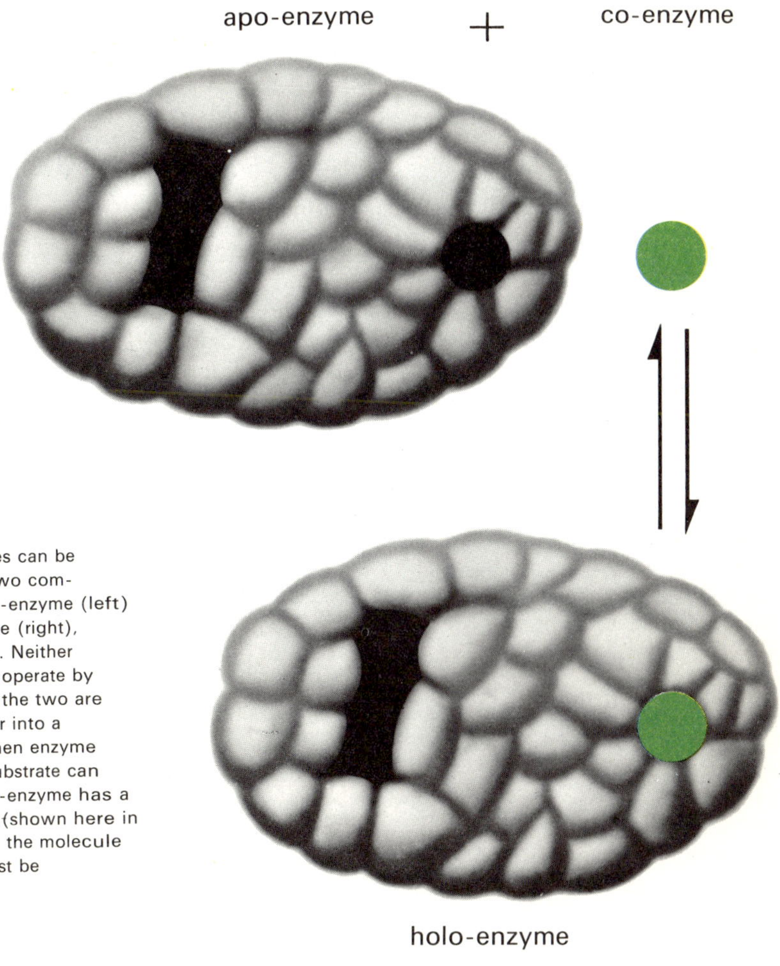

4. Most enzymes can be separated into two components, an apo-enzyme (left) and a co-enzyme (right), coloured green). Neither component can operate by itself, but when the two are brought together into a holo-enzyme, then enzyme action on the substrate can begin. The holo-enzyme has a particular place (shown here in black) to which the molecule of substrate must be 'matched.'

tion of the bonded substrate. The two functions are governed by different parts of the enzyme.

Most enzymes can be divided into an *apo-enzyme* and a *co-enzyme*. Either part by itself is incapable of operating on the substrate; it is only when the two are brought together into a *holo-enzyme* that the enzyme becomes activated.

The two components of an enzyme have quite different properties. The co-enzyme is a substance of low molecular weight, chemically well characterized; sometimes it is only a metallic ion such as copper or molybdenum. In contrast, the apo-enzyme has a very high molecular weight, and without exception, is albuminoid – it is a *protein*. Thousands of enzymes are known, and as a result, thousands of apo-enzymes. The number of co-enzymes, on the contrary, is exceptionally small – only a few dozen. Earlier (p. 20) we said that the enzymes that work by removing hydrogen atoms can be grouped together. If we investigate their co-enzymes, we find in many (though not all) of them the same phenomenon – namely, a chemical bond which exists between the different amides of nicotinic acid and phosphoric acid. Their structural formulae are known, but we do not need to learn them for our purposes. Anyone who has taken a vitamin preparation and has read the label on the bottle knows that the amides of nicotinic acid belong to the vitamin-B group. Not all hydrogen-extracting co-enzymes are derived from the molecule of the amide of nicotinic acid; others include riboflavin = vitamin B_2. It is no accident that riboflavin and the amide of nicotinic acid are vitamins. One group of enzymes, which splits off organically bonded CO_2, has as co-enzyme thiamine-pyrophosphate, a compound of thiamine with two molecules of phosphoric acid. Thiamine (aneurin) is familiar to everyone under the name Vitamin B_1. Those enzymes that contain the amino group (NH_2), have as their co-enzyme pyridoxalphosphate: pyridoxine (adermin) is vitamin B_6.

It is fascinating to follow out these very close relationships between vitamins, co-enzymes, and metabolism, but we must leave this subject and give our attention next to different matters, namely that often the co-enzymes are the bearers of *effect specificity*.

They may abstract hydrogen atoms, CO_2 groups, or NH_2 groups according to their chemical composition, which is individual, but never simple, and thereby alter the chemical substrate to which they are bonded. Thus laden, these co-enzymes may abandon the apo-enzyme to which they were hitherto attached, transfer themselves to another apo-enzyme with a different substrate specificity, and give up to it the hydrogen atom or the groups that they have picked up. The double decomposition has the character of a perfectly comprehensible chemical reaction.

The co-enzyme has apparently not the least influence on the choice of substrate with which it is to react. The choice of substrate, and the bonding together of the enzyme and substrate, is the province of the apo-enzyme, the albumin product. The enzyme protein is thereby bearer of the substrate specificity, and there this specificity can be very strongly imprinted – we have seen that an enzyme may be very competent at distinguishing between two 'optical isomers' (p. 20). This prompts the question, How does the enzyme 'know' that it must bond with this substrate and not with that one (since quite obviously the bonding is *not* left to chance); by what means is the enzyme able to detect such fine differences? Has this ability, this information been learned, or did the enzyme have it – implanted by heredity – at the time it came into existence?

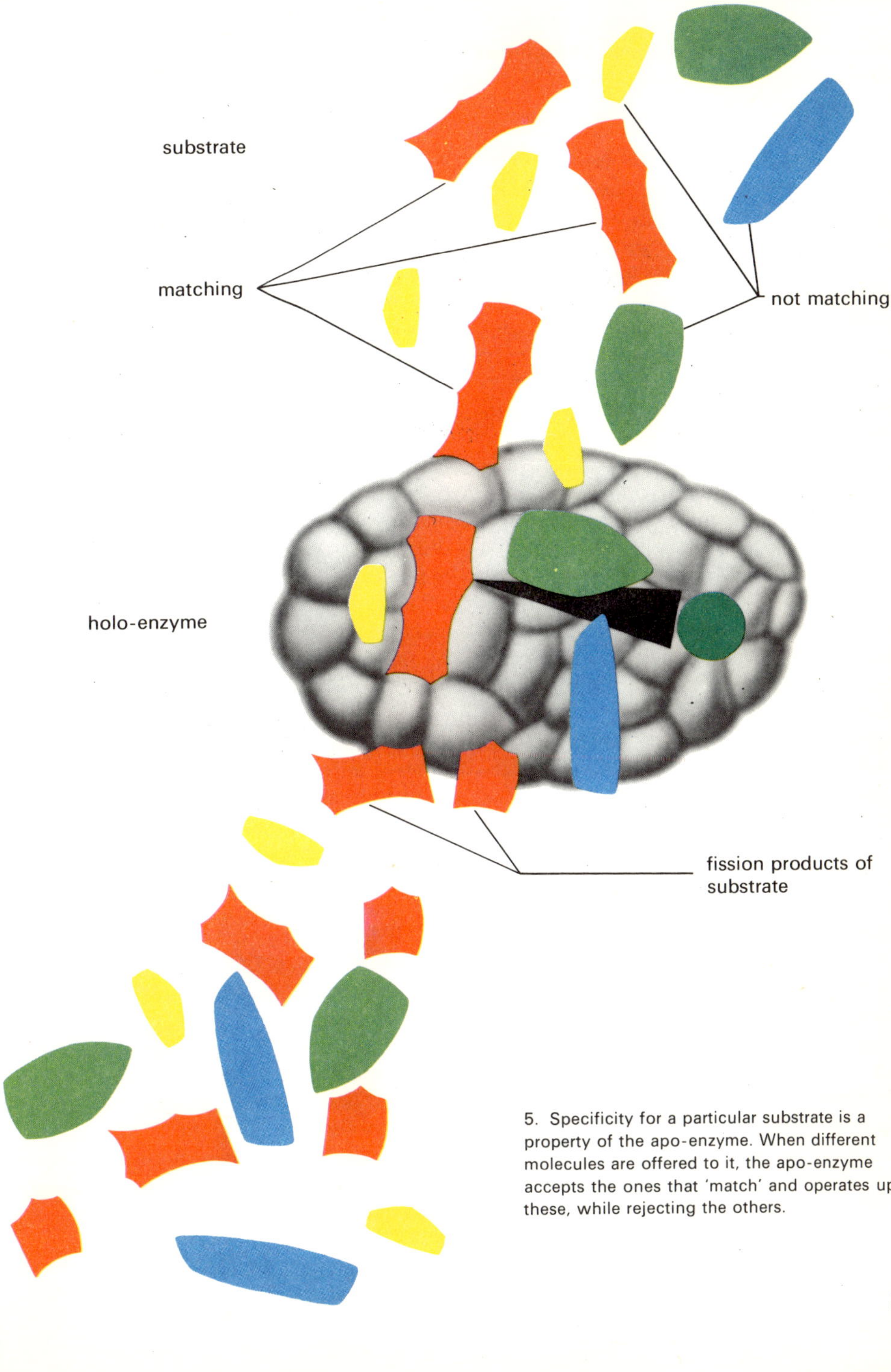

5. Specificity for a particular substrate is a property of the apo-enzyme. When different molecules are offered to it, the apo-enzyme accepts the ones that 'match' and operates upon these, while rejecting the others.

1.04 Old and New Facts about Albuminoid Compounds

Components and Plans

To answer this question we must concern ourselves, for better or worse, with the special properties of albumin. Albumin was originally the common name for the colourless part of a hen's egg, which when uncooked is an almost clear, viscous substance that can be twisted into threads, is somewhat sticky, and is not easy to hold. Later, albumin became the collective name for a whole class of substances (albumins) of similar appearance and behaviour. These substances are to be found in every living cell, where they make up the greater part of the protoplasm.

Changeable in its form, albumin is water-bound but not miscible with water; it is a mixture of many different albuminoid compounds, colloidal yet viscous like glue and therefore incapable of being crystallized. For many people albumin is fully qualified to be considered a 'living material.' Many people will still remember the reverential awe of many scientists and many teachers when they spoke of the 'living albumin.' If the *Ursleim*, in which Ernst Haeckel believed, was the mysterious life substance itself, which had been brought forth from the primeval sea and in its turn had arisen from a non-living jelly of silicic acid, there was a persistent idea that albumin was in some way connected with *Ursleim*, with somewhat formless primeval characteristics. From 1923 onwards Oscar Hertwig believed in the existence of cell elements rich in albumin which lie below the limit of vision under the microscope but which are sharply distinguished from the atoms and molecules of chemistry and physics by their living characteristics (assimilation, growth, and multiplication by fission). A famous biologist once said to the chemists, 'If you can only create genuine albumin, it will crawl away.'

Nowadays little of all this mystery is left. Indeed today we know certain elongate protein molecules which are able to contract and expand again (and thereby indeed 'crawl'); they do not do this of their own accord, but only if they are supplied with the necessary energy. This is supplied in chemical form and is measured out in very small doses. Nowadays, with more modern apparatus, and methods, not only can the different albumins be separated from each other, but also those of the same group can be so extensively purified and concentrated that they can be crystallized, making chemical and physical analyses possible.

The reverential awe of the unknown is dispelled, but in its place we have astonishment, and sensible respect for the 'architectonic' of the proteins, in which vast complexity and perplexing uniformity are endlessly combined without repetition. There are millions of different proteins. Many are well known; many have now been analyzed down to the last detail. All are built up from only twenty different components, and these belong ultimately to the same group of substances: the amino acids. What principle of construction underlies them?

All twenty amino acids which have been obtained from the breakdown of protein, are built to the same plan:

1. At one end, attached to the first carbon atom, lies an acid radical (carboxyl group, COOH).

2. Next to this, attached to the second carbon atom of the chain (i.e., in the alpha position) lies an amino group (basic, NH_2).

3. Then a residual group, or 'residue' R. In its simplest form it is an atom of hydrogen, but in complex amino acids it may be a complicated benzene-ring system.

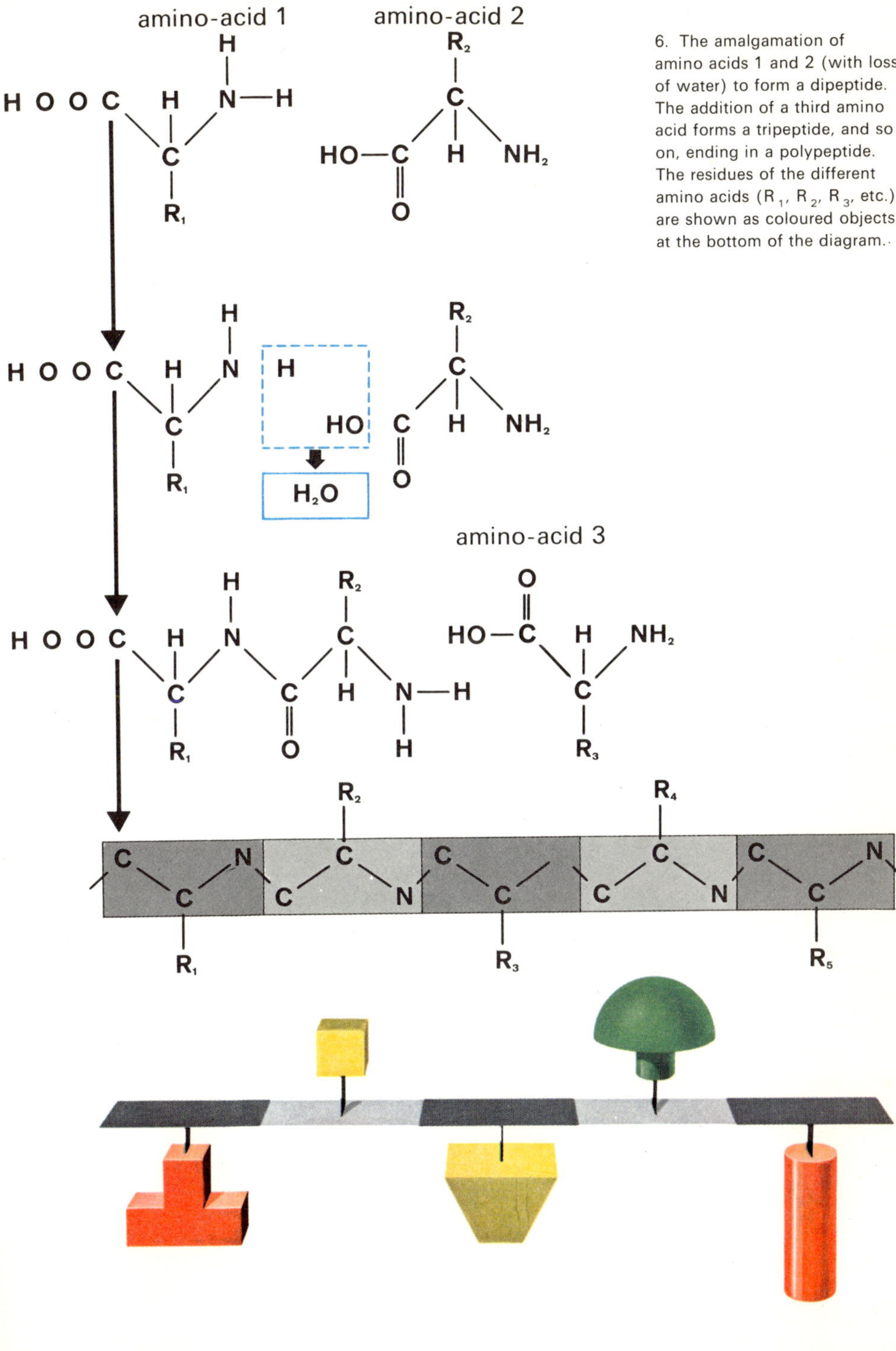

6. The amalgamation of amino acids 1 and 2 (with loss of water) to form a dipeptide. The addition of a third amino acid forms a tripeptide, and so on, ending in a polypeptide. The residues of the different amino acids (R_1, R_2, R_3, etc.) are shown as coloured objects at the bottom of the diagram.

$$\begin{array}{c} COOH \\ | \\ HCNH_2 \\ | \\ R_1 \text{ to } R_{20} \end{array}$$

NOTE 1: There are some amino acids in which the NH_2 group is attached not to the second carbon atom, but to one of the following ones and is thus in the beta or gamma or delta position, or in one of the positions after delta. Such amino acids are not normally found in proteins.

Some of the most important amino acids are alanine, glutamic acid, aspartic acid, proline, leucine, cysteine.

Two amino acids may bond themselves together into a *dipeptide*: a molecule of water is lost, and a link is formed between the amino group of the first acid and the carboxyl group of the second.

The living cell can repeat this reaction at will; from a dipeptide a tri-, tetra-, and ultimately a polypeptide can be made – with a short or longer chain in which the residual groups of the individual amino acids are arranged. Thus the proteins may conveniently be referred to as polyaminoacids. The characteristics of the acid and amino groups are lost when they are incorporated into such chains, but may be retained if they form part of a side-chain.

A chain includes between one hundred and thirty thousand segments – i.e., residual groups of amino acids; and its molecular weight may be from seventeen thousand to five million. (When dealing with large and very large molecules, like those in such organic products as cellulose and sheet plastic, the term *macromolecule* – from the Greek *makros*, long – is used.) Among those which are known and have been distinguished, about 29,900 different proteins may thus

7. Three different amino-acids 1, 2 and 3 can be grouped into 27 different 'triplets'. Those which are enclosed in red boxes and joined together by arrows are mirror-images of one another.

exist according to the length of the chain – and that is quite certainly an underestimate. In the search for some other way of distinguishing them, use was soon made of the fact that it is not a matter of indifference which amino acid is attached to which stalk; on the contrary, the sequence of the amino acids – which comes after which – is of the highest importance. To illustrate this, let us consider a simple tripeptide. It is composed of three different amino acids, which we will number 1, 2, and 3. There are, as will easily be seen, 27 different combinations ($3 \times 3 \times 3 = 3^3 = 27$) and therefore 27 different tripeptides. Of these, however, 2×9 differ only in being mirror images of one another. If the tripeptides can be 'read' from front or back as we please, then 9 mirror images drop out, and we have only 18 different possibilities. We know from the structure of the polypeptides (with which we have already been concerned) that many of them in fact can be 'read' only in one direction. In this respect, these polypeptides are like telephone numbers. If I ring number 1234, I get a different person than if I ring 4321. Every sequence of numbers is thus 'characteristic' for one definite subscriber; the sequences are non-interchangeable. Each amino acid, among tripeptides as well as among polypeptides, is or can be specific for one, quite definite quality.

Excursion into Permutations and Combinations: How Many Proteins Are Possible?

The number of possibilities of different proteins exceeds the imagination. There are twenty different amino acids. If we build from them polypeptide chains with only a hundred amino acid groups, then calculation shows 20^{100} or about 10^{130} different possibilities. 10^{130} is a 1 with 130 noughts – a number that would require more than three lines of print and that is more than one billion times greater than the number of atoms in the whole world.

Living creatures have thus, at least theoretically, more built-in possibilities than they can ever realize in practice. Of course there may be limiting circumstances, perhaps even powerful ones, but these reduce the possible number of variations by an insignificant amount. The further development of the living organism in future eras may still produce unexpected differentiation.

1.05 Chains, Coils, and Spirals

Protein Architecture as a Result of the Sequence of Amino-acids

The polypeptide chains of proteins (one of which is illustrated in fig. 6) by no means always lie in long, straight chains. On the contrary, they may be variously *coiled, rolled up,* or *folded,* either completely or only partially. One of the commonest arrangements is the spiral, or helix, to give it its original name, which the discoverers Pauling & Corey gave it in 1951. This type of polypeptide chain is wound round regularly like thread on a reel, although the reel itself does not actually exist. The thread is wound round a hollow space.

This very regular structure is stabilized by means of fixed bonds – the so-called hydrogen bridges – between strands. We do not need to concern ourselves with hydrogen bridges here. The important matter for us is simply the fact that the spirals, which are regarded as *secondary structures,* originate from *primary structures* which we know already as amino-acid chains. Moreover, if only a part of a polypeptide chain is in the spiral, then

8. Secondary spiral structure of polypeptide chains. Left, the α helix; right, the shorter and thicker β helix.

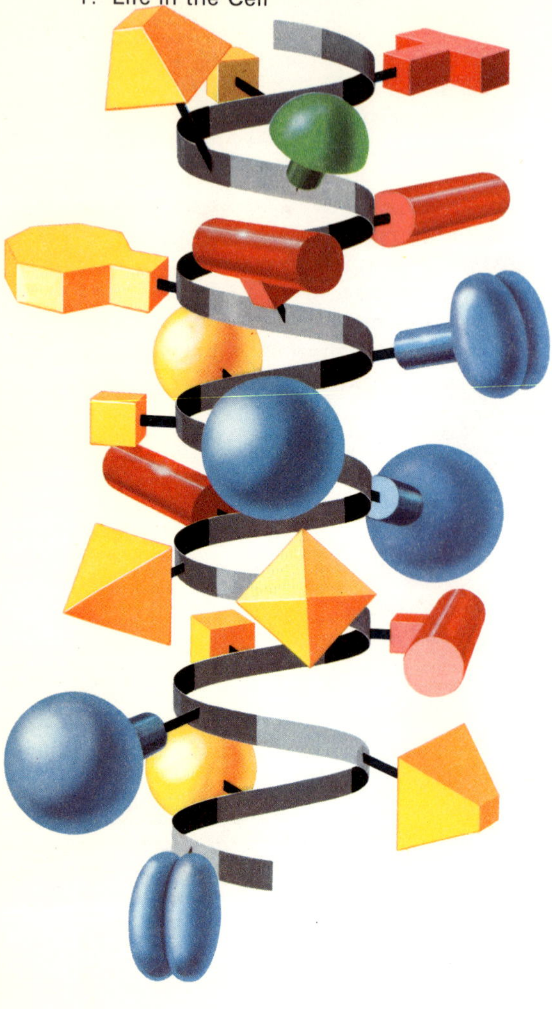

9. The α helix, shown this time with the symbolic coloured figures of the amino-acid residues (cf. Fig. 7) arranged in position and directed outwards.

another part remains extended and coils itself irregularly, and this, too, is a consequence of the primary structure. A second form of secondary structure is a thicker spiral with narrow coils. The first spiral (the alpha helix) has 37 amino-acid groups on 10 turns, or 3·7 per turn; whereas the beta helix has 51 on 10 turns, or 5·1 per turn. The two forms are convertible, one into the other. That means that in the change-over from the loose to the tight spiral – on and off the 'reel' – the chain is shortened and lengthened. It must be remarked here that such shortening of individual albuminoid (protein) molecules plays a major role in muscular contraction.

Let us take a look at the next Figure (fig. 9). We know that not all the amino-acid groups are arranged on the hollow cylinder; there may even be no space for them there. They are rather directed to the outside, and they form a spreading structure outside the 'reel.' The cylinder is prickly. All the same, these amino-acid groups do not stick out into the air, but into a watery medium. However, many amino acids can bind water molecules to themselves more or less closely, and so form a regular water sheath and many others can do this to some extent. As a result the upper surface of the spiral is somewhat smoothed, though it still remains uneven.

The upper surface of a straight polypeptide chain, just like the upper surface of a coil or helix, has a certain characteristic 'structure.' This 'structure' is by no means accidental; for it is decisively determined by the linear sequence of the amino acids in the chain, i.e., by the primary structure. For example, if an amino acid with a small residual group (perhaps a single CH_3 group) has to right and left a neighbour with a bigger residual group, this gives it a small, pit-like hollow. The hollow is broader if two 'small' amino acids stand next to each other, enclosed between bigger neighbours.

In this way there arise 'regular irregularities,' and it is easy to see that the pattern is of critical importance for reciprocal action between albuminoids and other chemical compounds: the latter, which also have a

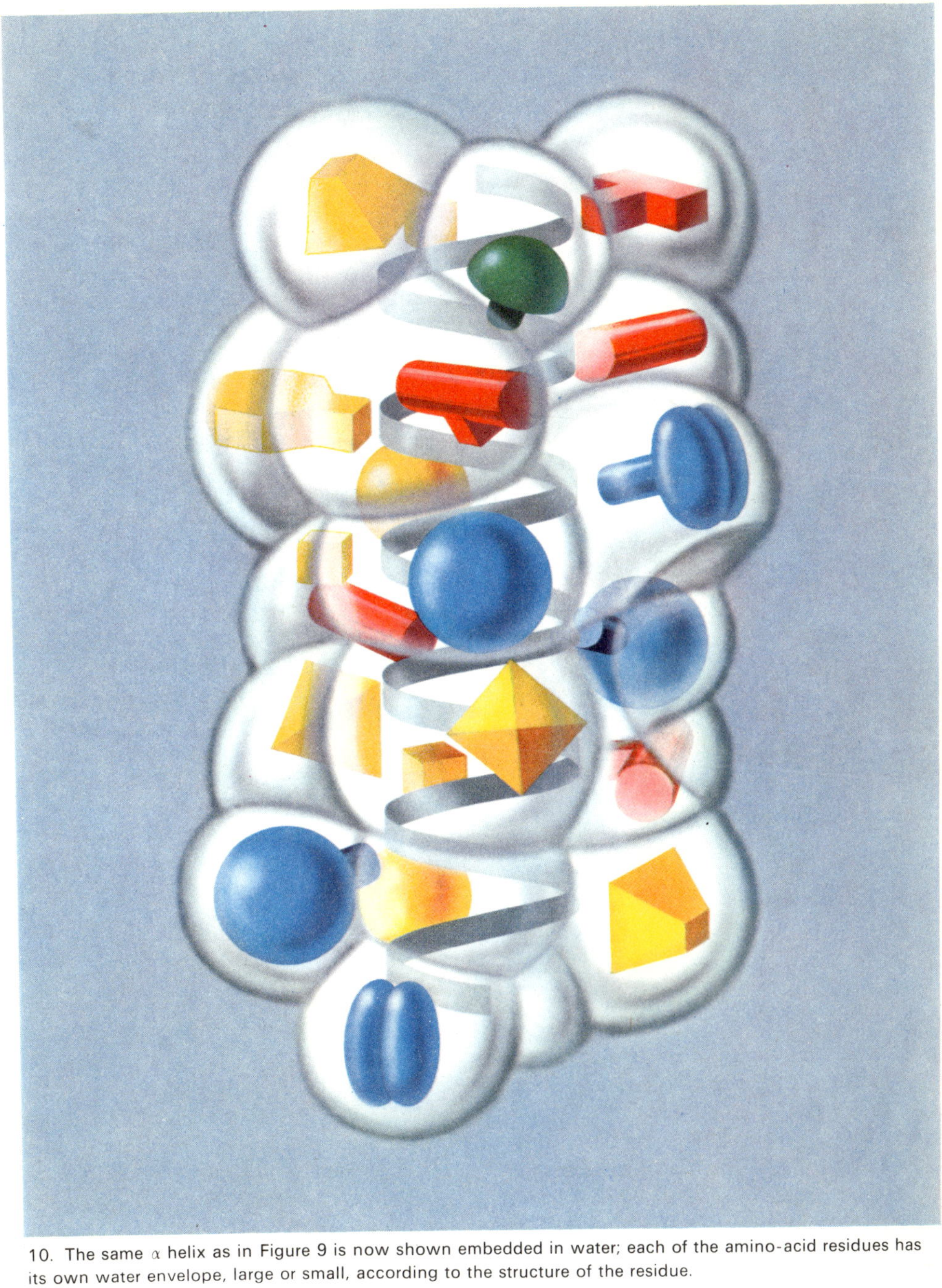

10. The same α helix as in Figure 9 is now shown embedded in water; each of the amino-acid residues has its own water envelope, large or small, according to the structure of the residue.

11. If the bends of the α helix all lie in one plane,r the 'tertiary structure' has the form of a disk, as can be seen by cutting a cross-section of the α helix at any point.

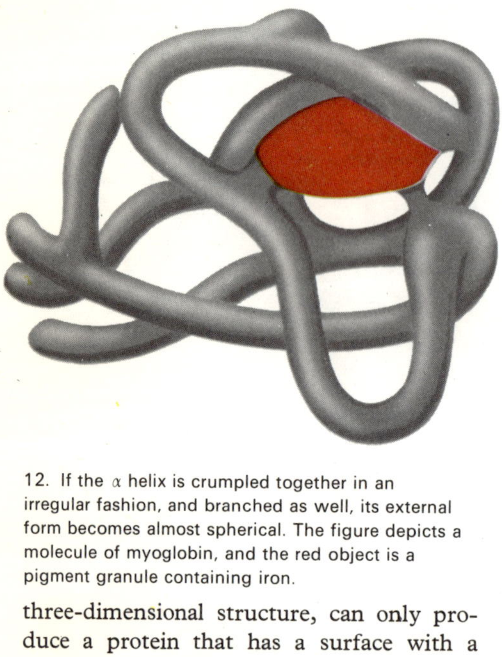

12. If the α helix is crumpled together in an irregular fashion, and branched as well, its external form becomes almost spherical. The figure depicts a molecule of myoglobin, and the red object is a pigment granule containing iron.

three-dimensional structure, can only produce a protein that has a surface with a corresponding pattern, and then only where the protein and its partner 'match' one another.

Now we need only say *apo-enzyme* instead of protein, and *substrate* instead of partner, and we have a pattern showing how and when a bond is possible between enzyme and substrate, and when not. Perhaps the outer surface of the albumin molecule is so shaped that only a particular kind of substrate molecule will fit. Before we accept this supposition as correct we must also consider *tertiary* and *quaternary* structures, which are much easier to explain than their names suggest. They refer to threefold and fourfold structures. Just as the polypeptide chain, the primary structure, is not always extended, so the helix (the secondary structure) need not always be a straight, hollow cylinder. It may be folded in one particular place, or in several. If all the 'kinks' lie in the same plane, then the tertiary structure has the shape of a disk; if not, then the structure is more like a coil. Finally, if a polypeptide chain, and with it a helix, branches again, then arise such peculiar bodies as, for example, myoglobin. Myoglobin is one of the substances that resemble hemoglobin, the colouring matter of blood (of which we shall have more to say later). It is a coloured substance which, like hemoglobin, contains iron; it is found in muscle tissue and is active in the supply of oxygen to the muscle. Its molecular weight is upwards of seventeen thousand.

We speak of a quaternary structure if several protein molecules combine into a great complex. In the simplest case several helices are linked together into a single rope. Naturally this is possible only if the amino acid groups that stick out sideways do not get in each other's way, and then the result is a long, extended complex, which we can call a *linear protein*.

The case of hemoglobin itself is different, and somewhat more complicated. Like an enzyme, hemoglobin comprises an albuminoid component (globin), and a colour compound ('hemo') which contains iron. The albuminoid component consists of four

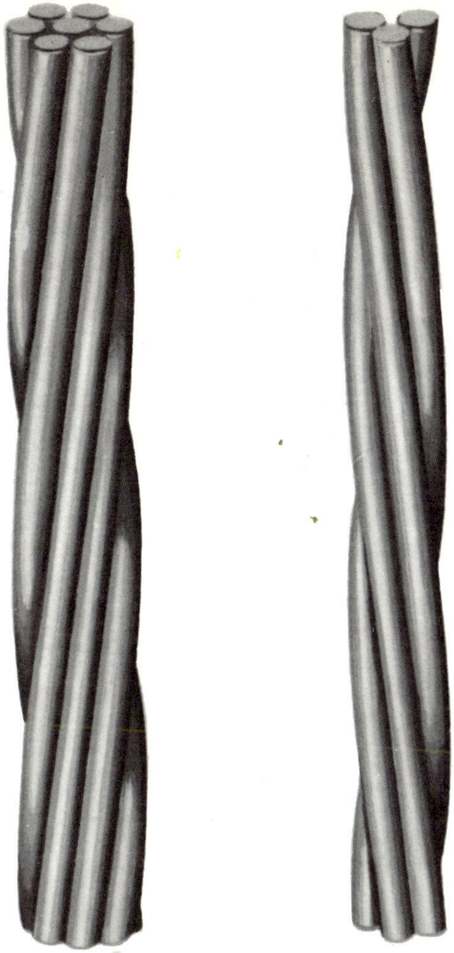

13. Several helixes (three are shown on the right; seven at left) may combine to form a 'cable.'

albumin molecules, each with a molecular weight of about 16,100. The complication lies in the fact that part of the polypeptide chain forms the secondary structure of the helix, while the intervening parts are unorganized. In addition, foldings of the spiral segments appear as tertiary structures, and four single molecules of this structure make up the complete complex (quaternary structure) of hemoglobin, with a molecular weight of 64,450. This is not a linear protein, but a *globular* one. Moreover, the four partners form two pairs of different lengths; two have 146 amino-acid residual groups in their chains, the other pair 141. In the hemoglobin molecule, besides the primary structure, we have secondary, tertiary, and quaternary structures. And on top of this we must say that, as far as we know, the spatial structure of the finished molecule has arisen spontaneously *from the specific sequence of amino acids of the primary structure*. In addition this primary structure decides ('commands' the molecule) if and where the spiral structure shall arise, if and where it shall be folded, and with what other molecules it shall finally combine.

Proteins – and thus also apo-enzymes – are for this reason highly specialized in their shape, and have a characteristic, non-interchangeable structure of the outer surface. They are specialized to a particular kind of substrate, and they do not react with other substrates. Our previous supposition has therefore been confirmed.

14. Model of a molecule of hemoglobin, consisting of four protein components, grouped together in two pairs. The four red objects represent the pigment granules.

The Information Lies in the Sequence of Amino Acids

The right answer to a scientific question leads on to another question, that follows logically from it. So it is here. In view of the enormous number of possible combinations of amino acids (20^{100}!), it is out of the question that a cell itself could experiment until it found the right combination. The chance involved would be impossibly long, very much longer at any rate than the chance for success in a lottery. This means that the sequence of amino acids must actually be 'dictated' by the structure of the polypeptide chain, and this means, further, that in turn the sequence of amino acids must contain the knowledge of *which substrate to react with*; the sequence of amino acids is, in fact, *programmed.* The programme is written with twenty symbols, those of the twenty different amino acids – it is written, that is, with an alphabet of twenty letters, so to speak. Since in this context the word 'alphabet' may give cause for confusion, it is better to speak of a set of twenty symbols by which the information is *coded,* or in other words, *concealed.* Anyone who understands the code, anyone who has the key, can extract the information, can decipher the report. If, however, information has already been made use of in protein synthesis, i.e., stored up, then the cardinal question arises, Where does this information come from?

What is the *first cause?* What orders which sequence of amino-acid groups is to be present in any particular instance and decrees whether any modification is necessary? Where is the information center, and how is the news transmitted from it?

At first sight it appears that these questions are the same questions that arose from the protein synthesis above, but appearances are deceptive. It has long been known how peptide bonds could be formed between amino acids, thus forming dipeptides. In 1912 Emil Fischer synthesized the first peptide. True, it was a test-tube synthesis, carried out with pure chemicals, which can never be achieved in the living cell – but it gave a genuine peptide, and was a first-rate achievement by Fischer. Today peptides can be synthesized in other ways, of which some approximate to what happens in the living cell. However, in these cases we are dealing only with peptides that are to some extent 'uncoded'; these consist either of only a single kind of amino acid (for example polyalanine) or only a few kinds, and then the sequence is more or less accidental.

Our cardinal question is aimed not at the chemistry of the links between amino acids, but towards the source of the information that directs the arranging of the amino acids and towards the mechanism of the extraction and evaluation of this information. Naturally this does not exclude the possibility that answering this question will throw a little light on the chemistry of the cell.

It is clear that the cardinal question cannot be answered by exclusively chemical methods. Moreover, nothing is known of where to look, and whether the answer will be a long time in coming. Hundreds of research workers have worked on it, with the most varied methods. Starting from widely different points, they have assembled brick upon brick. Many investigations that at first seemed hopeful led into dead ends; many more incidental, and previously unnoticed, discoveries opened up new angles – which would make a long and exciting story. We are not in the least interested in the trails and false trails that have been followed in the past: we want to know – preferably without too much trouble, and if possible even with pleasure – what is now possible and what may become so. So let us deny ourselves the stories and look for our goal, the source of information in the cell. For there must be, within the cell, some part, some region or some substance, by means of which the cell can synthesize albumin (protein) for itself.

The area in which we have to look will be narrowed down if it is considered that the albuminoids are really macromolecules. If these are to be codified, with completely specific, non-interchangeable sequences and structures, then this can best be done by a kind of 'negative,' as in photography, or by stamping, as in pressing a phonograph record – hence the negative may be called a *matrix,* or *template.* In fact, in early days of cell-biology a kind of 'template-theory' for protein synthesis had been worked out.

A template cannot, however, very well be smaller than the positive that is to be pressed from it. For this reason, chemical compounds with smaller molecules can be excluded; there remain only the three groups of substances with macromolecules, among which we can look for the template, first, carbohydrates, such as starch; second, nucleic acids; and third, proteins.

The proteins can quickly be struck from this list. If they were coded for the template of the protein positive, and thus the finished enzyme protein, then they themselves would also have to be coded – and so we would be back to the same question, What codes the template protein (or has already coded it)? Carbohydrates can also be excluded: first, because they are mostly uncoded; cellulose, for example, consists exclusively of glucose esters lined up on each other; second, it is certain that a protein synthesis can take place in the absence of carbohydrates. This is not so for the nucleic acids.

The study of protein synthesis in living cells, and of particular constituents of cells, has made it clear that this sythesis *is not possible* in the absence of nucleic acids. Research workers all over the world have therefore concentrated, with an amazing perseverence and consistency, on the nucleic acids, and it is still hardly ten years since the secrets of the storage and transmission of information were worked out in their essentials.

This means that we must concern ourselves to some extent with the structure of the nucleic acids. Unfortunately modern biology is not conceivable without chemistry, without a little knowledge of the structure and function of many kinds of molecules, especially those of the nucleic acids. The amount that is needed is comparatively small. Although many books and papers state that the structure of nucleic acids is even more complicated than that of the proteins, this is actually not true. In reality the nucleic acids, as we shall see very shortly, are amazingly simple. Moreover, even if they were in fact more complicated than all the substances that we have met so far, this must be accepted for better or worse, because this is where we meet the substance about which so much has been written in recent years, of such outstanding importance, and so difficult to understand: DNA, RNA, m-RNA, etc., are no more than particular kinds of nucleic acids, written as NA for short.

1.06 What, Therefore, Are Nucleic Acids, and What Do They Look Like?

A Preliminary Glimpse

Although nucleic acids were discovered and named almost a hundred years ago, little more was known about them than that they were compounds with the characteristics of an acid: since they readily combined with the 'basic' colour-compounds. Since they were found to be present in the nuclei of cells, they were named 'nucleic acids.' Nowadays they are also referred to as polynucleotides. At the moment all that we can say about them is this: with their own kind. Hence there are two kinds of polynucleotides, or nucleic acids, each determined by its type of sugar. The abbreviation for the sugar is D or R, and for the nucleic acid NA, so we have:

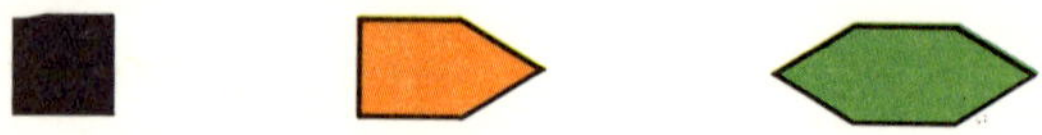

15. Structure of a nucleotide.

1 molecule of phosphoric acid + 1 molecule of sugar + 1 molecule of an organic base

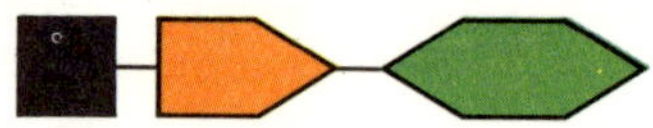

= 1 nucleotide.

just as the proteins arise through an arrangement of amino-acid units into rows, and thus, simplifying it somewhat we have polyamino-acids, in the same way nucleic acids are built up by the bonding together of 'nucleotides.'

In dealing with proteins we have to take account of twenty different amino acids, but there are only four, or sometimes eight, different nucleotides – a fact that simplifies things considerably. On the other hand, the nucleotides are certainly complex. Actually they are made up of phosphoric acid, sugar, and an organic base.

1 molecule of phosphoric acid + 1 molecule of sugar + 1 molecule of an organic base = 1 nucleotide.

Fortunately the phosphoric acid is always the same. For the sugar there are only two alternatives: ribose and deoxyribose. Ribose is a '5-sugar' – that is one that contains only five carbon atoms instead of the six found, for example in glucose. The other sugar has one oxygen atom fewer than ribose (hence the name *de-oxy-ribose*). The two sugars never stand next to each other in a polynucleotide, in a molecule of nucleic acid; ribonucleotides, and deoxyribonucleotides, are only linked

DNA = **D**eoxyribo-**N**ucleic-**A**cid
RNA = **R**ibo-**N**ucleic-**A**cid

At first it was assumed that the two types occurred separately in the cell, and in fact, that DNA was present exclusively in the nucleus, while RNA occurred only outside the nucleus, that is, in the cytoplasm. Nowadays it is known that while most DNA is indeed in the nucleus, and therefore in the chromosomes, DNA is also to be found, among other places, in the chloroplasts of green plants. RNA occurs not only in cytoplasm, but also in the so-called nuclear bodies, or nucleoli: one or several of these can be observed in every cell nucleus. Finally we may even have double molecules, which are 'bastard' molecules, so to speak; these are formed when a DNA chain links, temporarily, with a chain of RNA. Before we can make an exact study of the origin and effect of these remarkable fissions and reunions, we must first become acquainted with the third component of the nucleotide, the organic base.

This is called a *base* because it reacts like a weak alkali, such as soda lye, and so can form salts with acids. Its chemical constitution, which has been known for a long time, is not precisely simple, but we do not need to know it in every detail. For our purposes it is enough to know that one half of it belongs to the group of *pyrimidines,* the basal structure of which is a hexagon

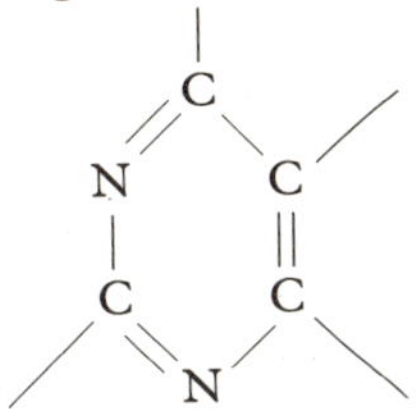

The other half belongs to the group of *purines,* which somewhat resemble the pyrimidines, combined with a ring of five:

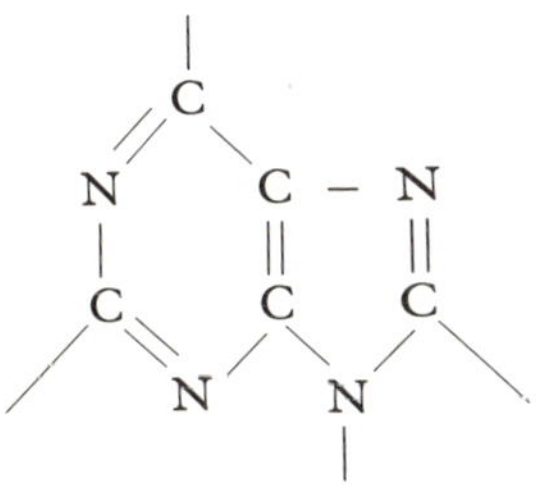

Next let us look at DNA. This comprises two purines, by name *adenine* (A) and *guanine* (G), as well as two pyrimidines, *cytosine* (C) and *thymine* (T).

In RNA we find three of these bases – namely, adenine, guanine, and cytosine; but in place of the pyrimidine thymine we have the pyrimidine uracil (U) – thus a total of five bases is involved, from which four are chosen.

Here already is the complete structural principle of both nucleic acids: DNA consists of deoxyribonucleotides placed one behind the other, with the four bases adenine, cytosine, guanine, thymine; RNA of ribonucleotides, one behind the other, with the four bases adenine, cytosine, guanine, uracil. The length of the chains ranges from seventy-seven to several million nucleotides.

From the standpoint of structural chemistry the mononucleotides are thus of quite simple construction, from which few surprises are to be expected. The same would apply to the polynucleotides if only they were not *macromolecules,* assembled from *different nucleotides.* Now, the nucleotides, of which there may be four different ones, are *approximately* evenly spaced along a large part of a chain of DNA or RNA, and at first glance the sequence shows no particular significance. Consequently, if the synthesis of a protein can even once be achieved without the use of nucleic acids, then the suspicion will arise that, on the analogy of the sequence of amino acids, the *sequence of bases* has something to do with the 'coding' of information.

In that case this sequence must:

1. Be definitely and finally proved.
2. Hold information and release it.
3. Be essential in the synthesis of proteins.

The implications of this were first discussed about fifteen years ago; during the last ten years or so mathematicians, physicists, chemists, biologists, and communications engineers have tried together to clarify this and to confirm it by experiment.

16. Part of a molecule of DNA, with twenty-four (deoxyribose) nucleotides. At first glance there seems to be no regularity in the sequence of these.

1.07 Communications Techniques within the Cell

The Discovery of the Triplets

In chains of polynucleotides such as DNA and RNA three successive bases follow each other along a polynucleotide chain, whether of DNA or RNA, and then constitute a so-called *triplet*. (And here again there may be confusion, since the term 'triplet' is also used for an important constituent in the structure of an atom. In a similar way the term 'quaternary' can refer either to the structure of a protein or to a geologic period. The term is derived from the Latin word for 'four,' but how this is relevant is known only to the initiate. If he is a geologist, he thinks of the quaternary period; if he is researching into proteins, then he understands by this term that on top of the third level of structure there is imposed a fourth. The word 'triplet,' from the Greek, implies that some kind of three-foldness is present, whether it is an arrangement within the atom, or among the nucleotides or bases). Exactly the same question can be asked about these triplets as about the proteins, how many combinations are possible? Among proteins, starting from 20 amino acids, and with a chain length of 100 amino-acid groups, this possible total is 20^{100}; among triplets with 4 bases, and a chain length of 3, the number is only 64. This is not very many, but still more than are necessary.

A triplet, of course, is more than just an accidental grouping of three nucleotides: a triplet regulates the assembly of one particular amino acid – and here at least we have found a link between nucleic acids and protein synthesis. This most important finding has a significance which has still not been appreciated in every detail – a large part of the present book will be concerned with it. In the following sections and chapters an attempt will be made to discuss a few consequences. However, in order to understand it properly, we need to get to know a little more about triplets.

The maximum number of sixty-four *different* triplets can be attained only if the sequence of nucleotides or bases can be read only in one order, like the digits in a telephone number. We could set out such nucleotide sequences diagrammatically, in the usual way, but that would not be particularly intelligible. So let us take advantage of the fact that – within DNA, or within RNA – the nucleotides can be distinguished only by their bases, if at any given time they have the same sugar and phosphoric acid groups. It is sufficient, moreover, if we specify the *bases*, and since the five bases happen to begin with different letters, we can simply use these initials as convenient abbreviations.

In RNA we have:

adenine nucleotide = A
cytosine ,, = C
guanine ,, = G
uracil ,, = U

In DNA we have:

adenine nucleotide = A
cytosine ,, = C
guanine ,, = G
thymine ,, = T

Some of the triplets derived from RNA will then look like this:

GAU will stand for the amino acid aspartic acid
GCU will stand for the amino acid alanine
CCU will stand for the amino acid proline
UUU will stand for the amino acid phenylalanine

GAU is, so to speak, the key word, the code word, for aspartic acid, UUU for phenylalanine, and so on. The basic sequence GAU–GCU–CCU–UUU . . . is thus the written order to the cell, that is, to the albumin from which the cell is made up in the first instance: build up an albuminoid compound, and attach to the first position a molecule of aspartic acid; to the second, one of alanine; to the third, proline; and to the fourth phenylalanine, etc., etc. This can be stated more generally: the sequence of bases contains the information for the sequence of amino acids, it is a fount of information, but at the same time it is the 'negative' or 'template' that is wanted. The sequence of amino acids follows from the sequence of the triplets.

Each triplet is thus a signal, a unit of information, a code word, or as it is termed today, a *codon*. Codon sequences and amino-acid sequences keep in step, so that they are 'co-linear.'

We have just written four triplets (codons) side by side, but separated by dashes. In one respect this is wrong: because the degree of separation between two bases within one triplet is not in fact less than that between two bases belonging to adjacent triplets, the sequence is more correctly written GAUG CUCCUUUU . . . or G-A-U-G-C-U-C-C-U-U-U-U. . . .

The protein-building agency can correctly understand (or interpret) the information only if it starts with the first mononucleotide, takes the first three mononucleotides together as a triplet, and accepts the amino-acid groups appropriate to these as exit points from the polypeptide chain

G-A-U-G-C-U-C-C-U-U-U-U. . . .

If the first nucleotide (G) happens to be missing, or is overlooked and forgotten, then the following codons are read:

(G) A-U-G-C-U-C-C-U-U-U-U. . . .

The reading of the amino acids is now as follows: the first triplet becomes methionine, the second leucine, the third leucine again, and so on – a faulty sequence, resulting in a faulty protein. (See Table 1.) Moreover, the protein-building agency must obviously read the codons only in one order, in this case left to right.

The (enzyme) protein synthesis goes still further in matching the codon series; it stops when the RNA molecule comes to an end – or if it comes to a signal 'protein-synthesis stop'! For the moment we need not bother about 'punctuation marks' of this kind, but later, when we come to the function of DNA, this will no longer be so; we have introduced the subject of 'punctuation marks,' at this point for another reason.

Of the sixty-four triplets that are possible, only twenty – i.e., one for each of the twenty different amino acids that are present in albumin – are needed. What about the other forty-four, 'surplus,' triplets? A few may certainly serve as punctuation marks; the rest are possibly meaningless, and thus not code words for any amino acid – in which case they ought not to participate in the codon sequence, since otherwise a hole, a gap, would appear in the chain of amino acids. In all probability the code should be called a 'degenerate' one. There is not just a single codon for every amino acid, as in the simplest system; there are several. (It may have occurred to the reader that in the above false sequence of triplets, [G]-A-U-G-C-U-C-C-U-U-U-U . . ., the triplet CUU answers for leucine, and so does the previous triplet, CUC). It must therefore be looked upon in the same way as if we 'describe' the sound mmmm when we can not only use just one m but two or three more, without changing the meaning. At the present time it appears that there are sixty-one or sixty-three triplets that can be used for codons, and that the remaining three, or one, are still (so far) 'surplus.' Table

Table 1

First Position ↓	Second Position U	Second Position C	Second Position A	Second Position G	Third Position ↓
	U	C	A	G	
	phenylalanine	serine	tyrosine	cysteine	U
	phenylalanine	serine	tyrosine	cysteine	C
U	leucine	serine	leucine	?	A
	leucine	serine	leucine	trytophane	G
	leucine	proline	histidine	arginine	U
	leucine	proline	histidine	arginine	C
C	leucine	proline	glutamine	arginine	A
	leucine?	proline	glutamine	arginine	G
	isoleucine	threonine	asparagine	serine	U
	alanine	threonine	asparagine	serine	C
A	methionine	threonine	lysine	arginine	A
	methionine	threonine	lysine	arginine	G
	valine	alanine	aspartic acid	glycine	U
	valine	alanine	aspartic acid	glycine	C
G	valine	alanine?	glutamic acid	glycine	A
	valine	alanine	glutamic acid	glycine	G

1 shows the present state of knowledge of the triplets of RNA.

It cannot be denied that a codon, the meaning of which has been discovered by experiment – as, for example, that UUU = phenylalanine – 'occasionally' may incorporate a different amino acid. This seems seriously to weaken the authenticity of the code system. Thus something must be said about this – but later on, after we have taken a look into the laboratory and seen how researchers carried out the investigations that led to the deciphering of the code.

Information at Firsthand

Let us go over once again what we know so far about RNA and its role in the synthesis of proteins: they are produced by the combination of mononucleotides; at any given time 3 mononucleotides make 1 triplet = 1 unit of information = 1 codon; a codon directs the incorporation of one specific amino acid into protein; codon sequences and amino-acid sequences are co-linear; RNA thus carries the information for the amino-acid series, and indeed is keyed with a code of four symbols; it translates the information into the twenty-symbol code of the amino acids. From this is derived a very simple picture of the mechanism of protein synthesis. The RNA is the original source of the information; an elongate chain of RNA with, for example, 146 codons (this number is that of hemoglobin), lies in contact with the corresponding 146 amino acids, which either by chance or by positive movement come to lie on its outer surface. The peptide bonds between the amino acids become linked, and the finished, coded protein detaches itself from the template. Linked together, it folds itself in the way prescribed by its primary structure, and already it is a complete apo-enzyme. Almost all of this picture is inaccurate; the operation is considerably more complicated, and we shall try to approach it step by step.

The first step is comparatively simple; but little by little we shall learn that it has almost incalculable consequences. This is the realization that the information implicit in RNA must be regarded as *secondhand information.* The real, original source of information is not RNA, but DNA, and this real, original information is *genetical* (in other words it has something to do with *genes*: thus we are now, for the first time, touching on the subject of heredity, a fuller discussion of which will start in the next chapter). Once again we are confronted with the problem of the transmission of information, this time from DNA to RNA. A step that is technically so simple cannot be hard to realize in the living cell. In the event it works so simply that everyone was astounded – if it is compared with such familiar devices as the Morse code, the telephone, and television, and taking into account the molecular scale of it all.

Transmission of Information: the Complementary Principle

The transmission of information is based upon a principle that is much simpler than the name 'complementary principle' would suggest. The four bases associated with DNA and the four associated with RNA, which have now been familiar to us for some time, always go together in pairs:

In DNA, adenine and thymine; likewise cytosine and guanine.

In RNA, adenine and uracil; likewise cytosine and guanine.

If the structure of these is recalled (see p. 35), it will be remembered that at times a pyrimidine and a purine are paired.

Each of the four bases has also its 'complementary base,' much as every colour has its complementary colour: green to red, blue to yellow, and vice versa. Using the accepted abbreviations, we get the following list:

DNA	A . . . . . T	RNA	A . . . . . U
	C . . . . . G		C . . . . . G
	G C		G C
	T . . . . . A		U . . . . . A

In practice, 'pairs' are of course found between nucleotides, together with their bases.

Next consider RNA and its nucleotides grouped together into triplets = codons. For every codon there is a complementary triplet: for the codon GAU, which produces aspartic acid, there is CUA; for the codon GCU, which produces alanine, there is CGA; and so on. In order to be able to refer to these 'complementary triplets' more conveniently, and to underline their complementary character, they are called *anticodons.* One molecule of RNA of 146 codons (as, for example, is needed for one of the hemoglobin chains with 146 amino acids) may be matched against 146 anticodons, which cannot be interchanged any more than the codons themselves. The anticodons attach themselves to RNA, *not* to an 'anti-RNA.' It is enough that the anticodons are arranged in an 'anti'-specific order. Later on we shall see that the anticodons bear the amino acids corresponding to the codons, and that it is *these* amino acids that join together.

In DNA it is different. Here, too, each triplet has its complementary anti-triplet. It is confusing now to learn that the DNA triplet is not also called a codon, but a *codogen* (meaning something which produces a codon); there is a good reason for this, as will be seen later. As in RNA, the anti-triplets, the codogens, lie in a sequence, which is specified by the triplets, but they are mutually bound together to a second DNA molecule which is exactly the same length as, and complementary to, the first. The two DNA molecules can remain together for a long time; DNA exists, though not invariably, as a *double molecule,* or two-strand system. The two strands are complementary to each other, rather like the two halves of a closed zip-fastener.

The comparison with a zip-fastener is not too far-fetched: a zip-fastener can be opened, and so can the two strands of DNA be separated from each other. Each of the two strands now consists of a sufficient number of nucleotides to satisfy its opposite number, the enzymes of the other half. Once again this fulfils the complementary principle, for when the reaction is over, there are present two double strands, which are completely similar in the sequence of triplets in each strand. Using the letters *X* and *Y*, this may be set out as follows:

As a starting point, let *X* be a single strand of DNA. This makes for itself a complementary strand, *Y*, and remains in union with it as an *XY* double strand (zipper closed). The zip-fastener now opens, and *X* and *Y* are separated; *X* completes itself with its complementary strand, *Y*, to give *XY* again; *Y* also completes itself, with complementary strand *X*, also producing *XY*. So now, in place of one strand of *XY*, we have two. The figure may make this clear. Here the lower part of the zipper is still not opened, while above, directly to right and left, the complementary strands are completed (fig. 17).

It is easy to see that the two 'new' double strands are fully equivalent, fully 'identical.' An exact duplication has taken place, giving an 'identical copy' of the DNA. Since the DNA carries the 'genetical information,' this information is now present in two identical double strands. If the cell now begins to divide, the two daughter cells each receive one double strand; as a result, each receives not only the same information as the other but also the same information the mother cell had.

Unintentionally we have again come round to the subject of heredity, and we are very close to its kernel, the gene, but once again we must retrace our steps. Although we certainly want to find out how information is passed on, we are inquiring into the information in protein synthesis, and how it is passed from the original source to the secondhand information implicit in the RNA. Now, what is mutually possible between nucleotides of DNA, and mutually possible between nucleotides of RNA, is also possible between nucleotides of DNA on the one hand and nucleotides of RNA on the other. This process of 'bastardizing' or 'hybridizing,' is an essential mechanism in the progress from genetical information to amino-acid sequence. The four bases, or rather mononucleotides, of DNA (namely, A, C, G, and T) can pair with the four bases of RNA (A, C, G, U) in the following ways:

DNA	RNA
A	U
C	G
G	C
T	A

Alongside every triplet of DNA there is a complementary triplet of ribonucleotides. The RNA triplets are the ones we call codons,

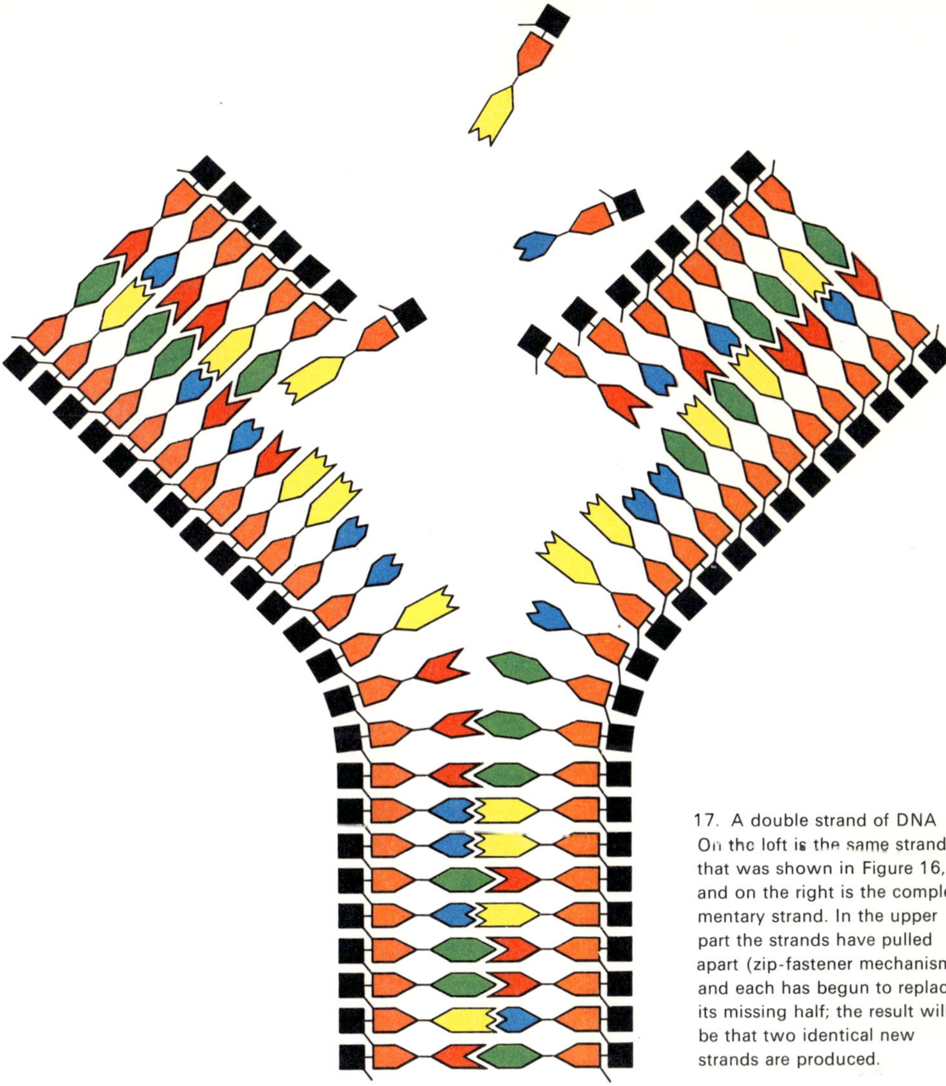

17. A double strand of DNA On the left is the same strand that was shown in Figure 16, and on the right is the complementary strand. In the upper part the strands have pulled apart (zip-fastener mechanism), and each has begun to replace its missing half; the result will be that two identical new strands are produced.

and we know that the codon sequence is the information for the sequence of amino acids in the proteins – in fact, *information at second-hand. Firsthand information* exists – written down in the sequence of codogens within the DNA. This is, so to speak, the *positive*, and on this positive the complementary RNA codons are set out, and linked together into a chain. The codon sequence, with its complementary bases, obviously constitutes the *negative,* and this is no other than the *matrix* (or *template*) that we were looking for. We have said already that in this negative there was an arrangement of triplets, complementary again. These are the *anticodons,* which are not bonded to each other. These are 'off-prints' from the negative (the codons), and so are positive in their turn. A diagram (fig. 18) will make this clear. On the left is a part of a single strand of DNA, and on this the codogens are indicated by letters, corresponding to their bases. On the right is the strand of RNA (RNA_1) with the complementary codons (i.e., the 'negative'), and at the lower left are the anticodens, of RNA_2, which lie alongside the codons, and which are again complementary (i.e., 'positive'). It can clearly be seen that the letters in DNA and RNA_2 correspond – naturally after allowing for the fact that the place occupied in DNA by thymine (T) is occupied in RNA by uracil (U). The operation that takes place between DNA (left) and RNA (right) can be described as 'printing off': for every positive there must be a negative. But this is more than just a technical printing process. It involves a process of 'snatching away' the information from its source, the DNA. The information from now on resides as a negative in the RNA, and moreover in an RNA with molecules considerably shorter than the DNA chain.

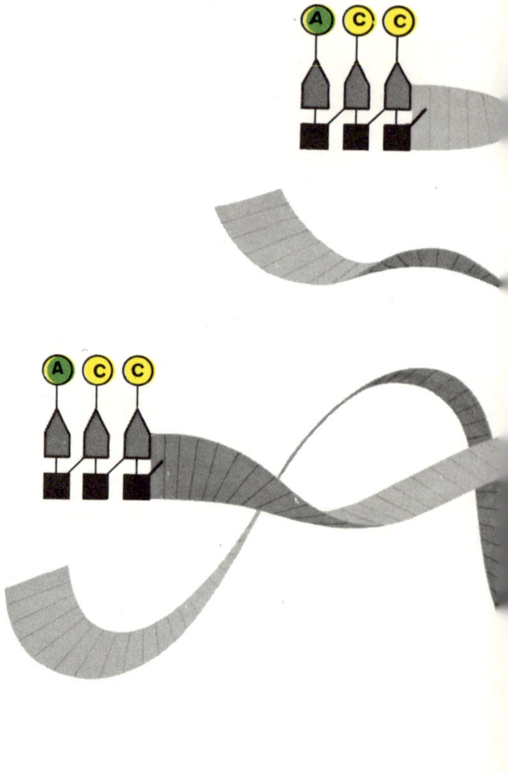

1.08 Messenger, Factory, and Transporter
Three Kinds of Nucleic Acids for the Manufacture of Proteins

There are organisms – we still have to consider them in detail – in which the total information for the enzyme proteins (and there may be hundreds of these) may lie in a

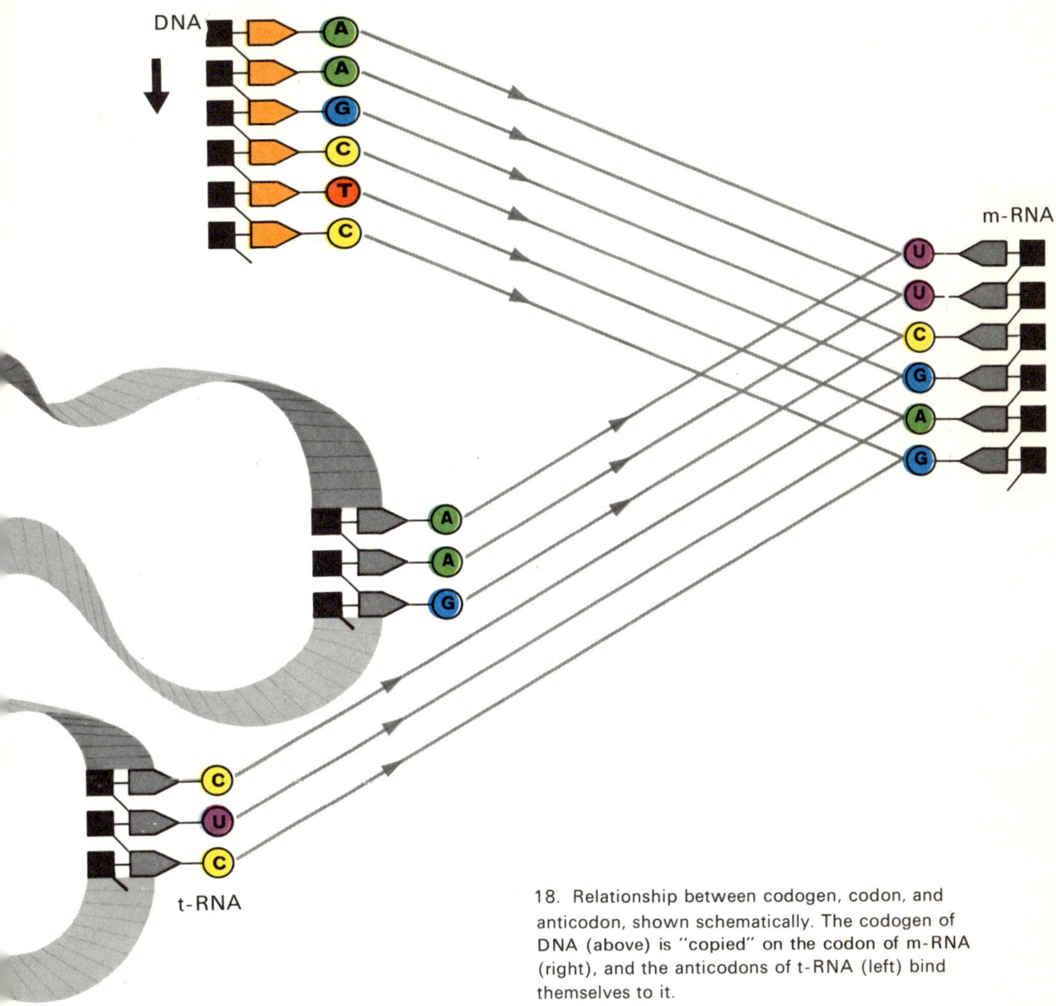

18. Relationship between codogen, codon, and anticodon, shown schematically. The codogen of DNA (above) is "copied" on the codon of m-RNA (right), and the anticodons of t-RNA (left) bind themselves to it.

single DNA chain, with ten thousand triplets. This giant molecule does not lie in the cell nucleus, since some organisms do not have any nuclei, but the giant molecule may be called a chromosome, the only one that the organism possesses. This mass of information is not held in one piece, but in sections. Each section comprises one enzyme-protein, and includes between one and several hundred triplets; in hemoglobin, for example, there are 146. This RNA_1 is thus much shorter; for hemoglobin it has 146 codons. For this reason it is decidedly more mobile, and so it can move inside the cell, and with the DNA as a source of information, carry out some kind of protein synthesis. The

RNA_1 thus contributes to the protein manufacture – or so it would appear – the information that commands, Let this one, specific enzyme-protein be made. The RNA_1 is thus acting as a kind of messenger, and since the discovery was made in an English-speaking country, the English word 'messenger' has become part of the international term 'messenger RNA,' abbreviated to m-RNA.

Let us summarize what we have learned so far about m-RNA in practice. Information, which is present in the DNA in the form of a sequence of codogens, is extracted by the complementary pairs of bases, copied by the m-RNA, and in this form (negative), is used as an information source for a kind of protein synthesis. Hence the protein synthesis, as we already know, is not based on the *DNA* itself as the source of information; nor is it based directly on the m-RNA – as we have already established, this is equally too simple. Where, then, is the 'protein factory' – where is this information stored that is delivered by the 'messenger' and 'translated' into a sequence of amino acids?

In a single cell there are many 'protein factories,' and as might be expected, for the greater part they are to be found outside the nucleus, and usually in the cytoplasm. For the present let us consider only those in the cytoplasm. This time it is not a matter of chemical compounds alone, and the ones that do take part are not unknown to us. Proteins are synthesized in particles that have the form of little granules. In fact they are too small to be visible under the light-microscope, but they can be seen in cell sections examined under the electron microscope. Approximately spherical in shape, they have a diameter of about 200 ångstrom units. (One ångstrom unit, (Å) = one ten-millionth of one millimeter. The ångstrom unit, Å, is used not only because it is easier to pronounce but also because it is more suited to the scale of the world of molecules than is the systematic unit nm = nanometer = one millionth of a millimeter.) The granules are sometimes distributed at random in the cytoplasm, without any arrangement; at other times, however, they are in rows like those in a pearl necklace, on both sides of certain layers in the cytoplasm. In the chapter headed 'A Trip into the Cell' we shall come across them again.

For the present, however, we must dispense with electron-microscope shots, because even with the electron microscope, protein synthesis cannot actually be seen. Meanwhile cells can be cut up with rapidly rotating knives, or smashed to pieces by using very high-speed vibrators filled with tiny glass beads, or destroyed with ultrasonic vibrations; each method results in a uniform mixture – a so-called homogenate – of all the components of the cell, including the protein-building particles. This homogenate is a 'cell-free system,' or more precisely, a 'cell-less' system. Since all the cell components are randomly mixed in the homogenate, the original basic order of the cell has been lost. As far as its metabolism has been investigated, the homogenate is not completely dead, since all the possible chemical reactions between its components can still go on, but these will be uncoordinated and uncontrolled; and the reaction is often uncontrollable, because the chemical changes that go on in the intact cell will now take place either quite differently or not at all. As a result of the mixing together of the components of the cell, the basic materials will either be widely separated from one another, or kept apart by some impenetrable layer.

If the chemical exchanges which go on inside the cell are to be made to operate outside the cell, in a test tube, *in vitro,* then the necessary components of the cell must be separated from each other, isolated, and as far as possible 'cleansed' of all unwanted impurities. Today it is indeed comparatively simple to do this, since the cell fragments are of different sizes and different weights. Modern centrifuges can separate the components of a homogenate again and again; a low speed of rotation throws out the coarsest particles, e.g., bits of the cell walls and incompletely disintegrated cells; these form a basement layer, a sediment at the end of the tubes of the centrifuge, while all the other components remain in the fluid above. By raising the speed of the centrifuge step by step, the homogenate can be divided into a large number of parts, or 'fractions.' The individual fractions may already be to a large extent uniform, i.e., consisting of one single kind of cell component, such as nuclei. If not, then the centrifuging must be refined still further, and this is especially necessary if the 'fine' particles are to be separated from the 'finest.' For this purpose, ultracentrifuges have been built, which develop particularly high centrifugal forces, up to 200,000 g, and which reach 300,000 r.p.m.

One of these fractions will therefore contain our granules, which were wanted for use in protein synthesis, in high concentration, and comparatively 'clean.' Later on we shall see how protein synthesis can be carried out with the aid of these granules, outside the cell, in a cell-free system, but first we want to know where these particles come from. The chemical analysis is not difficult, provided that a sufficiently large amount of material can be isolated. It transpires that about half comes from protein and half from RNA; the percentage varies between 40 and 60. Thus we do not need to learn about any new groups of chemical compounds at this stage. Talking about RNA gives an opportunity to mention the particles called *ribosomes* (in Greek *soma* = body), i.e., bodies which carry RNA. Their RNA component is called ribosome RNA, or ribosomal RNA, or r-RNA. With m-RNA this makes the second ribonucleic acid that we have met.

Polysomes

When the particle is combined with its corresponding protein (in the proportion of 60–40 per cent) its molecular weight lies in the neighbourhood of six million, or in the same order of magnitude as the biggest proteins, and of RNA, with which we are familiar. It is remarkable that, in contrast to these, the molecular weight of r-RNA pretty uniformly reaches six million, and is thus not as variable as that of the proteins, which have molecular weights ranging from five million down to 6,000.

Indeed the maximum molecular weight of six million is attained only if the fluid, the medium in which the experiment is carried out, and in which the ribosomes are to be found, contains large numbers of magnesium ions. When the magnesium content falls, smaller molecular weights (three, two, and one million, in round figures) are recorded. The ribosomes are then present in smaller fragments.

To be precise it should be said that accurate estimates of molecular weights could never be made by the classical methods, which were very wasteful. The 'particle size' of the ribosomes could be characterized much more simply by the so-called sedimentation constant. This, which is somewhat difficult to calculate, is a measure of the speed of sinking of particles in the ultracentrifuge. The bigger (i.e., heavier) the particles are, the quicker they sink. The sedimentation constant is expressed in terms of Svedberg units (S), after the research worker who did the scientific groundwork for the technique of ultracentrifuging and made possible the building of the first ultracentrifuge.

The 'complete' ribosomes with a molecular weight of six million are 100-S-ribosomes (i.e., those with a sedimentation constant of 100 Svedberg units). Their fragments are units of 70 S, 50 S, and 30 S. It is an interesting fact that one 100-S ribosome can break into two 70-S units; and each of these 70-S units can then break into one 50-S unit and one 30-S (but the Svedberg units cannot, like molecular weights, be *added together*). The pattern of events is as follows:

As a result of increasing quantities of magnesium ions, the fragmentation can be reversed, and complete ribosomes reassembled. It follows from this that the 100-S and 70-S

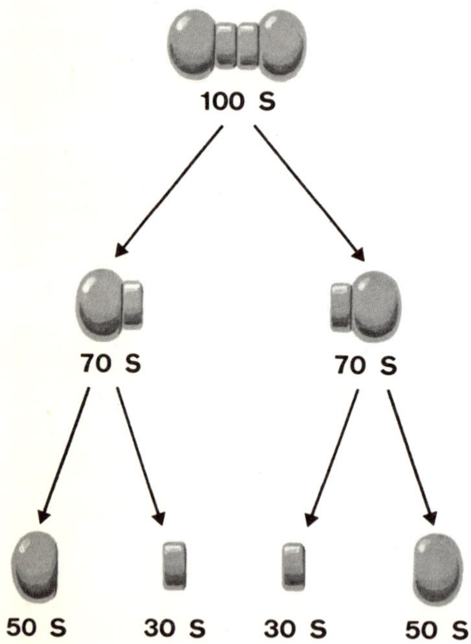

19. Fission of a ribosome of 100 sub-units into two components of 70 S-units each, with each of these then dividing into a 50- and a 30-S-unit component.

particles must be just as competent at protein synthesis as 50-S and 30-S units. The aforesaid plan can only reproduce the larger masses from the separate units, *not* out of general disorder. Neither process is sufficiently well understood at the present moment. Meanwhile the findings of the electron microscope are so ambiguous that we have to be content with pictures of one 70-S fraction. An electron-microscope transparency that was made when this fraction was spread out in a thin film shows the ribosomes as very nearly spherical bodies, most or all of which lie close to each other. We shall very soon know whether the aggregation of the ribosomes is accidental or not. (Incidentally, the photographs show how 'clean' a ribosome fraction can be obtained with painstaking preparation.) The ribosomes' similar form and uniform molecular weight – this contrasts strongly with the differences in form and molecular weight between proteins – leads to the assumption that ribosomes are not, or are only to a very small extent, *directly* concerned with the coding. In fact today it is generally accepted that the ribosomes are *non-specific* particles as far as their RNA and protein content are concerned. It follows that in certain bacteria only a very small part of the r-RNA, namely 0·1 or 0·2 per cent is complementary to the DNA base sequence (i.e., to the sequence of codogens). To all appearances the release of ribosomes is dependent upon the m-RNA on the outer surface of the molecule. The RNA molecule is held fast in the extended shape, so that the codons of the m-RNA lie parallel to each other. Only in this form can the m-RNA perform its function as a template; when it swims freely in the aqueous medium (in a cell-free system) or in the cytoplasm (inside the cell), on thermodynamic grounds the molecule is probably so knotted that many codons are completely inaccessible.

20. An electron-microscope picture of a purified fraction of ribosomes at the 70-S-unit stage. Magnification about 150,000.

Now the anticodons have their turn. On the outstretched codons of the n-RNA lie, not the corresponding amino acids, but the appropriate *anticodons*. Already these have extracted the right amino acids from the store that the cell has produced, and carried them along to the outer surface of the ribosomes. This RNA functions as a *transporter*; for this reason, it is given the title *transfer RNA*, or t-RNA (an early name was soluble RNA, or s-RNA, because its molecule being comparatively small, it remains in solution during the chemical reaction between DNA and m-RNA). This is the third, and for the moment the last, kind of RNA that has so far been discovered and given a name.

m-RNA r-RNA t-RNA

The molecules of t-RNA are never so small that they consist only of one anticodon and the corresponding amino acid. The facts are much more complicated. The molecule of t-RNA – of which twenty, probably fifty-six, and possibly sixty-four different versions exist (embodying the code) – has at times seventy-seven nucleotides in a chain, one behind the other. All versions are alike in one respect: at one end of the molecule there is always placed the triplet CCA, and on this triplet the amino acid is fixed. In addition, somewhere along the chain there must naturally be the operative anticodon.

Until very recently no more than this was known about the *structure* of t-RNA, but on the other hand its *function* was fairly well known by experiment. One molecule of t-RNA (No. 1) withdraws from the amino-acid store the amino acid corresponding to its anticodon (amino acid No. 1), and attaches it to the CCA end of itself (the fact that energy for this operation must be found somewhere does not concern us at the moment). Thus laden, the molecule moves towards a ribosome, with an m-RNA molecule stretched alongside it, locates the codon No. 1 appropriate to its own anticodon No. 1 (the complementary principle!), and clings there. Immediately a t-RNA molecule with the anticodon No. 2 attaches itself to the adjoining codon No. 2; this carries on its CCA end an amino acid, but naturally a different one, belonging to anticodon No. 2. The two amino acids now lie next to each other. By this time the peptide bond between them is tied, the first t-RNA has fulfilled its function, released itself from the ribosome, and is ready to pick up another amino acid (a similar one). Meanwhile the third molecule of t-RNA, with anticodon No. 3 and amino acid No. 3, has arrived at codon No. 3, and thus the peptide bond between amino acid No. 2 and amino acid No. 3 is formed.

Thus the polypeptide chain grows out from one end at top speed, until the last codon is in position. The last step in protein synthesis, the transfer (also called 'adaption') is thus explained, at least in its essentials. In conjunction with the first step, the operation of copying, it produces a result that corresponds with the pattern of the protein structure.

It will not escape the reader that this picture, though it may be basically true, is still obscure in a few particulars. Thus the few facts that we know, for example, about the structure of t-RNA are not sufficient to make its *function* clear to us. Experiment shows us how t-RNA actually operates, but the chemist cannot tell us *why* it works in this way and not differently. There are other obscure points. What really happens on the outer surface of the ribosomes? What happens to the m-RNA and the t-RNA when they have played their part?

One thing, however, is obvious: the phenomena of protein synthesis are not

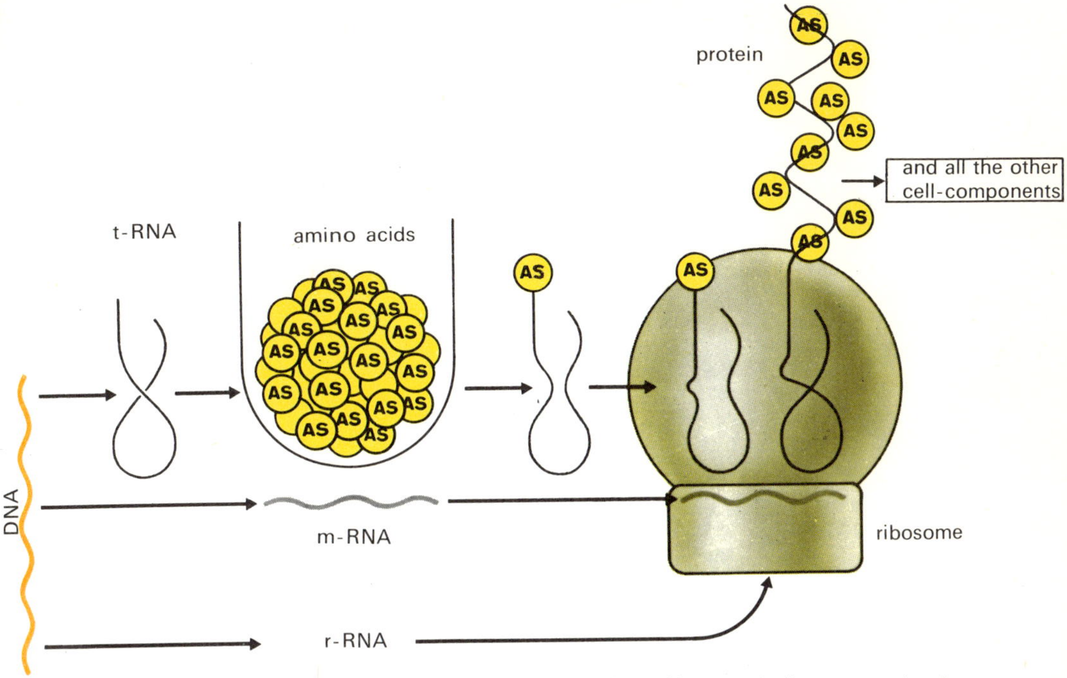

21. The construction of a protein molecule, shown schematically. The DNA gives rise to three kinds of RNA: the r-RNA produces the ribosomes; the m-RNA carries the information over to these ribosomes; the t-RNA, loaded with amino acids out of "store," places these alongside the m-RNA in the ribosomes. As the amino acids are gradually transported to the ribosomes, their peptide bonds unite, and they form specific enzyme proteins, which in their turn go on to construct all the other components that are necessary for the operation of the cell.

simply of academic interest, and they have not been studied out of idle curiosity. How precisely must the copying and transfer be carried out if a living cell is to be built up in the orderly way that it usually is? A single mistake in copying or transfer can have permanent effects: a 'wrong' amino acid is incorporated, resulting in a 'wrong' protein, a 'wrong' apo-enzyme, which then sets in train a 'wrong' chemical reaction. And what if the chemists ever succeed in making artificial m-RNA, insinuating it into a cell, and releasing it there, and so making an enzyme at the whim of the experimenter? Shall we then be able to make artificial genes? We shall learn more about this in the next chapter, but before that let us look for answers to the questions asked above.

Let us look again at t-RNA. Recently substantial progress has been made by an American research team, which has demonstrated the complete structural plan of a t-RNA, and an alanine-binding t-RNA at that.

This is how it appears (Fig. 22). The picture confuses almost more than it enlightens. The molecule seems to consist of two halves, which have laid themselves alongside each other, but a 'double strand' has not been formed; a true pairing of bases (naturally this can also take place between complementary bases of one and the same chain) is obviously possible only along quite short stretches of the chain. Between these are visible larger and smaller 'blisters,' in which the bases confronting each other are not complementary.

In addition to this we notice new symbols in the chain: 1, Ψ (psi), Me, DiHU. This need not depress us, since they are usually concerned with easily variable bases, which are closely related to, and perhaps interchangeable with, those that we already know. DiHU, for example, is none other than uracil with two more hydrogen atoms. Me simply lacks the ring, and has a molecule of a methyl group. Finally it must be admitted that the conformation shown in the figure is only one of three possibilities.

All this is not easy to follow. Nevertheless, by means of the colours in the diagram, the triplet CCA standing at the end, and the distinguishing group GGC can be recognized – this is indeed complementary to the codon GCC, and is released in reverse order. The Americans indeed express themselves cautiously: they are of the opinion that the nearly related IGC series may be a possibility as a coding triplet. Whether this is so or not, the anticodon actually lies in the middle of the chain, not at the end which is opposite to the CCA triplet. So far there is no answer to the question of how the anticodon, lying in the middle of the chain, can ensure from that position that the CCA end, which is the same in every t-RNA molecule, shall attach itself to alanine, and to no other amino acid. In the present state of research it remains a problem. Many questions are still to be

alanine-specific t-RNA

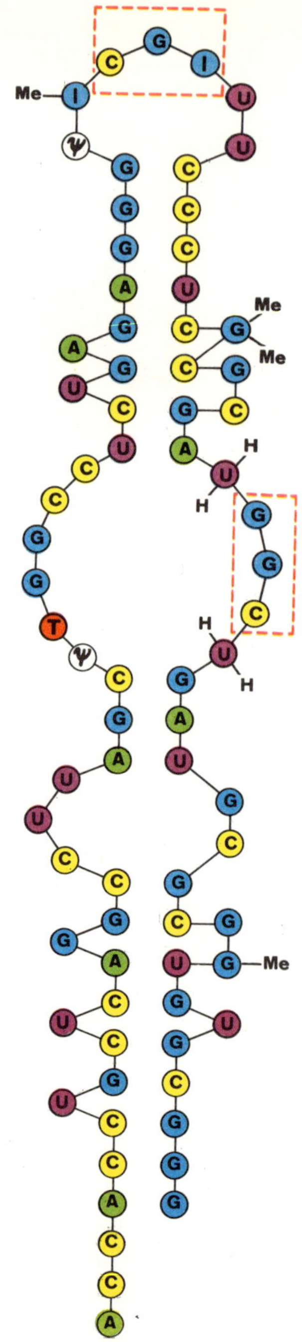

22. Molecular structure of an alanine-specific t-RNA, one of three possibilities. Formation of groups of bases is shown at various places in the molecule; the groups enclosed in the red dotted lines are identification groups (anticodons).

asked, but nevertheless we are standing, so to speak, right in the research laboratory itself, in the front line – this explains why there are so many sentences qualified by 'maybe,' 'it is possible,' and 'probably.' Nevertheless we must not fail to give credit to the scientific achievement that has led to structural formulas, and to the deductions that have followed directly from these formulae: all the more, since it concerns the first molecule of a nucleic acid of which the nucleotide sequence has been discovered.

There is no particular difficulty in understanding the formation of a bond between amino acids that stand in the CCA ends of the t-RNA molecule; this takes place, as might be expected, through the operation of enzymes. Large amounts of the enzymes involved are obviously present in the vicinity of the ribosomes.

On the other hand, it is still not quite clear what is going on on the outer surface of the ribosome, where codon and anticodon come together. It is certain that by no means every molecule of RNA is in contact throughout its length with the outer surface of a ribosome. Fig. 26 illustrates the comparative sizes of ribosome, and m-RNA molecules fairly accurately, and shows very well that only a small part of the larger can be attached to the stalk. As far as present knowledge goes, it would seem that generally only two codons are attached at the same time, and for this reason two adjacent points of attachment are provided. These are not associated with *any specific coding groups* – we recall that only an insignificant percentage of the r-RNA is present as 'specific' triplets, and for the rest the significance of the various attachments is not yet known.

As soon as they are attached to two adjoining codons of the m-RNA, two molecules of t-RNA, equipped with amino acids, place

Ala-tRNA Ser-tRNA I + II Tyr-tRNA

23. 'Cloverleaf' structure of three different molecules of t-RNA.

themselves with their anticodons alongside. At the same moment the linking of the two amino acids takes place. The t-RNA molecule No. 1 is then without its amino acid; it detaches itself and goes to fetch another amino acid. At this moment the molecule of m-RNA which has its No. 1 codon free moves along one step, so that codon 2 arrives at position 1. In the same way position 2 is occupied by codon 3, with which anticodon 3 makes a complementary pair. The dipeptide 1–2 becomes attached to the tripeptide 1–2–3 with the assistance of t-RNA_2, and all is set for the m-RNA molecule to move on again by one position. This works almost like a sewing machine.

The RNA molecule slides step by step alongside the outer face of the ribosome, and the peptide chain comes into being at the same speed. When the thread of m-RNA reaches the end, when the last codon moves into position on the outer surface of the ribosome, then the primary structure of the polypeptide is 'stitched together.' In other respects there is no resemblance between this process and the operation of a sewing machine; the thread of m-RNA is not really like a thread of cotton because adjacent sections of it are tied together by the amino acids. Moreover, it is very short-lived, and is

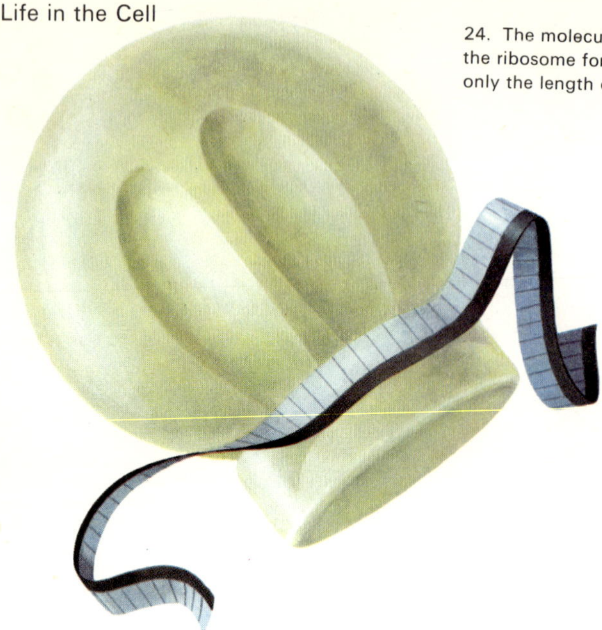

24. The molecule of m-RNA (blue) is attached to the ribosome for only a very short distance, probably only the length of two codons.

soon dismantled. The polypeptide chain, on the other hand, in all probability remains on the ribosome until it is complete, and then detaches itself for the first time, folds itself, twists itself into the secondary and tertiary structure, and attaches itself to the quaternary structure, if necessary with other partners – and thus for the first time the enzyme protein is complete. What has just been described step by step, in detail, is carried out by the cell competently and continuously, i.e., not in a series of steps. Estimates of speed, arrived at by different methods, have brought astonishing figures: in one second something like five thousand peptide bonds are joined; an enzyme protein with 146 amino acids needs only about one thirtieth of a second to come into being. At molecular level many things take place at an even faster rate, as will be realized from what has gone before. One of the reasons is unquestionably that the distances apart of atoms and molecules are immeasurably smaller than the distances that separate objects in our own macroscopic world.

Polysomes

Here are the facts about short-lived molecules.

Many proteins have a so-called *half-life* of about two days. The term 'half-life' is one that has been taken from physics. To say that a protein has a half-life of two days means that within two days one half of any given material (which may be molecules, or atoms, or nuclei) is expended, dissipated, broken up. This is not to say that in twice that time, four days in this case, the other half will have suffered the same fate. On the contrary, in the second two days only *half of those remaining* (i.e., one quarter of the original number) will be affected. For example, uranium isotopes (atomic weights 235–238)

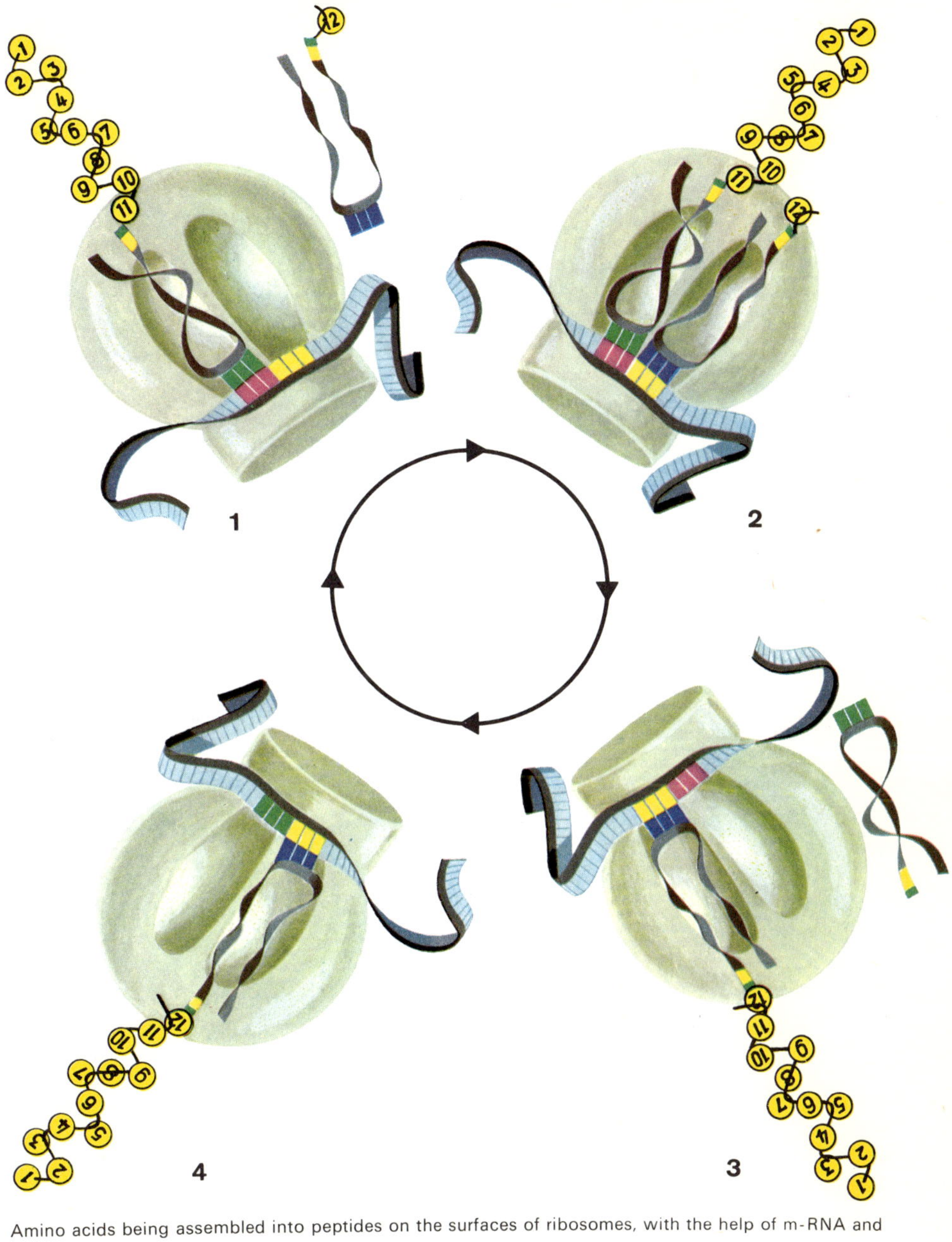

25. Amino acids being assembled into peptides on the surfaces of ribosomes, with the help of m-RNA and t-RNA.

have half-lives ranging from 710 million years to 4·5 milliard (thousand million) years; the half-life of radioactive carbon, C^{14}, is 5,760 years.

m-RNA, with a half-life of about thirty minutes, is very short-lived compared with proteins that have a half-life of two days. When one considers however, that the production of a single protein molecule takes only a thirtieth of a second, it is evident that a molecule of m-RNA is able to perform its coding function many times during its existence, and thus to take part in the synthesis of many protein molecules – similar ones! So it works like this. A messenger molecule that has already moved forward one place over the ribosome, and thus has a free place 'up in front,' can attach itself to a free position on another ribosome molecule and thus set in motion the production of another, identical, polypeptide chain. And since the second ribosome is soon 'overrun,' the process can be repeated. The number of ribosomes that at any one time can be attached to the same messenger molecule, depends naturally on the length of the molecule itself, and on the distance apart of the ribosomes, which are nearly equal; from an average length corresponding to 150 codons come about five to seven ribosomes. The expression 'polysome,' or 'polyribosome,' is applied to the temporary association of one RNA molecule and five ribosomes. It would need several of these to become visible in the electron microscope (see Fig. 20). It is possible that some of the clumps that are to be seen in an electron-microscope picture of the ribosome fraction may be polysomes.

The presence of polysomes explains one thing that was a problem previously: namely, that out of one complete fraction of ribosomes, only about 20 per cent, i.e., one fifth, are active in protein synthesis. In a watery suspension of ribosomes and molecules of m-RNA, the probability that a piece of the messenger that has already been in action and has been released again will attach itself to a second ribosome is very small, because the distance to be travelled is too great. In the living cell, on the other hand, we have a high degree of orderliness; distances to be travelled are small, and so the messenger can very quickly find a second, a third, and fourth, and even a fifth ribosome. Under these conditions 100 per cent of the ribosomes are active.

It is not very likely that a molecule of m-RNA would be capable of acting with a total of five ribosomes and then would disintegrate. It is not as short-lived as that. Presumably it can act on twenty, thirty, or more ribosomes, and thus code that number of polypeptide chains, before it is exhausted. In this case of course, it cannot attach itself to more than five ribosomes at the same time.

1.09 A Glance into the Laboratory

How Are Experiments on This Subject Carried Out?

Here is a point that has so far not been explicitly stated: our knowledge of the process of protein synthesis comes from experiments which biochemists have carried out *outside the living cell*; it is a practical possibility to produce proteins, or at any rate, protein-like substances, in cell-free systems, *in vitro*. This does not mean – for the present anyway – that we can make lifelike proteins artificially. The biochemist as a general rule works with substances and particles that he has by no means made himself, but has taken from living cells. Nevertheless, inherent in biochemical work is the possibility of being able to replace the natural materials of the cell one by one with com-

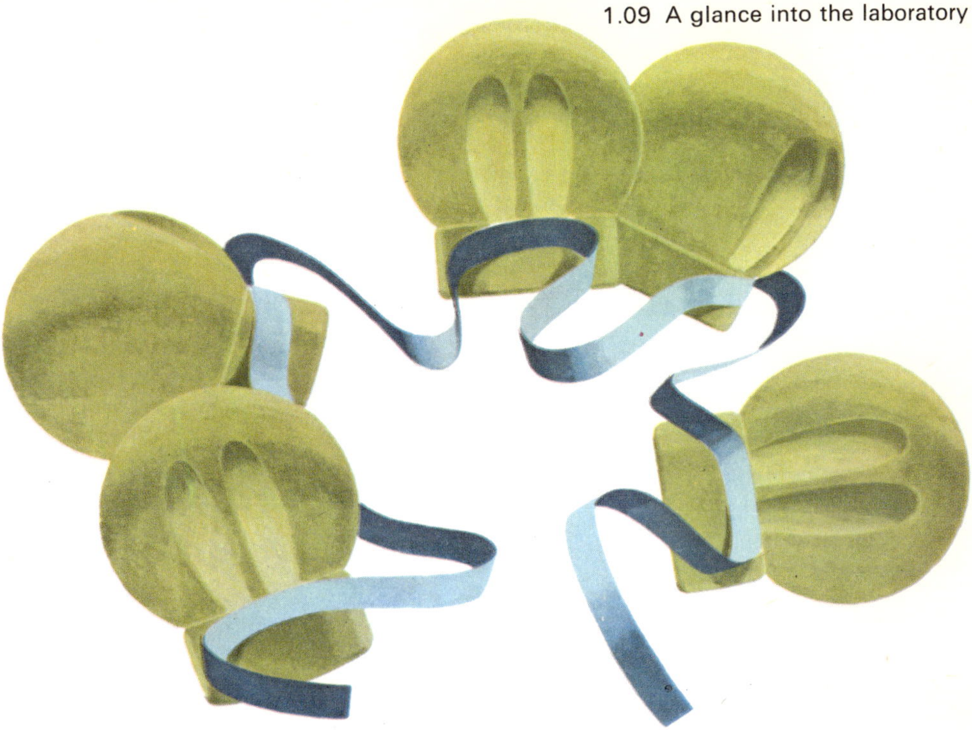

26. One molecule of m-RNA may be attached simultaneously to five or more ribosomes. Such ribosome complexes are called *polysomes*.

pletely synthetic substances. A beginning has been made already. Artificial molecules of m-RNA can be introduced into a cell-free system in place of m-RNA from a cell, and can be used to code proteins.

In journal articles one comes across forecasts that man will soon be able to make artificial protein, and perhaps even artificial cells; few readers ever ask themselves whether such possibilities of meddling with the workings of the living cell are disturbing or even frightening. Perhaps it would be frightening if we were to see such experiments in progress, and to see which deductions have proved to be correct, and what things may or may not be done in the future.

We will try to do this.

Among the substances that must be synthesized are, naturally, our three kinds of RNA – messenger RNA, ribosome RNA (preferably 70-S-RNA), and transfer RNA – and also a sufficient store of amino acids, which must include all the twenty different ones.

We have already seen how ribosomes can be extracted as a pure fraction from a cell-homogenate by making use of the ultra-centrifuge, but naturally this is not as easy as it is made to appear on p. 44. Rotating a centrifuge at a high speed generates a great deal of heat, and heat would soon change the ribosomes and make them useless. So the centrifuge must be cooled enough to ensure that the temperature does not rise above 5°C; consequently, the first step, homogenization, must take place at 5°C.

After centrifuging, and after the ribosome fraction has settled, the m-RNA and t-RNA remain in the 'supernant' liquid. Molecules of m-RNA are bigger than those of t-RNA; the messenger molecule often has 150 triplets (450 mononucleotides), whereas the t-RNA has only 77. Thus m-RNA and t-RNA can in turn, be separated from each other by centrifuging at a higher speed. In addition, there are other methods of separation – for example, 'gel filtration.' The 'gel filter' is somewhat similar in operation to an 'ion exchanger' that is used for softening drinking water, but in place of an artificial plastic material it makes use of carefully prepared cellulose. This is packed into a vertical glass tube, from ten centimeters to six meters long, as required, and correspondingly narrow or broad. Into the top end of this is poured the liquid from the last centrifuging, which may contain other bodies besides the molecules of m-RNA and t-RNA mentioned above. The liquid soaks in, and remains while the processes of sedimentation and differential solubility deposit the substances in different parts of the cellulose pile. After this the cylinder is flushed out with suitable liquids, e.g., salt solutions of high concentrations. This detaches the various substances one by one, according to their molecular weights and solubility – a process known as 'elutriation.' The elutriate runs out of the bottom of the cylinder and is collected in fractions.

This method of fractionating often gives better results than centrifuging. It often happens that two different substances have the same molecular weight, and the same particle size. The centrifuge cannot separate them, but the cellulose pile can do so, since it is only rarely that they will have the same molecular charge and the same solubility.

By this method m-RNA and t-RNA have been separated from each other, and from the other substances mixed with them. The twenty different amino acids can be bought in the purest possible state from a chemical works, and all mixed together: ribosomes, t-RNA, m-RNA, and amino acids, all dissolved or suspended in water.

All the same, this cell-free system is not yet capable of producing protein; several factors are still missing. The list of all the components that must be present at the same time for this synthesis is as follows: (1) ribosomes, (2) m-RNA, (3) t-RNA, (4) all twenty different amino acids, (5) the appropriate enzyme, (6) ATP, (7) a system for regenerating ATP, (8) magnesium ions, (9) GTP.

Probably the reader will not yet know what items 5 through 9 have to do with it. He may come across one of these factors later on, and ask what it is, and then he can at least turn back to this page.

Item 5: the appropriate enzyme. The attachment of an amino acid to the CCA end of 'its' t-RNA molecule is looked after by an enzyme. Its correct name is aminoacyl-t-RNA-synthetase, because it synthesizes the complex made up from t-RNA and the residual group of the amino acid (aminoacyl). Probably there are the same number of these complexes as there are different ways of making t-RNA.

Enzymes are necessary in order to detach an amino acid from its t-RNA, and its linkage with the next amino acid in the peptide chain. This step is comparatively simple. It only involves breaking one bond (amino acid to t-RNA) and forming another one (amino acid to the corresponding amino acid in the half-completed peptide chain). This requires hardly any energy, and the appropriate enzyme can bring it about with comparatively little difficulty. It is quite different if it involves attaching a 'free' amino acid (i.e., one from the store of amino acids within the cell) to the end triplet, CCA, of a molecule of t-RNA. This has a definite price, a definite expenditure of energy, of chemical energy in fact. But what do we mean by 'chemical energy,' and where is to to be found?

Strictly speaking, energy is present in every bond of an organic compound; in every bond, that is, attached to a carbon atom (with the exception of CO_2). For example, if starch or cellulose is burned, CO_2 and water are produced, with a great deal of heat. This is even more obvious when benzene, which is more inflammable, is burned. Simplifying it a little, we might say that the chemical energy that was latent in the substance has been changed into thermic, or heat, energy. The human body, like a single cell, makes use of foods and reserves of substances containing carbon, such as fats, carbohydrates, as stores of energy, and most of the body's chemical reactions make this energy available for conversion into other forms. Naturally no actual burning, no naked flames, occur in the living cell – if heat were developed on that scale, the cell would be destroyed. Instead, the release of energy takes place in small steps: currency of large denomination (starch or fat) is changed into small coins, and these still hold their energy in the form of chemical energy.

A store of small coins is very handy for the cell. If the cell had to break down one molecule of starch every time it carried out one step in protein synthesis – such as the attachment of a free amino-acid molecule on to a molecule of t-RNA – this would be very unreasonable. For one thing, it would take up rather a lot of time, and for another, the demolition of even one molecule of starch would release more chemical energy than was needed for the single step in synthesis.

This brings us to item 6, ATP.

ATP

The Universal 'Small Change' of All Cells is Called ATP

ATP is another abbreviation, one that stands for a compound with the name of **A**denosine **TriP**hosphate. This means nothing to us at first, but careful examination of the following equations will make it clear:

Adenosine triphosphate = adenine + ribose + phosphoric acid + phosphoric acid + phosphoric acid.

Adenine + ribose = adenosine.

Adenosine monophosphate (abbreviated AMP) = adenine ribose-nucleotide. Adenine diphosphate is abbreviated ADP; adenine triphosphate, ATP.

ATP is nothing other than an adenine ribose-nucleotide to which two more phosphoric acid radicals have become attached, so that the complete molecule has three phosphate groups. The bonding of second and third phosphate groups introduces chemical energy into the molecule – these are 'energy-rich' bonds, and 'energy-rich' phosphates. The chemical energy of the phosphate bond can easily lead to the making or breaking of other chemical bonds. In the example we had, when a free amino acid became attached to the t-RNA, the necessary chemical energy was derived from the ATP. The ATP releases one phosphoric acid radical and reverts to ADP (adenosine diphosphate), which is less energy-rich. The chemical energy from the third phosphate bond is absorbed by the bonding of the amino acid to the t-RNA.

Adenine + Ribose + Phosphoric acid + Phosphoric acid + Phosphoric acid

↓

Adenosine

↓

Adenosine Monophosphate (AMP)
= Adenine-Ribose-Nucleotide

↓

Adenosine Diphosphate (ADP)

↓

Adenosine Triphosphate (ATP)

During this reaction, in addition, a small amount of energy is lost as 'fall-out.'

We can now understand that if we want our cell-free system to build up proteins, we must supply it with ATP, since for every bonding of one amino acid we shall need one molecule of ATP. For a polypeptide chain of 146 amino acids we shall need 146 molecules of ATP.

Item 7: a system for regenerating ATP. Our methods of analysis are not yet refined enough for us to be able to identify just a single such polypeptide chain. Thus our cell-free system must be allowed to work away for a long time, until it has produced a detectable quantity of protein. In the process it would be highly impracticable to allow the large amounts of ATP that would be required to be supplied as ATP, because this would bring our entire protein synthesis to a stop. Instead of this we make use of item 7, an ATP-regenerating system. This is a small trick, which fortunately we need not discuss in detail, since it does not affect the main process. All we need to say now is that it has to do with a substance called phospho-enol-racemic acid, which is a product of the breakdown of carbohydrates during respiration, and which, in co-operation with an appropriate enzyme, 'regenerates' ATP from ADP plus phosphoric acid.

Item 8: magnesium ions. It is understandable that our cell-free system will also need magnesium ions, since of course the sub-units of the ribosomes are held together by means of them (see p. 46). However, we also need these ions for another reason. Every transfer of energy by ATP requires the presence of magnesium ions, an experimental fact that cannot yet be explained. Just as little is known about the reason why magnesium ions should facilitate the cohesion of sub-units of the ribosomes.

Item 9 (the final factor): GTP. GTP is an abbreviation for **G**uanosine **TriP**hosphate, also a nucleotid. It is very little different from ATP. Instead of an H atom, the guanine contains an OH group, and it has its NH_2 group in a different position from that of adenine (A).

GTP has a most interesting part to play. It takes care of the transport of codons to the outer face of the ribosome, and simultaneously the movement of the m-RNA. It is, so to speak, the motor of the ribosome sewing machine. Unfortunately we still have no idea how GTP works. Certainly the energy-rich phosphate bonds, which clearly exist in

ATP, play some part, but what particular function they have we do not know. It has invariably been found that without GTP neither a beginning nor a continuation of the protein synthesis is possible.

At last we have been introduced to all the factors involved in protein synthesis. Let us bring them together in their proper proportions, and let us 'incubate' this brew at 37°C, so that we obtain – outside the cell, in a test tube – real, genuine proteins. It is true that there are no very long chains, and it is a comparatively variegated mixture of different chains, but this is not surprising, since even the m-RNA, which carries the specificity of the polypeptide, has had to be obtained from a living cell, and then only as a mixture. All the same, there is no doubt that it is genuine peptide, which can be further studied. It was – and is – indeed sad that the output of protein is modest. We still cannot produce *in vitro* more than 1 per cent of what an intact cell can produce in the same length of time. There may be many reasons for this. In the highly organized system within the cell everything goes more quickly, and with less trouble, than it does in the confusion that is produced by mixing together the nine factors. Moreover it is possible that one or more unknown factors are still missing; the addition of these might multiply the output. Possibly, too – though this cannot yet be said with certainty – there may exist not only the methods of protein synthesis that we have described above but several others as well.

Disappointment over the small output is less, however, if we think how much simpler is the cell-free system than the intact cell, with its tens of thousands of different substances, enzymes and particles. So nobody need be discouraged, particularly since the relatively simple cell-free system offers very real possibilities of new experimental investigations which will give a deeper insight into the processes.

When a start was made on varying individual factors and interchanging them with each other a most exciting discovery was made. A cell-free system was made from some particular group of organisms or cells such as a species of bacterium, and this was complete except for the m-RNA, which was omitted. The system was then completed with an m-RNA which had been extracted from some other organism, such as a cell of yeast, and protein synthesis began immediately. This shows that the cell-free system did not know that it was working with a 'strange' m-RNA; it made no distinction between it and the 'proper' m-RNA and used it for its own purposes.

This can only be understood if the mechanism of protein synthesis as well as the process of coding are not species specific, but *universal*. If that is so, then experimental work will be very decisively simplified. Above all, this would show us that the method of working of the cell-free system is of no particular importance, and its effects are more or less limited, if not entirely accidental; yet we have here *a basic process of life at a molecular level*. The universality is such an important fact that we must devote a special section to it forthwith. First, however, we must struggle on for a while if we wish to know how a cell-free system can help us to decipher the code.

1.10 Breaking the Code

The most important discovery in sixty years

If a 'strange' m-RNA can be substituted for the 'proper' m-RNA in a cell-free system at will, this suggests that we might try an artificial, synthetic m-RNA – if such a thing exists.

So far, completely synthetic m-RNA cannot be made, but this is not strictly necessary for the purpose suggested above. It is sufficient if natural m-RNA is split into mononucleotides and these are recombined into new, 'desirable' combinations. Since 1955 an enzyme has been known that works in this way: the so-called *polynucleotide-phosphorylase.* As fig. 26 shows it separates one mononucleotide from the other, and, with the addition of a second molecule of phosphoric acid, sets it free as a nucleotide-diphosphate. Such a process – i.e., detaching of one bond, with the taking up of phosphoric acid – is called a *phosphorolysis.* This is analogous to the process of hydrolysis, which, as we already know, is the detaching of one bond, with the taking up of water.

After complete phosphorolysis we have a mixture of four different substances, the A-, G-, C-, and U-nucleotide-diphosphates. We have already met one of these, adenine diphosphate, ADP, this is derived from ATP by the loss of phosphoric acid. Now, it is comparatively simple to isolate the different diphosphates and to obtain, for instance, pure U-diphosphate, UDP (correctly known as uracil-ribose diphosphate = uradine diphosphate). From this the same phosphorylase, if it works in the reverse direction, it is possible to synthesize a polynucleotide that consists only of uracil mononucleotides: poly-U = poly-uracil-ribose-phosphate = poly-uridilic acid. Sometimes the second phosphate may split off again. Such polymers, made from only one kind of nucleotide, are called *homopolynucleotides.*

Now, in poly-U we have the sequence of bases. U-U-U-U-U-U . . .; only UUU can act as a triplet codon, though this can be indefinitely repeated. If our cell-free system accepts poly-U as messenger, and with its help synthesies polypeptide, then this ought to consist of only one kind of amino acids, those for which UUU is the code word. What no one really dared to hope for, in fact, came about: the cell-free system made a *poly-phenylalanine.*

Admittedly the amount was not very great, and it required a very great deal of trouble to prove that exclusively phenylalanine, and not any other amino acid, had been made. To prove that this was so, the method was very cunningly altered. Instead of one experiment, twenty similar experiments were set up. At the start every experiment included all the twenty possible amino acids, but each time one of them (naturally a different one in each experiment) was 'marked' in a special way. Radioactive carbon C^{14}, served as a marker (this is the C^{14}, the half-life of which was discussed on p. 52). The atoms of C^{14} are not very stable, but gradually disintegrate; in the process of disintegration they release radiation, which can be measured with a Geiger counter. If one of the normal carbon atoms in one of the amino acids is replaced by a radioactive isotope, then the radiation that can be measured is proportionate to the number of molecules of that amino acid that are present. The store of amino acids that is essential in each of the twenty experiments now contains nineteen normal and one radioactive ('hot') amino acid.

In series 1 amino acid no. 1 is marked; nos. 2–20 are not marked. In series 2 amino acid no. 2 is marked; nos. 1 and 3–20 not marked; and so on until series 20, in which amino acid no. 20 is marked, and nos. 1–19 are not marked.

If our 'artificial' poly-U is actually operating as a messenger, and if UUU is the code word for phenylalanine, then naturally phenylalanine will be made in all the experiments. But it should be *radioactive* phenylalanine only in that series in which the phenylalanine has been marked and the nineteen other amino acids have remained unmarked. In all the other series no radioactivity should appear in the protein that is produced; if it does, then UUU must be the code word for another, different amino acid.

The following table shows how these experiments worked out: In this table, the

27. Splitting and reconstitution of m-RNA. 'Natural' m-RNA is treated with the polynucleotide phosphorylase (centre) and divided into four different mononucleotide diphosphates (shown on the right, with the phosphoric acid represented by black squares). Working in the reverse direction, the reaction recombines such mononucleotide diphosphates into a polynucleotide (i.e., m-RNA, left), with the separation of phosphoric acid. If the system is supplied not with all four kinds of nucleotides, but with only one, in this case uridine diphosphate, then it produces 'artificial' polyuridilic acid (poly-U).

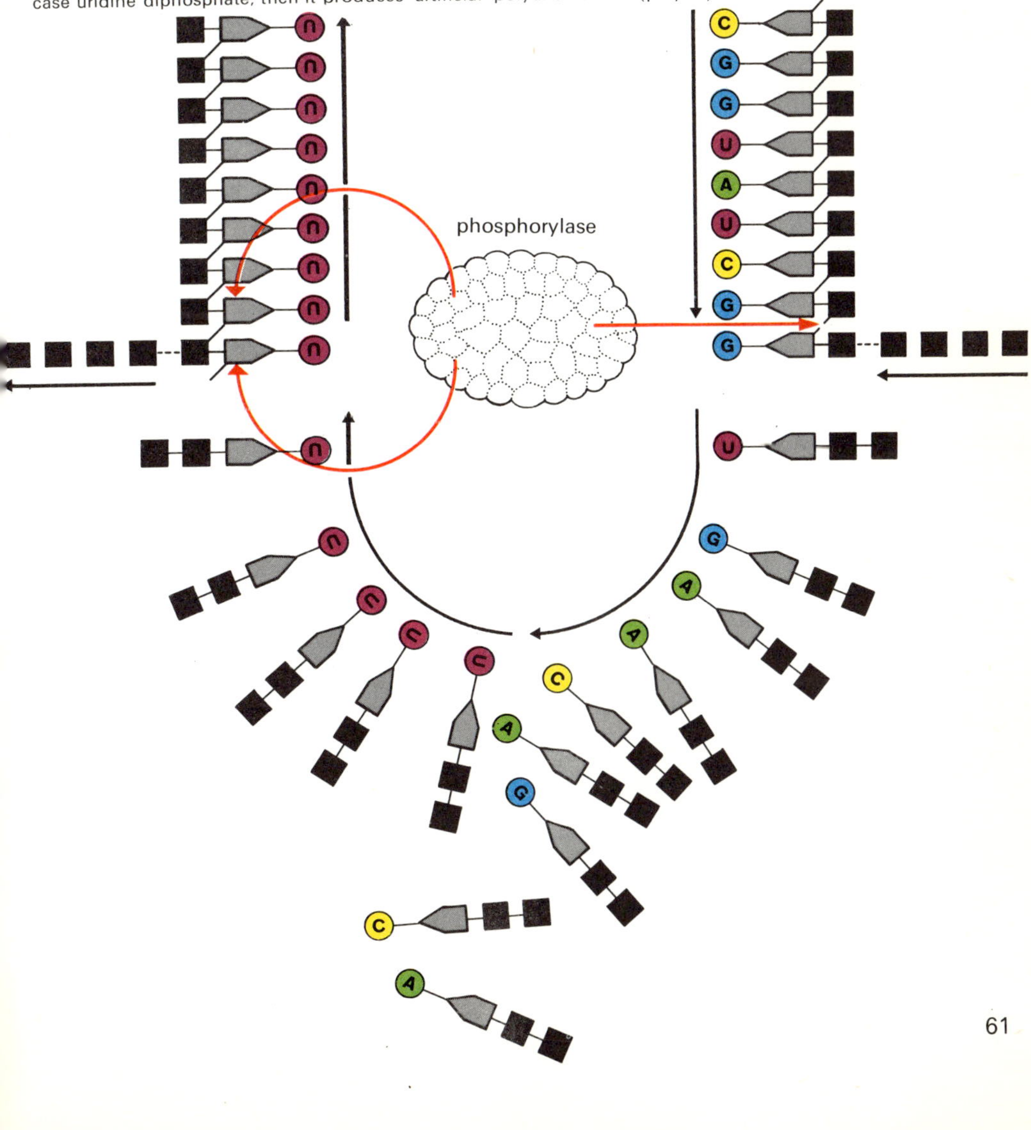

Table 2

Radioactively Marked Amino Acid in the Mixture	Function of Messenger RNA is Served by			
	Poly-A	Poly-C	Poly-U	Yeast RNA
1. alanine	0	0	0	28
2. arginine	2	0	3	93
3. asparagine	0	–	–	–
4. aspartic acid	0	0	3	77
5. cysteine	0	0	2	12
6. glutamic acid	0	0	0	76
7. glutamine	0	–	–	–
8. glycine	0	0	0	126
9. histidine	6	0	0	154
10. isoleucine	1	3	12	90
11. leucine	1	0	93	51
12. lysine	*1179*	0	6	58
13. methionine	3	1	6	27
14. phenylalanine	0	0	*2700*	57
15. proline	0	*157*	3	48
16. serine	1	0	16	219
17. threonine	2	1	1	42
18. tryptophane	0	0	9	135
19. tyrosine	0	0	2	23
20. valine	0	0	18	28

left-hand column shows which particular amino acid was marked out of the mixture of twenty, the right, the number of units of radiation. This indicates how many of the amino acids marked in this way have been incorporated into the protein that has been made. The same information is given for the artificial messengers poly-A (poly-adenine), poly-C (poly-cytosine), and poly-U, and also, for comparison, for natural messenger-RNA from yeast cells.

It is easy to see that with the help of poly-A as messenger, a protein has been synthesized which incorporates *lysine* as its particular amino acid, whereas poly-C is responsible for proline, and poly-U for phenylalanine. In the case of poly-U, it can certainly happen that in a concentrated mass *leucine* can also be made, and then on line 11 of the table, under poly-U will appear the number 93. This is very small compared with the phenylalanine number, 2,700, but nevertheless well outside the margin of error (numbers less than 10 may confidently be put down to impurities). The dossier for this story – 'Poly-U Also Makes Leucine' – is not yet closed, and without doubt the immediate future will see the matter cleared up.

However, we can say with certainty at the present time that we know the meaning of three code words. The triplets AAA, CCC, and UUU stand for the synthesis of lysine, proline, and phenylalanine. The first penetration into the mystery of the amino acid coding has gone well.

Further progress was made when the production of homopolymers (homopolynucleotides) was succeeded by that of heteropolymers. Instead of using only one kind of base-ribose-diphosphate, such as UDP, two kinds were mixed together, e.g., UDP and GDP, in the proportions of five to one. In this way it was possible to synthesize the polynucleotide phosphorylase, which is a compound polynucleotide. This substance contains U and G all mixed up together in a completely random ('statistical') order. This is more important than it seems at first glance.

At the top of this table stands the mixed polymer poly-UG, with bases in the proportion 5:1. In the second column are all the possible combinations (triplets). Now, it is a simple matter for the statistician to work out what is the probability with which each of the possible triplets will occur in the random chain. We are interested less in the absolute values of these probabilities than in the relative values. The triplet UUU must be the most frequent, as a matter of course; after all, we have included five times as many U as G. So let us fix the probability of UUU as 100 and express the other probabilities as percentages of this: 20, 4, and 0·8 respectively (column 3).

Now this heteropolynucleotide, in which the relative probability of the eight possible triplets is known, is again offered as messenger to the cell-free system, once again in twenty experiments, but this time another amino acid is radioactively marked. When the incorporation of the amino acids is complete (usually this state is reached in thirty minutes)

Table 3

Polynucleotide	Triplets	Relative Abundance of Each Triplet	Relative Proportions of Amino Acids
Poly-UG U:G = 5:1	UUU	100	phenylalanine 100
	UUG, UGU, GUU	20	cysteine 20, valine 20
	UGG, GUG, GGU	4	glycine 4, tryptophane 5
	GGG	0·8	
Poly-AC A:C = 5:1	AAA	100	lysine 100
	AAC, ACA, CAA	20	asparagine 30, threonine 23, glutamic acid 44
	ACC, CAC, CCA	4	proline 5
	CCC	0·8	
Poly-CG C:G = 5:1	CCC	100	proline 100
	CCG, CGC, GCC	20	alanine 22, arginine 19
	CGG, GCG, GGC	4	glycine 5
	GGG	0·8	

the radioactivity is measured, and the readings for the various amino acids are compared with each other. The reading for UUU (= phenylalanine), which we already know from our experiments with poly-U, is taken as 100. And behold, the relative figures of 20 for cysteine and valine, 4 for glycine, and 5 for tryptophane are just as the theory predicted. The artificial messenger has built certain amino acids into the substance, corresponding to the orders of the respective triplets, and it is therefore permissible to associate the triplets with their particular amino acids.

Admittedly the experiments just described lead to a conclusion only about what bases are assembled by the codon triplets, and not about the order in which they are arranged (cf. pp. 35–6). This problem can be approached step by step, however, and it is a question of time rather than of method before the code has been completely deciphered. The mystery of the 'breakdown' of the code, that is the fact that several triplets may stand for one and the same amino acid, whereas previously they have each stood for one, or at the most two bases, will soon be completely elucidated.

As unqualified people, we shall not meddle in the investigations that are in progress, the results of which are sometimes not yet completely confirmed. We shall again leave the laboratory, and turn to the question that we set aside on p. 59.

1.11 Is the Code Universal?

Do All Living Organisms Speak the Same Language?

Is the code, that is, the system of coding, really of such universal application as we supposed it to be? If this question is to be answered with absolute certainty, very many more species of animals, plants, and bacteria must be studied, more than hitherto were possible, on technical grounds alone. It is true that we do not need to know the operation of coding in *every* species of living thing, but we ought to have checked something close to a representative cross-section; we are still a long way from that however. On the other hand, all the organisms that have so far been tested code their messages according to the same principle and make use of the same code words; their t-RNA, m-RNA, and ribosomes, in a cell-free system, can apparently be exchanged for each other at will. Moreover, this is equally true for artificial messenger RNA. Poly-U regulates the incorporation of phenylalanine into cell-free systems that have been obtained from the following organisms: *Escherichia coli* (a colic bacillus from the flora of the intestine), *Salmonella typhimurium* (which produces mouse typhus), yeast, the unicellular green alga *Chlamydomonas,* sea urchins, the fruit fly *Drosophila,* rats, rabbits, and man.

Just as positive as the experiments made with the homopolymer poly-U are those with heteropolymers, either in bacteria or in mammals. Still more impressive, however, is an experiment with natural RNA. A t-RNA obtained from *Escherichia coli,* with radioactively marked leucine present (and thus a 'coli-* C-leucyl-t-RNA'), which in the normal way incorporates leucine into coli-protein, operated also in a cell-free system made from reticulocytes from rabbits. Such reticulocytes produce, in particular, rabbit hemoglobin. Ribosomes and m-RNA only are extracted from this, and the coli-t-RNA is added; the coli-anticodons 'fit' on to the rabbit codons, and the newly-made rabbit hemoglobin contains radioactively marked leucine. Even the code of the viruses, e.g., that of the tobacco-mosaic virus, which produces the so-called mosaic disease on tobacco plants, seems to be identical with that of

Escherichia

So long as no contradictory discovery is made, we can say the probability is very high that the code is universal. When talking about protein synthesis all living creatures speak the same language. In fact, we are on the track of a basic phenomenon of life, which is common to all living organisms. At least as important as its universality is the fact that it *takes place at molecular level,* that individual molecules of well-known substances direct important functions of living systems – we are landed right in the middle of molecular biology.

If the molecular groundwork of protein synthesis can be controlled in this way – if, because of the universal coding system, real and artificial code words can be interchanged at will – are not the publicists right if they express misgivings, if they are afraid that man will soon be able to manufacture every living protein, and eventually manufacture cells and fearful organisms?

It does not seem quite as bad as that, at least for the moment:

1. We cannot actually make even one protein. We can only take parts of natural cells, mix them together, and in this cell-free system *allow* protein synthesis to take place. The cell-free system is doing on one occasion only what the living cell has been doing for millions and millions of years. It cannot be denied that this process gives rise to a 'new' protein, e.g., polyphenylalanine, with which the cell probably could never have operated. Nor can the possibility be excluded that the 'new' protein might give rise to a very dangerous, possibly a very poisonous, albumin. So long, however, as we still cannot forecast in what way any one particular amino acid is going to attach itself to one particular place in a polypeptide chain, and so to give a particular characteristic to the finished protein, we must go on probing at random – so long shall we remain unsatisfied. It would take millions of years before all the 10^{130} possibilities could be investigated.

2. Up to now it has not been possible to introduce artificial m-RNA into an intact cell. The molecules are too big to pass through the cell walls. Admittedly, work is already going on at the present time with so-called 'naked protoplasts,' i.e., cell bodies from which the walls have been dissolved away with special enzymes. Such naked cells will accept substances with big molecules more readily. This development must be kept in view.

3. Even if we succeeded in getting m-RNA into the intact cell, and there making 'non-specific' protein, by far the greater part of this would very quickly be broken down again. Even so, the possibility cannot be excluded that one pair of molecules might be left over; this pair might possibly be capable of changing all the contents of the cell. The example of virus infections, which will be dealt with in the next chapter, shows only too well that this lies completely within the bounds of possibility, and that not all operations, speaking therapeutically, lead to healing. Care is thus imperative, but it is useless to be afraid. Nervousness arises mostly from lack of knowledge and uncertainty. So long as we do not know what heredity actually is, so long as we do not know if and how the hereditary structure can be altered, there will always remain the fear that one day artificial RNA will be used to make artificial animals and people. We should therefore be doing well to find out, in the next chapter, what knowledge has helped, and will help, to explain the molecular basis of genetics.

2. MODERN GENETICS

Why Do Children Always Differ?

Is heredity really the most important aspect of biology, as we are often told today? Is it more important than, for example, metabolism? This is a matter for argument, but the question of which is the more important is not really a scientific one. Just as metabolism is scarcely possible without a mechanism of heredity, so heredity cannot take place without metabolism. These two manifestations of life (like all the others, among which are growth, development, irritability, etc.) are mutually dependent on each other, and it is rather pointless to work out the percentage 'importance' to be attached to each. In this respect it is similar to the matter of nutrition: man cannot live *without* vitamins, but, equally, he cannot live on vitamins *alone*. Are vitamins less important than fats, carbohydrates, or proteins, just because they are present in minute amounts? All such disputes about precedence are irrelevant to the study of genetics, the science of heredity. From what has gone before, it will be firmly grasped that heredity, like metabolism, is tied up with living individuals, since it takes place only in living cells. This has the very obvious result that children resemble their parents. Hardly anyone is surprised when a grain of rye gives rise to a rye plant and not a stem of bananas, or that frog spawn does not develop into rabbits and mushroom spores do not turn into insects. This follows from such unchanging, and clearly ancient, laws, that it would seem a hopeless prospect to change the unchangeable, and to probe into its mystery. And yet the degree of resemblance between parents and children often suggests that inheritance plays a small part. Not only are children sometimes quite unlike their parents, but children of the same parent may show no more than a slight resemblance to each other. Visiting relatives want to find family similarities, and say that one child has her father's eyes, and the other has his nose – and in these respects there is a sort of family character. These differences are certainly the basis of Mendel's Laws, but yet they are by no means as straightforward as they are in the pure strain of pea plants with which the father of the science of heredity experimented about a century ago. From this point it is only a short step to a resigned and facetious definition: *Vererbung ist, wenn's mendelt*.

At best this is a half-truth (literally, as we shall see later, p. 71), and we do not need to make ourselves miserable about it just yet; but we shall not resign ourselves to it either. This much is obvious though: neither noses, nor chins, nor eyes can be inherited without the existence of a 'pattern' which causes the daughter organism to develop and to form the hereditary character in question.

We have the advantage of knowing that such patterns, called *genes*, are none other than the units of information that we learned about in the previous chapter, *units of genetic information*.

Hence the essential features of heredity are (1) transmission of information to the next generation, and (2) utilization of this information in developing the appropriate character.

The first question that arises is, By what mechanism is genetic information passed from the parent individual to the individual of the next generation?

We already know the material basis of information: it is the DNA, the 'original source' of the preceding chapter, in which the information is written down with the help of the code of four bases. DNA certainly accounts for an important part of the genetic information, but it must be pointed out that research is also going on into the possibility that yet another kind of genetic information may exist. Here at any rate we shall be concerned only with that information that is carried in DNA. The sum total of all the information units (genes) will be referred to as a *genome*, and inheritance by this means as *genomatic inheritance*.

2.01 On the Oldest Methods of Begetting Children

How does the individual offspring arise? Let us consider the oldest, most primitive organisms in the hope that they will also be the simplest. Undoubtedly these organisms are the ones that remain single celled throughout their life: bacteria, flagellates (single cells with a whip-like flagellum and a nature that is part plant, part animal), amoebae, siliceous algae (diatoms), ornamental algae (desmidiaceae), and many more. These have developed a very simple method of propagation: they divide their single cell into two halves, and each half grows into a new cell. Many single-celled yeasts follow a somewhat different path, i.e., they form little buds, which when they have reached a certain size, detach themselves from the mother cell, and continue to live and grow as independent daughter cells. Still, this comes to the same thing in the end: out of one cell, two come into being. Since the daughter cells are identical

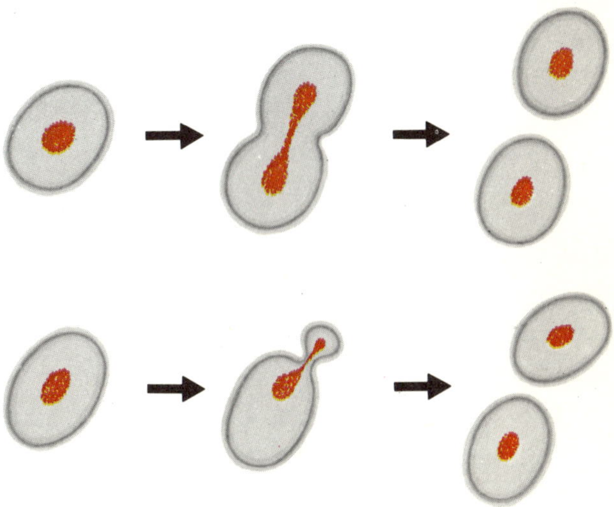

28. Two methods of reproduction by single cells: (top) constriction in the middle until the cell divides into two; (bottom) budding, the bud breaking off when it has reached a certain size—as occurs in many yeasts, including bakers' yeast.

with each other and with the mother cell, they must contain the same information, the same genomes. Hence the genomes of the mother cell must be doubled; one of these remains in the mother cell, the other becomes part of the daughter cell, either by splitting or by budding. *Each of the daughter cells receives and retains one genome.*

This is easy to understand. In the preceding chapter we have seen that DNA is present as a twin-stranded 'double molecule' in which the two strands are complementary to each other. By a process of separating the two halves (zip-fastener mechanism), and complementary completion of the missing 'halves,' two double strands come into being. Cell division must be achieved in such a way that each of the two halves acquires one of the two double strands – but apparently this presents no difficulty.

Light and Heavy DNA: Semiconservative Transmission

It can be shown experimentally how well this mechanism works. Normal DNA can be extracted from coli-bacteria (which we have met already), and can be centrifuged. This then lies in a particular place along the centrifuge tube, according to its weight. Then the coli-bacterium is allowed to multiply in a nutrient solution, which as a source of nitrogen, contains an ammonium salt of a special kind. In place of normal nitrogen, with an atomic weight of fourteen (N^{14}), we have the 'heavy' isotope N^{15}. This heavy nitrogen, repeatedly offered, is taken up by the reduplication of the DNA for many successive generations (the bacteria are able to live with N^{15}), and so eventually the entire DNA contains only heavy nitrogen. As a result, it is itself heavier than N^{14}-DNA, and so it lies farther along the centrifuge cylinder.

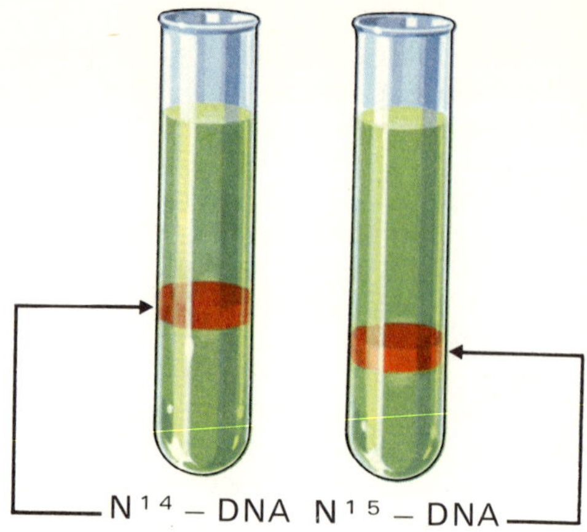

29. 'Light' DNA (containing N^{14}) and 'heavy' DNA (containing N^{15}) come to lie at different levels in the tube of the centrifuge.

The experiment proper begins when a culture of bacteria with heavy DNA is placed in a nutrient solution which still contains only normal, 'light' nitrogen. Any DNA that is synthesized from then on will be 'light' DNA. Samples are then taken at different times and examined for the weight of their DNA:

1. After about twenty minutes. At this time most of the cells have only divided once. The whole of the DNA lies in the centrifuge tube neither in the position for N^{14} nor that for N^{15}, but exactly in the middle: the DNA has become 'half heavy.'

2. After twice as long, i.e., when the daughter cells of the first generation have themselves divided. The DNA is no longer uniform. On the contrary, two fractions have appeared, a 'half-heavy' fraction, as in case 1, and in addition, a 'light' fraction – both fractions in equal parts.

3. After three times as long. The 'grand-daughter' cells from case 2 have divided for the first time. Here, too, we find 'half-heavy' and 'light' DNA side by side, but the quantity of latter is three times that of the former.

These apparently perplexing results immediately become clear if figure 30 is studied. There the strands of N^{15}-DNA are shown as thick lines; the N^{14} strands appear as thinner ones. The DNA double strands of the original culture (with N^{15}) have divided themselves; each separate strand has made for itself a complementary strand, with N^{14}, and combined with it to make a new double strand of $N^{15} + N^{14}$. After the next cell division each daughter cell contains a complete double strand (the complete genomatic information), *but this contains only one of the original strands; the second strand is new*. This type of transmission of information is called 'semiconservative.' It would be called 'conservative' if alongside the old double strand a new double strand were to be formed and if then one daughter cell received only original DNA and the other daughter cell only new DNA.

The semiconservative transmission guarantees that the two daughter cells shall receive the same genomatic information.

Photographing a Genome

Under the right conditions, the separating of the double helix into its two separate strands, the opening of the zip-fastener, can even be made visible, and once again in *Escherichia coli*. This time the DNA is marked not with heavy nitrogen, but with radioactive hydrogen (tritium). Tritium can easily be incorporated into thymine. If a particular strain of *E. coli* is used, which cannot produce thymine for itself, then this must be delivered ready-made, so that this radioactive thymidine can be incorporated directly into the DNA. (RNA has uracil in the place of thymine, and so cannot be marked by means of radioactive thymine!)

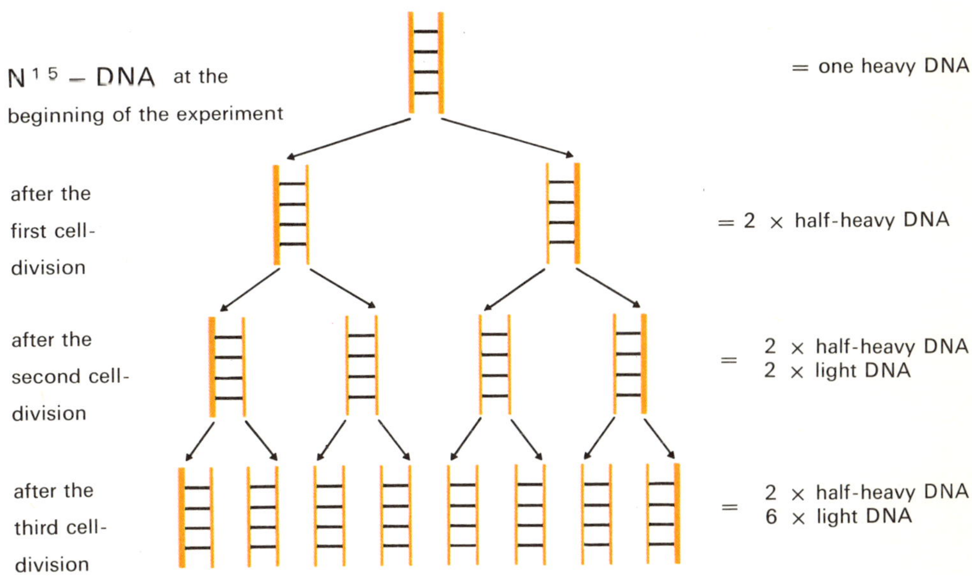

30. 'Semiconservative' transmission of heavy DNA to the first, second, and third generations of offspring. Whether a single strand of DNA is 'light' or 'heavy,' it is completed with a strand of 'light.' The resulting proportions of 'heavy,' 'half-heavy,' and 'light' DNA are shown on the right.

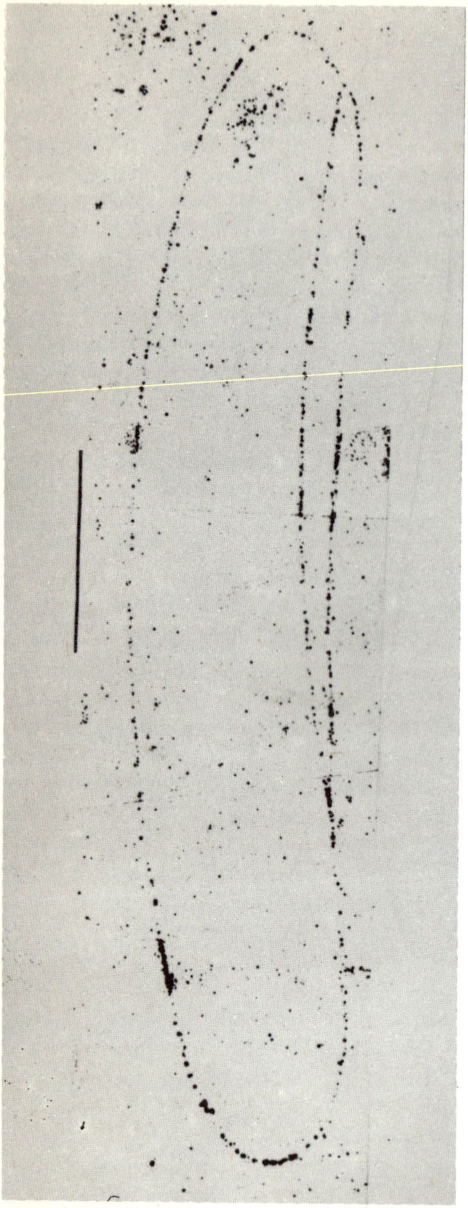

31. A genome of *Escherichia coli* spread out into a thin film and 'marked' with radioactive tritium. The genome is oval or circular, and on the right can be seen a length that is duplicated. The short line on the left of the genome represents a length of 100 microns.

By this method the newly formed DNA is radioactively marked.

Tritium sends out β-particles at a low energy level. Consequently the β rays do not travel very far, but they penetrate sufficiently to darken a photographic plate. If now the *coli* cells, in which the DNA is radioactively marked, are carefully broken up, it is possible to extract the DNA, to spread it out, and cover it with a photographic emulsion. If it is left covered long enough – say, several weeks – in the dark, and then developed, it shows a long chain with black spots. The emulsion is darkened only in the immediate vicinity of the DNA. In some examples it can quite clearly be seen how one part of the DNA is split into a loop, with the strands separated, each of which is exactly complementary to the other (shown clearly in fig. 31).

What Goes for the Protein Goes for the DNA
The DNA Helix

Thus the mechanism of splitting and doubling of the individual strands is apparently quite simple. The double strands were held together 'merely' by hydrogen bonds, and these are comparatively easy to break, since the chemical energy involved is small. Admittedly we are still a long way from knowing what gives the order to operate the zip-fastener and separate the two strands. This much is certain: the order must already have been formulated before the cell division or budding started.

If the double strands of DNA were merely like ladders, of which individual strands were the sides and the hydrogen bonds were the rungs, then in practice the separation of the two sides of the ladder would be very simple. However, the molecules of DNA have their peculiarities. Like proteins, which on top of their primary structure (sequence of amino acids) have at least one secondary structure superimposed, i.e., the helix, so in the DNA the long-drawn-out chain is coiled into a helix. This is not a simple spiral, but a double one: two molecules of DNA combined into a double strand are so to speak screwed into each other. Fig. 32 shows this schematically, while Fig. 33 shows a model of coloured balls. From this it is evident that the two individual strands cannot be separated without further consequences. In the first place, they must be untwisted, i.e., the double strand must rotate about its longitudinal axis. (It might happen to every coil in the strand, or only to every tenth coil. If it breaks off, it may come together again if the gap is closed. This simple picture is no longer accepted today.)

A twisting together of this kind, which must occur even in semiconservative transmission, is physically entirely possible. To understand the physiology and the microbiology of it, some unusual assumptions must be made. From biochemical data (the time needed for the duplication of the DNA, and the estimated mass of a DNA double helix) it has been calculated that the rotation about the longitudinal axis might take place at 10,000 r.p.m. We have no alternative but to accept this figure; for an explanation of such fantastic speeds we shall have to wait.

32. Two molecules of DNA are twisted together into a double helix.

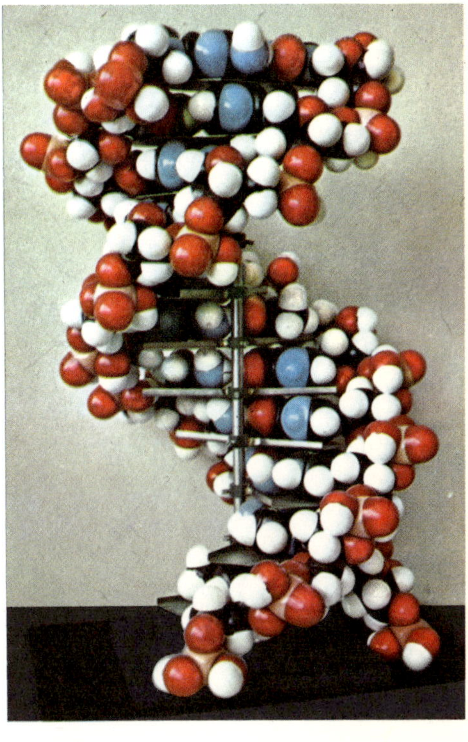

33. A model of such a double helix of DNA, comprising one complete rotation of the double spiral.

The problem of twisting together has become further complicated by a fact that has not previously been taken into consideration. During the preparation of the coli-DNA for photography it was said, reasonably enough, that it must be carefully extracted and *spread out*. So what we are seeing in the photograph is not the natural shape of the double strands of DNA when they are in the bacterium cell, but rather a flattened helix. If the length of the helix is measured, taking into consideration the magnification factor, we arrive at a figure of just under one millimeter (900 μ). One cell of a bacterium, however, is only 1 to 1·5 μ long, i.e., only one thousandth of a millimeter. The double helix, with its genetic information, must therefore be folded a great deal before it can be fitted into a bacterium cell. We do not know whether this is a regular folding, incorporating the tertiary structure of the protein, or whether the helix is bundled together at random. In either case the internal twisting and the reduplication of the DNA are not interfered with.

We shall come back to this question presently, when we turn our attention to the higher plants and the animals. Up to now we have been talking only about bacteria, and bacteria have no nuclei, and in the strict sense no chromosomes either (even though when using simple language we may talk about the chromosomes of bacteria). In reality, bacterial cells have only a single, long and convoluted double helix of DNA, which carries the complete set of genomes and thus contains all the genetic information. After cell division, each of the daughter cells possesses *exactly the same information*, the same double helix of DNA, with exactly the same series of triplets. The daughter cells are identical with each other and with their mother cell.

This type of propagation is called asexual, or vegetative; what takes place may be called 'heredity par excellence,' since it gives rise not merely to similar offspring, but to *identical* ones. We should call it 'inheritance by splitting of a double helix of DNA.' This expression is cumbersome, but correct, and will make it easier for us later on to compare with what happens in plants and animals.

2.02 Nuclei and Their Component Parts

An Advantage Possessed by Plants and Animals over More Lowly Organisms

On the time scale of evolution, plants and animals are undoubtedly not as old as bacteria, and appeared on earth at a later date. They are also more highly developed. From more or less coiled strands they eventually build up multicellular organisms, with characteristic outer form and inner specialization. This higher state of development is implicit in the individual cells. Whereas in the bacterium we can find at most one or two hanks of DNA in the cell body, i.e., inside the cell walls, in plants and animals it is normal to find a *nucleus*. This is surrounded by a nuclear membrane and embodies, as we already know, DNA in the form of one or more nuclear bodies (nucleoli), which in turn are completed with RNA.

Here, as in bacteria, the DNA carries the genetic information. If plant or animal cells divide, as they do very frequently, then they must have developed some mechanism by which genetic information can be passed on. Now, the DNA in the cells of the higher organisms does not lie in a single, tangled, double helix, but in several 'parcels,' which are known as chromosomes (from the Greek *chroma*, colour, and *soma*, body) because they are coloured with basic elements attached to the nucleic acids.

Admittedly these chromosomes cannot be made visible in a 'resting' cell, i.e., one that is not in the process of dividing. The colouring matter is distributed more or less uniformly throughout the nucleus. At most some trace of a nucleolar structure, or a kind of network, can be seen, and round this a nuclear covering, or nuclear membrane. We shall see later, in the chapter headed 'A Glance into the Cell', that this nuclear membrane actually belongs not to the nucleus, but to the surrounding substance, the so-called cytoplasm.

It is different when a cell begins to divide. The first step in this direction is *normally* a division of the nucleus (we can disregard the exceptions). This time it is by no means a simple halving of the contents of the nucleus, as if by a thread; this would not guarantee that the genetic information was passed on exactly to the two daughter cells. Instead of this, the contents of the nucleus thicken up by concentration and contraction to a fixed number of chromosomes, which is constant for each species. These now contain, collectively, the complete hereditary material, the DNA. If they are examined in detail, it will be seen that the chromosomes, which are usually elongated, are actually split longitudinally, and thus consist of two longitudinal halves – half chromosomes, or *chromatidia*.

Meanwhile the nucleus, or rather the nuclear space, has usually become drawn out in length. The two halves of each chromosome migrate to the two poles of the nucleus, the one half to the 'front,' or 'upper' end, the other half to the opposite pole. In this way the half chromosomes are collected together – naturally there are the same number of them as there were of the original complete chromosomes. They become surrounded by a new membrane, draw away from each other, lose their outlines, so that the entire contents of the nucleus again become uniform in colour; and now within the one cell there are two nuclei, and moreover two identical nuclei. The next step is the appearance of a new cell wall between these two nuclei. This divides the rest of the contents of the cell, in particular the cytoplasm, into two halves, and finally we have two identical cells, which carry in their two nuclei genetic information that is identical between the two of them and identical with that of the mother cell.

This process, which is of such decisive importance for the cell, is called *mitosis*. We must study it a little further, because here we have something similar to, though not identical with, the 'inheritance par excellence' that we saw in bacteria: an inheritance on the basis of mitosis. The analogous phrase for the bacteria was 'inheritance by splitting of a double helix of DNA.' The above explanation of mitosis has been simplified as far as possible, to make the principle clear. To understand it more fully, we must take it in a series of steps. These steps are naturally rather arbitrary, since the process of mitosis itself is a continuous one. However, the division into steps has proved a very practical one, and we shall take advantage of it.

Mitosis Analyzed

Prophase (opening phase): The contents of the nucleus can be divided into a strongly coloured and an uncoloured part. The coloured part evidently consists of granules or strands, randomly distributed, but presently with a progressive thickening; it can be seen that the granules are part of strands or bands that are twisted round each other. These are the chromosomes when they are not quite completely formed. *Even at this early stage the chromosomes are already longitudinally split into chromatidia.* Thus the doubling of the information units is already achieved.

Other things take place during prophase. The nuclear membrane disappears, 'dissolved' into the same area of the cytoplasm, from which it had originated. In addition, the nucleoli detach themselves and migrate to the opposite pole of the cell, to become the so-called polar bodies (in plant cells) or centrosomes (usually in animal cells). Fibrous strands grow out from them and join them to each other and to the chromatidia.

Metaphase (assembly): The elongate chromosomes that result from prophase become rolled like a screw (the process is usually called 'false spiralization'), and thereby become shorter and thicker. This is on the same principle as the rolling up of a protein molecule into a helix, but this time we are not operating at molecular level. Whereas the protein helix has an overall size of about 25Å, the size of a chromosome is about 500 times greater, i.e., more than 10,000Å $= 1\mu =$ one thousandth of a millimeter. When the rolling up is complete the chromosomes – and naturally the chromatidia too – are at their maximum shortening, usually they have become stout little rods; sometimes they are almost spherical. These rods are often constricted or notched, though the notch is not always placed in the middle. All the chromosomes now lie fairly precisely in one plane, halfway between the two poles and at right angles to the polar axis. This plane is known appropriately as the 'equatorial plane,' since it runs through the equator, so to speak, of a nuclear sphere – albeit a somewhat drawn out one – with its two poles.

Now threads grow out from the polar bodies, or centrosomes, and extend 'deliberately' towards the chromosomes. Since all these fibers together look like a spindle, they are known as *spindle fibers*. They are attached to the kink of the chromatidia (the commissure).

Anaphase (migration): Next the compact and mobile chromatidia migrate to the poles of the cell, one chromatidium to the upper pole, the other to the lower. To all appearances they are not themselves mobile during this stage, but seem to be drawn along by the spindle fibers, which shrink continuously. Possibly other, 'stem fibers' come into action, which lie between the two chromatidia of one chromosome, and slide them apart.

Telophase: The aim of the mitosis has now been achieved, the chromatidia have been brought to the poles. Now the prophase takes place in the opposite direction, so to speak. The chromatidia which are rolled up like a screw now extend and become unrolled; then their outlines become completely blurred. The nucleoli appear again, and the cytoplasm forms two new nuclear membranes. The result is two daughter cells, in which the chromatidia must now be referred to as chromosomes. During the *interphase*, between now and the next cell division, the new chromosomes must once again split longitudinally into two chromatidia.

Between the two daughter nuclei there still remain the remnants of the spindle apparatus, also called the *phragmoplast*, which is still recognizable for a time. In the equatorial plane there are blister-like objects, which join up into a kind of cellular disk. At first this is still more or less fluid, but it quickly solidifies and extends up to the 'old' cell walls. By this time the rest of the cell contents are divided into two, and from one original cell we have two new, completely identical fragments. Not only is the information precisely transmitted, but the continuity of living organisms is preserved.

The aims and objects of nuclear and cell division have now been set out – or at least

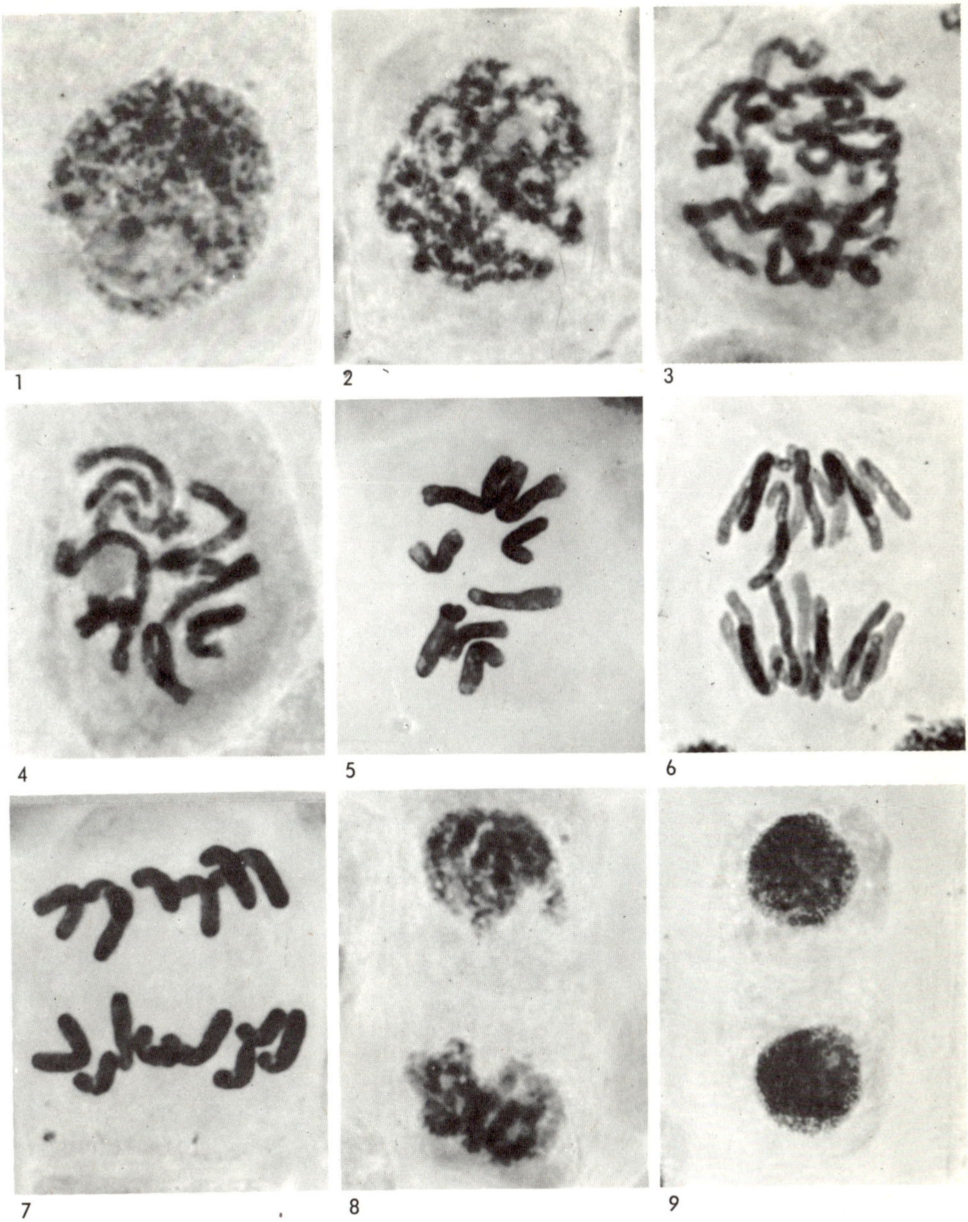

34. Photomicrographs of successive stages in cell division: 1 and 2, prophase; 3 and 4, metaphase; 5, the equatorial plate; 6 and 7, anaphase; 8 and 9, telophase.

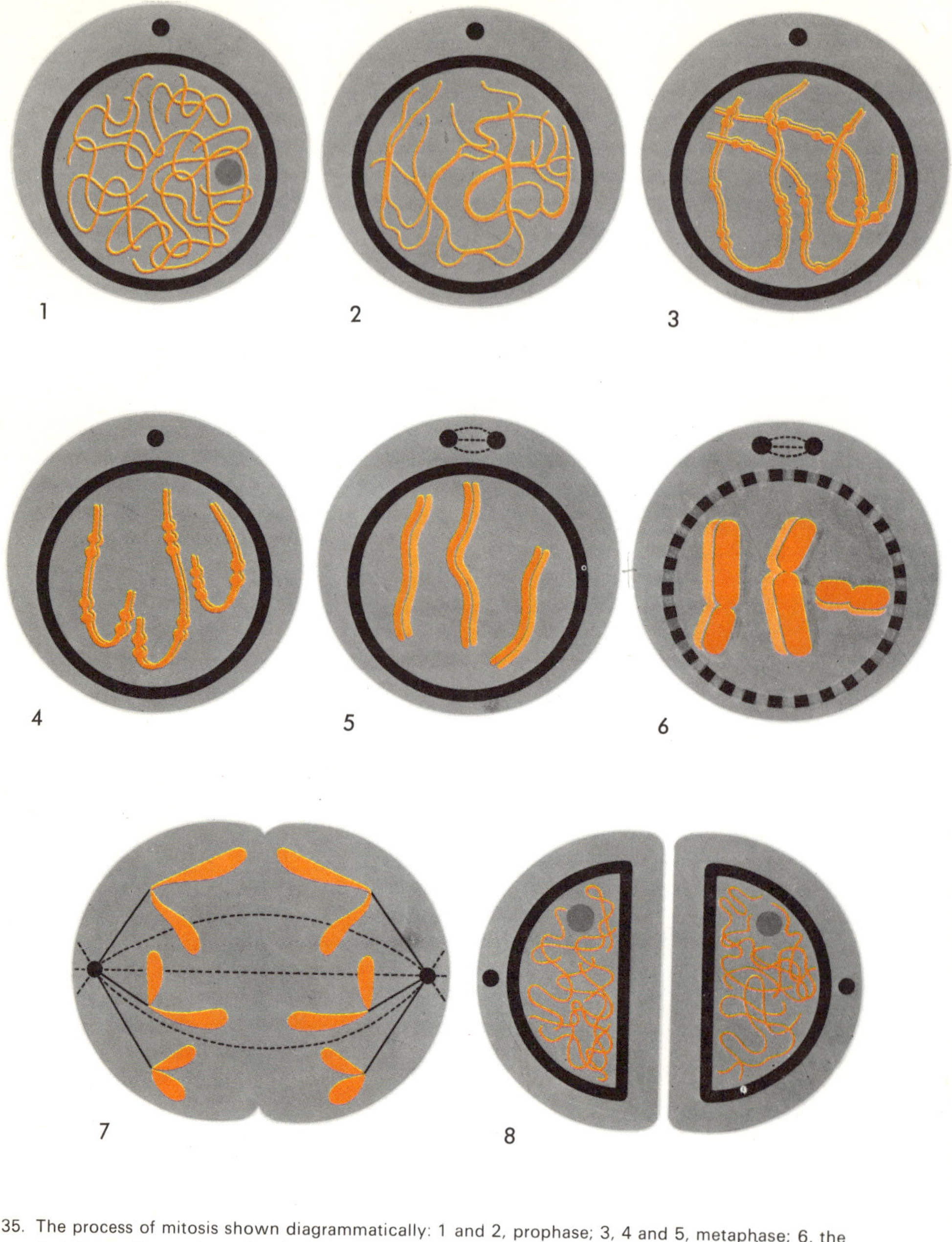

35. The process of mitosis shown diagrammatically: 1 and 2, prophase; 3, 4 and 5, metaphase; 6, the equatorial plate; 7, anaphase; 8, telophase.

some of them have. So we know fairly accurately what purpose mitosis serves, and we are aware that the mechanism described above is outstandingly effective, and indeed may be the only way possible. But we still do not know very well how this process originated, and where it came from. A description is only very rarely an explanation. A minute description of the sea battle of Salamis tells us nothing about how the Persian Wars, and that particular battle, came to take place.

The Starting Signal for Mitosis

It must not be forgotten that although mitosis can be followed even under the light microscope (its normal duration is about one and a half to two hours) the processes which start it, regulate it, and determine its direction are undoubtedly going on at molecular level, more or less as they do in protein synthesis. If one asks, What gives the signal for prophase to begin? – or, What orders the chromosomes to unroll and arrange themselves in the equatorial plane? What makes the draw-strings contract and the supporting fibers stretch out? What dissolves the nuclear membrane and the nucleoli, and then builds them up again? – then answers are still lacking in spite of the fact that mitosis has been known for many decades. Naturally there has been no shortage of attempts to find out. Many years ago it was thought that the answers had been found in the so-called nucleo-plasma relationship, which stimulated mitosis. Something like the following was believed to be the explanation. A 'young' cell, which has only just finished telophase, begins to grow, but the nucleus grows more slowly than the cytoplasm. Hence in terms of volume, weight, and surface area the relative balance between nucleus and cytoplasm is altered, to the disadvantage of the nucleus. It was supposed that when this process reached a certain limit the process of metabolic exchange, or exchange of chemical substances, between the nucleus and the plasma would be so greatly altered that the nucleus would be caused to start dividing. This sounds plausible, and is quite possibly even true, yet decisive experimental evidence is still lacking. Moreover, as long as we do not know which exchanges of which substances are retarded, which (different?) substances are released by mitosis, and how they are liberated, we have not clarified anything fundamental. It seems to be even more difficult to answer another question that has not concerned us up to now. There can be no doubt that the replication of identical genetic information goes on in nucleated organisms by the zip-fastener mechanism of DNA. How, then, is the double helix arranged in the chromosomes when they are five-hundred times as thick? Are there complete bundles of double molecules (of the same kind?) lying together, or are these molecules rolled up as individuals, or in parcels? If, as is probable, the last is the case, then how is the twisting of the chromosomes arranged in such a way that the entwining of the DNA double helix (something that we have already met as a problem on p. 71) can take place as easily as the longitudinal splitting of the compact chromosomes into two chromatidia?

This makes it all sound so remarkable that this particularly difficult question leads straight on to others, and eventually to the question of the fine structure of the chromo- in general. If we are to answer this question (even approximately), we shall obviously have to learn a lot of new facts, which for the present were better left for later consideration: they concern the possibility of exploring structure not only by optical means such as the light microscope and by the electron microscope, but also by genetic methods. For the time being, we can safely postpone these questions; though in the appropriate chapters we shall certainly return to them.

2.03 Asexual Propagation Has Its Advantages . . .

Cuttings and Tissue-cultures

Instead of this, let us consider once more the effect of mitosis. Anyone who has begun to understand the process, even if not very well, will recognize this as the separation of a double helix. Asexual inheritance, or propagation, is typical of the lower organisms. The most highly developed plants reproduce sexually, and so sexual reproduction could be regarded as more 'important' than asexual. However, this is only partly true. Certainly it is not without justification that higher plants have given up asexual reproduction; they are able to maintain their specific identity without it. However, just because the lower, and simpler organisms make such a great use of asexual propagation, and to some extent use it exclusively, and renounce sexuality, asexual propagation must be regarded as one of the oldest basic phenomena of life.

Moreover, asexual reproduction and inheritance are by no means so rare even among higher organisms. If a multicellular (diploid) organism grows out of a fertilized (diploid) egg cell, then this will undergo only mitotic division. Mitosis is thus one of the basic steps in overall growth – growth here in the sense of the multiplication and not the enlargement of cells.

But, after all, cannot a starfish regenerate a lost arm, and a lizard a lost tail? Cannot certain worms, when their bodies are cut into halves, develop each half into a complete worm? These powers of regeneration are particularly highly developed in plants, so much so that great use is made of this property in artificially breeding plants. The reader has probably taken cuttings from begonias or African violets, and raised these to complete, full-sized plants. For many plants, even, that are so inbred that they produce few offspring, or only sickly ones by sexual reproduction, cuttings are the only means of survival and inheritance. This is certainly an artificial, but nevertheless a very productive way of propagation. And what about potatoes? No farmer extracts the seed (we shall see later on why not); instead of seed, farmers lay out potato tubers, which are parts of the mother plant which have formed spontaneously (whereas other plants have bulbs). Common to all these offspring is the fact that their cells originate from cells of the mother plant, vegetatively, asexually, and that they resemble each other just as much as they resemble the mother plant, because they have exactly the same genetic information.

Such identical descent is called a *clone,* and clones are not just a figure of speech, or – as in the multiplication of cuttings – a useful trick. They are also very important both scientifically and in practice. If, for example, we want to study the influence of climatic factors, or the effect of manuring or pesticides, then we must have many plants or animals which are genetically completely alike. This we can never achieve by sexual propagation and multiplication.

The possibilities of vegetative – i.e., asexual – reproduction are made use of in yet another direction, which is mentioned here in anticipation. We mean the so-called tissue culture. In biology a tissue is a formation of many similar cells; there may be, for example, skin tissue or connective tissue or glandular tissue, and in plants, assimilation tissue, storage tissue, and many others (from which it will be seen that tissues are generally called after their functions, and not after the form or structure of the cells of which they are composed).

If a few cells are isolated from an organism – plant, animal, or human – carefully kept sterile, and brought into a fluid medium or on to a more or less solid nutritive medium, then the cells will divide mitotically and grow into a bulky complex of cells, i.e., into a tissue, assuming that there is a sufficient variety of nutritive substances present. To this extent we have a means to hand of finding out the nutritive requirements of different tissues. But tissue culture has become indispensable not only for such physiological problems, but also for problems of development and differentiation for the problems of cancer, and for the study of biological protective reactions (i.e., immunology). More will be said about this, too, in later chapters. It may be just mentioned here in order to indicate that asexual propagation is not quite so unimportant as it appeared at first glance.

The sceptic will pick on one point here: in all that we have said about inheritance above, we have said virtually nothing about mitosis. A study of heredity cannot be based on this. If mitosis is impeded, perhaps by poisoning, then the formation of clones will not take place; if mitosis is allowed to go on without interruption, then the offspring will resemble each other more even than one egg resembles another. (Hens' eggs come from different races of hens, as well as being products of sexual processes.) So the possibilities of experimentally influencing the process of mitosis, and hence of fully analyzing it are virtually nil.

And something further. Since mitosis (in the same way as the splitting of the double strand of DNA in bacteria) always produces nothing but completely identical offspring, all or almost all of the individuals of which are completely similar, *evolution is not possible*. The principle of mitosis is *archconservative*; never by this means could the highly differentiated organisms have developed from the lower, uncomplicated ones. Life would have stood still at the level of the bacteria, or the flagellates or the amoebae, and a genealogical history which finally led down to man could not, and need not, ever have been written. The groups of information units (genes) which must at some time have had an origin, would remain forever unchanging, if only . . .

2.04 . . . But Mutations Ensure That New Forms Arise

Mutations Are Molecular Phenomena

Mutations are means by which hereditary changes can be brought about. The word 'mutation' comes from the Latin *mutare,* to alter. Those offspring that are produced after a mutation are 'mutated,' and are called 'mutants.' They have become new organisms. This is not to say that they are better; merely that they are different. In this way the conservative principle can be circumvented.

Mutations are not all that common. It has been reckoned that even in one species, only one out of 100,000 to 10 million individuals mutates. The mutation rate is thus 1:10,000,000 to 1:100,000 – that is, $1:10^7$ to $1:10^5$, or 10^{-7} to 10^{-5}. This refers to the so-called spontaneous mutation rate – 'spontaneous' in this context meaning without recognizable external origin. Nevertheless, when it is remembered that a single spore of bacterium, on a suitable nutrient medium, may grow in twenty-four hours into a colony containing many millions of individuals, the likelihood of a mutation occurring is no longer so remote.

By artificial means, mutation rates can be quite considerably increased, for example one-hundred-fold. Nitric acid, for instance, is an exceptionally effective 'mutagen' (mutation-releasing substance). It is known to react directly with the bases of the nucleotides, which arrange the DNA information units into a sequence.

Thus adenine (A), by the splitting off of one amino group, changes into hypoxanthine (HX). This hypoxanthine can no longer pair with thymine (as did the original adenine), but pairs with cytosine. A triplet with the sequence TAC is changed to THXC, and in the zip-fastener mechanism this reproduces the complementary triplet ACG. In the next splitting of this double helix the triplet ACG pairs, as we know, with TGC. So from an original TAC we have arrived at TGC. TAC is the code for tyrosine; TGC that for cysteine; that is, in the place of tyrosine the protein now contains cysteine, and so we have a different enzyme. It may happen, of course, that the 'wrong' amino acid is attached to an 'unimportant' position on the polypeptide chain, and then the enzyme will not change in its behaviour. It is equally likely that an 'important' position may be hit upon; then the activity of the enzyme is degraded, sometimes reduced to nil – and in certain circumstances this may be fatal, if the enzyme is absolutely essential to life.

The nitric acid deaminates (deamination is the process of removing amino groups) guanine and cytosine as well; the fourth base of DNA, thymine, contains no amino groups and so is not affected. The practical effect on guanine is nil, because the resulting xanthine again forms a pair with cytosine. Cytosine when deaminated forms uracil, which pairs with A instead of G. In this way nitric acid can be used to make decisive changes in the sequence of bases and nucleotides as well as in amino acids in protein. Similar effects can be produced by treatment with hydroxylamine, which changes the cytosine of the DNA. This experiment has a further and very significant result: that in practice the alteration of a single molecule of cytosine is sufficient to lead to a mutation.

This should hold good for all the other mutagens as well.

If we want to form a correct estimate of mutations and their results, we must take two other things into consideration:

1. If mutation takes place in *only one* of the two complementary strands of DNA, then the reduplication of the normal half that follows will result in only half of the offspring being mutants.

2. The deamination of A to HX (for example) has not yet achieved a mutation, but only a *pre-mutation*. A mutation proper may appear for the first time as a 'manifestation' at the next reduplication. But this may not happen. The pre-mutation may be 'normalized' again, healed, or otherwise done away with. If cells are irradiated with ultra-violet light, this produces a particularly large number of pre-mutations (although admittedly many cells are also killed). Moreover the mutations then also affect thymine; so that where two molecules of thymine lie next to each other in a strand of DNA, they attach themselves mutually to a so-called *dimer* in a double molecule, resulting in a false pairing of bases. If, however, the cells that have been treated with ultra-violet are afterwards placed in visible light, then the pre-mutations are again restored to normal, and fewer mutations result. The death rate, too, is reduced.

Moreover, the reversal of the ultra-violet

effect is brought about by a special enzyme, which eliminates the dimers of thymine, and may properly be known as a reparational enzyme; we shall have more to say about these later.

The reader need not bother with all the many methods that are known by which mutations may be released. Only two of these are important, and these because their operation has been very well studied, and because they produce mutations at a molecular level. They are

(1) the so-called base-analogues, and
(2) the acridines.

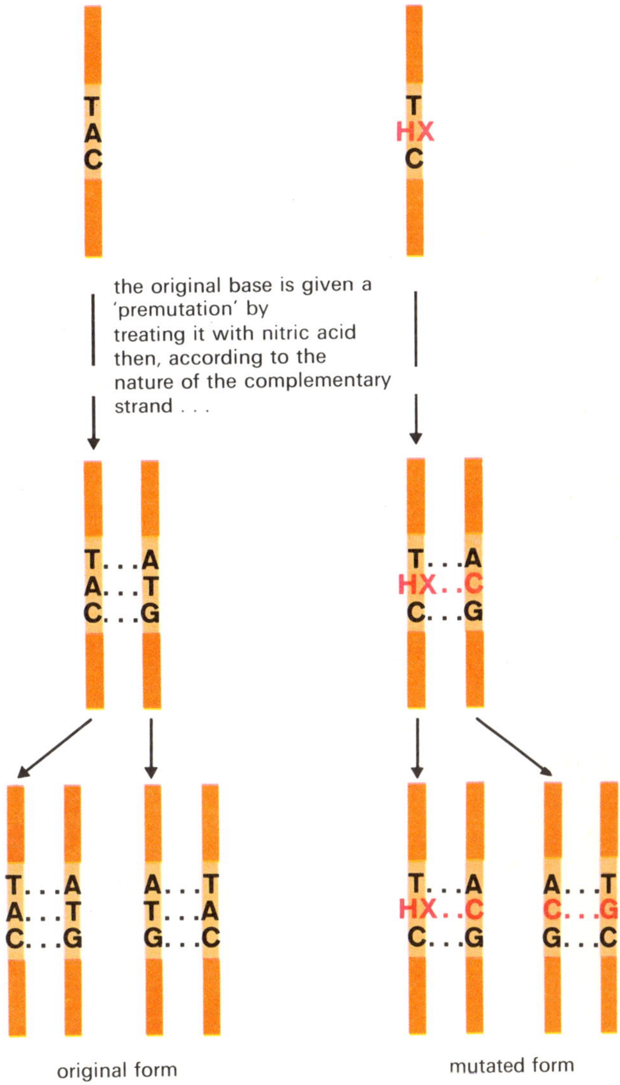

36. If a single base is artificially changed (e.g., adenine to hypoxanthine), this may lead to a different triplet (e.g., TAC becoming TGC), and may cause a 'wrong' amino acid to be incorporated at this point.

1. About base-analogues. These are bases which in the normal way are so similar to the bases that participate in the making of DNA that the cell and its enzymes confuse them, and incorporate them in place of the normal bases. They differ from the normal bases a little in the arrangement and stability of the hydrogen atoms, and in consequence, are inclined to fail to pair.

As an example we may take bromuracil (BU) which can be confused with thymine. Compared with the normal uracil, which forms part of RNA, it has an atom of bromine instead of a hydrogen atom attached to the fifth carbon atom. (Details are not important for our purposes.) As a result, the other hydrogen atoms are more loosely attached. For the most part this is of no consequence, since the normal BU (the so-called ketoform) makes a pairing in accordance with the sequence with adenine – exactly as if we were dealing with thymine. T → A has the same results as BU → A. Occasionally however the keto-BU changes into the 'enol form' (BU +), and then it forms a pair with guanine. The result: instead of T → A, BU^+ → G is replicated, and now G stands in place of A everywhere in the DNA – and this results in a completely different sequence of codons.

Acridines are substances all of which have a molecular structure based upon three benzene rings lying side by side. They differ from each other only in the side groups which are hung on to these rings. Among acridines are colouring materials such as acriflavine and acridine-orange, but also cell poisons and drugs such as trypoflavine, atebrine, and rivanol. Acridines react with phosphates including those in DNA, and they place themselves preferably between two phosphate groups, which are thereby pushed apart. The separation of these is increased from 3·4Å to 6·8Å, exactly double. Two things may happen:

Either the acriflavine settles on to a template strand of the DNA, which immediately begins to construct its complementary strand. At this moment the sequence of bases in the template strand has a gap, which leaves a clear place for a base. In the complementary strand this place is filled by an additional (or 'extra') base. Thus:

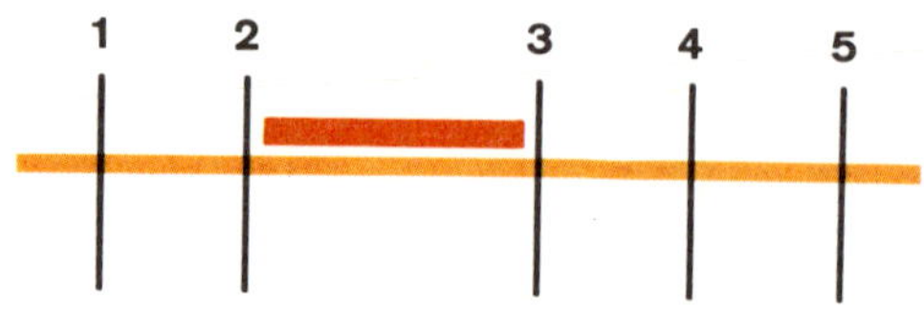

37. Above, a strand of DNA ('template') with acriflavine built into it. Below, the newly formed complementary strand, which has a gap opposite the acriflavin, into which an 'extra' base is fitted. This is the form that is subsequently replicated.

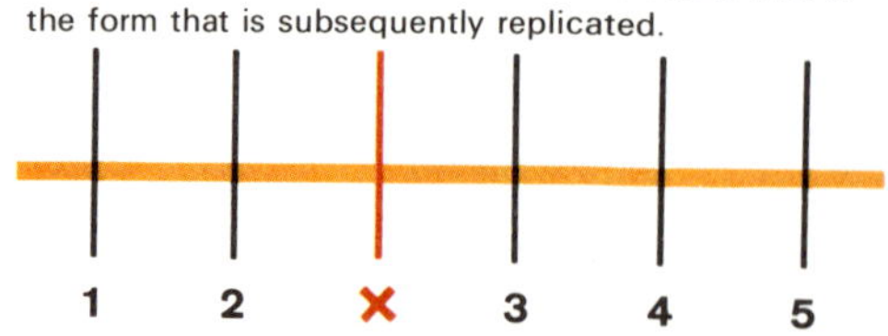

Or the acriflavine is not located in the template strand, but in the complementary strand, during the time this is being formed. Since it cannot now pair with any base it 'masks' the corresponding base of the template strand. In the subsequent replication (the second) its partner is therefore missing: a base has been lost. Thus:

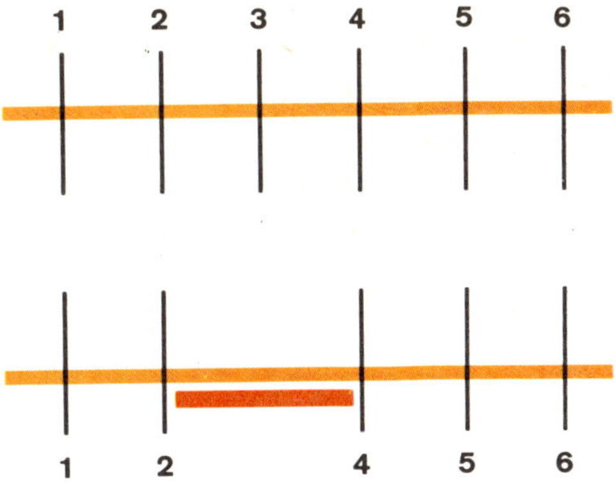

38. Above, the acriflavine is now not in the template strand, but in the complementary strand of the DNA. Below, the effect is that replication proceeds with base 3 omitted altogether.

In the second replication the following occurs:

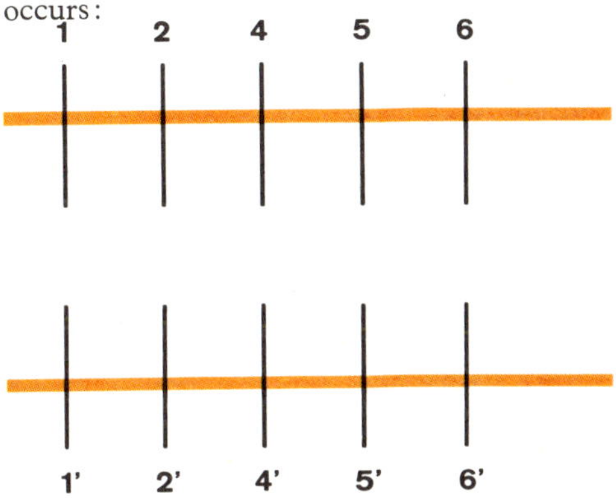

Either way the consequences are more serious than with bromuracil, or with irradiation with ultra-violet light, since whether there is an extra base or a position is lost, from then onwards all the triplets are misnumbered, and wrongly read.

To distinguish the bases of the triplets, and to emphasize the triple rhythm of their numbering, we will call them A, B, C. Then the undisturbed sequence is

ABC ABC ABC ABC

Case 1. If an extra base is formed, for example base *B*, then we have

AB*B* CAB CAB CAB

that is, after the triplet with the extra *B* we no longer read the triplets as ABC, but as CAB.

Case 2. If a base is lost, then we have

ABA BCA BCA BCA

Instead of ABC we now read BCA.

Both kinds of protein are completely undesirable for the cell.

These examples show that mutations can be understood as events taking place at molecular level. Often a very insignificant change in molecular structure leads to altered and even undesirable enzyme proteins. They may be fairly accurately called spot mutations.

All the same these can cause considerable disturbance. If higher cells, that is those of nucleated organisms, are irradiated very intensively with X-rays or with ultra-violet light, whole chromosomes may break up, or small pieces may be 'discharged,' which are then lost at the next mitosis. The breach may heal, or the pieces may become attached to faulty chromosomes (sometimes even 'hawked around'). In such cases – which are called *chromosome mutations* – the changes naturally extend over a much greater part of the DNA of a chromosome than in the case of a point mutation. But in the long run even here the changes are at molecular level.

Let us pause for a moment to think what we have learned about mutations. Mutations are molecular phenomena – good, then they can be investigated and explained, but why bother? The effects of experimental mutations are harmful ones, such as reducing the activity of enzymes, or even a complete breakdown – if not outright death, which is glossed over by calling it a *lethal mutation*. If the same is true of spontaneous mutations – as many indications suggest – then does this mean that any break of the 'conservative principle' can only lead to damage or failure?

If this is believed to be the case, and that reparation enzymes are now needed, then mutations appear as misfortunes, and they most probably ought to be avoided or done away with. Should we then speak generally of a *principle*, that the conservative pattern *ought to* be breached? Here, too, appearances can be deceptive. Certainly, an artificial mutation usually results in the loss of some attribute or characteristic. If we allow a bacterium – e.g., *Escherichia coli* – to mutate, then in the right circumstances, we obtain a form which can no longer ferment milk sugar (lactose). It has become lac-minus, or lac-negative whereas the original or 'wild' form was lac-positive. This is hardly an advantage for *E. coli.* The same is true of many characteristics, and many kinds of spontaneous mutation, but still we should not be surprised to hear that far-seeing and successful research workers are of the opinion that there really are *enzymes of mutation* – these must attack DNA directly and bring about mutations by this means. If mutations always lead to nothing but failures, then the possession of a tendency to mutate is a positive danger to life. Organisms that possessed such a suicidal enzyme would soon be extinct.

Not Everv Mutant Is a Failure

To take an example. We all know about antibiotics such as penicillin and streptomycin, and know that they kill bacteria. The normal wild forms of the bacterium, such as *E. coli,* are thus streptomycin sensitive, which can be written as S^s. If such a sensitive bacterium is treated with radiation, or with substances that induce mutation, and then implanted into a nutritive medium which contains streptomycin, no colonies at all of the bacterium will develop, since the streptomycin present has killed all the cells. If a very small residue of protein is left in the culture it will eventually grow into a colony. If a detailed investigation is made it is found that the cause is not an insufficiency of streptomycin round the centers of growth, but that the cells in these places *had a natural resistance to streptomycin* (S^r). This is not uncommon, and doctors and hospitals complain that as a result of unrestrained use of antibiotic remedies resistant strains are actually being 'cultivated' – in these cases the best thing is to change to another, different antibiotic. Naturally resistant strains are not cultivated deliberately, but they arise by mutation, and in the hospital itself, as a result of their reduced sensitivity to streptomycin, etc., they have a much higher chance of survival. Thus, while such mutations are most unpleasant for patient and doctor, they are advantageous for the bacterium.

Here we may disregard the fact that the origin of the resistance to streptomycin is rather more complicated than the mutations that have been described so far – in this context only the effect is interesting, and this is positive enough. How narrow is the division between advantage and disadvantage will be seen if the appearance of streptomycin resistance is rechristened the loss of streptomycin sensitivity, which is what it is often called in formal language.

Mutations can sometimes be advantageous, but only in the right set of environmental conditions. If there is no streptomycin present, streptomycin resistance is superfluous and may in certain circumstances even be a disadvantage. However, if streptomycin is present, then only the resistant cells survive, while the sensitive ones are killed. Hence it is to the advantage of the organism to override the strict conservative principle and to provoke mutations. No doubt some of these may be of no value, damaging, or even fatal – there are always a few advantageous ones among them. The original constitution is changed, new varieties have arisen, and new possibilities of improvement have come into being.

This being so, the existence of enzymes of mutation, even if they are only hypothetical ones at present, is both convenient and beneficial. And possibly the existence of the so-called 'hot-spots' is also understandable. Hot-spots are positions in a gene or in a double helix of DNA which are especially prone to mutation. Here mutations happen much more frequently than in other areas of the same gene or the same DNA molecule (see p. 101). If it were as dangerous as we have suggested to mutate or to have mutated, then these hot-spots would have been 'selected out' in the course of evolution, a long time ago.

Darwin, Mutants, and Selection

Selection is the key word under which the meaning of mutation can best be properly understood. An evolution along lines of descent, a phylogeny, is not consistent with the conservation principle, or with the idea of constant species, but only with the idea of selection of mutations, which conflicts with the conservation principle. Spontaneous or artificially induced mutations (where is the line between them?) form the material for higher evolution. The less viable mutants, those which are less well adapted, perhaps those in which the original structure was only inadequately formed, are crushed out of existence by the cumulative disadvantages brought upon them by the 'mutation spectrum.' They are selected out, while the better adapted forms survive them, and lead to progressively improved adaptation, to greater variety, and higher organization.

This is true for the bacteria, and it is also true for man. The geneticist Kaplan wrote in 1963, 'Man is a result of the appearance and selection of mutations over a period of many millions of years.' Today, more than a century since Darwin put it forward as a theory, the principle of selection of mutations has proved to be valid for every group of organisms. The only advantage we have over Darwin is a more exact knowledge of the basic phenomena, which we are more and more learning to control.

Variation and selection find detailed expression in *ontogenesis,* the individual life story of each separate living organism, especially of man. We all have experience of this in ourselves, in our children and in our pets. As soon as we meet with a problem, we explore every possibility to try to find the answer. If a mother puts a bowl of oranges on a high shelf out of reach of the children, naturally the children will try to reach them. An outstretched arm is too short – no good! They cannot be reached by jumping – still no good. Even reaching with a stick is not enough, because the bowl is too far from the edge. One of the children, a natural leader, notices that the orange on top is wobbly, so next time he tries to get that one. He has studied the different methods, varied them, and confidently chosen the one that looks most promising.

This commonplace example of sweet-toothed children holds good in all circumstances and for every situation that demands original thought. Variation, examination, selection of the most successful – these are the decisive steps in individual progress. Happy is he who has enough imagination to be able always to think of new variations!

2.05 Sexual Inheritance

Fertilization and Meiosis

Mutations play a decisive part in the production of organisms with new characteristics, but they are not effective by themselves. A mutant can pass on its new characteristic only through the operation of the conservative principle, i.e., by the separation of a double helix of DNA. Considering the low rate of spontaneous mutation, and the fact that it is just a matter of luck whether a mutation is good or bad, it is not unreasonably long, statistically speaking, until a cell that has mutated once mutates again; but it may be a *very long time* before a second *favourable* mutation comes along, that is until we have the making of two new characters together upon which selection can work.

Any device, therefore, that causes hereditary factors, genes, information units to mix and to form new combinations will be an advantage.

Mixing and Recombination In a few words, this is the discovery of sexuality, or the setting up of a sexual system (of which there are several).

Among the highest organisms, as well as in single cells, a sexual system works in two steps. The first step is to bring together the hereditary material from two different individuals, and the second step is to reduce this double quantity of material to the normal unit. The second is made necessary by the first, because the two identical contributions from the two individuals do not fuse together but lie alongside each other. If such 'doublets' then proceeded to regenerate by merging again, without reduction, the hereditary material would then become fourfold, and soon would build up to astronomical proportions. The reduction of the hereditary material can be achieved in different ways. Let us consider first the mechanism that is the most strictly organized.

This is the so-called *reduction division* of the older textbooks, which nowadays is called *meiosis*. The sexual system of which it forms a part is correspondingly called a meiotic system. It is significant that the system is named after the *second step*, not after the first, the union of the partners, the sex-cells, i.e., fertilization, which is what 'sexual' refers to in the everyday sense. A sexual process is only complete when it has fertilization *plus* meiosis, and from a phylogenetic point of view the meiosis is the more important step.

We must get to know this process in more precise detail, probably at a molecular level, and the outcome will be *sexual inheritance*. From this process arise, not identical offspring, as were produced by asexual reproduction (based on mitosis), but offspring that are slightly different, as a result of the recombination of characters that we have just mentioned, but which are still quite similar both to each other and their parents. Meiosis and meiotic systems do not exist among bacteria and other organisms without nuclei, in which the genome consists of a single double-helix of DNA, but only among organisms with chromosomes which are enclosed in a genuine nucleus. Fertilization, the first step in a meiotic system, usually consists of the merging of two different cells (the sex cells).

They locate each other:

1. By just wandering about.

2. Often by releasing an attractive substance; without which, encounters would be a matter of chance.

Such sex cells, or *gametes* (from the Greek *gametes*, spouse), may be exactly alike externally – that is, *isogametes*; nevertheless they usually have small differences in their hereditary material; hence we speak of *plus*-gametes (+ gametes) and *minus*-gametes (– gametes). In other groups of animals and plants the gametes are already different externally – larger, *macrogametes,* already marked down as female, and smaller, *microgametes,* recognizable as males. Most often, however, the cells of one sex have become very large and immobile: these are the egg cells, which are now with more justification referred to as the female cells, whereas the others remain small and are called male gametes, spermatozoa, or sperm cells. The majority of flowering plants, as well as many fungi, have gone further and abolished the power of either of the sex cells to move independently. In these organisms the egg cells are fertilized not by a mobile sperm cell, but by a (male) nucleus only. This makes little or no difference, however, since the practical result of a fertilization – at least as far as genomatic inheritance is concerned – is the fusing of two nuclei. The product, a fertilized egg or zygote, is the result of the fusion of its own nucleus with the male nucleus that has blundered into it.

The expression 'fuse' is actually misleading, since inheritance involves the chromosomes, the bearers of the hereditary factors, and these do not fuse with each other by any means. On the contrary, the male and female chromosomes remain clearly separated from each other inside the same nuclear membrane. Consequently the zygote has twice as many chromosomes as the unfertilized egg; it has two sets of chromosomes, and is *diploid,* whereas sperms and unfertilized egg cells, temporarily provided with only one set of chromosomes, were *haploid.*

2.06 Meiosis Is Not Pointless

The Old Is Newly Arranged

Many organisms are content with a diploid state in the zygote, which means that the first division, which splits this zygote, is a reduction division. In this way the haploid condition is restored. If, as we have previously assumed, the effect of meiosis consisted merely of the reduction of the number of chromosomes (whence it gets the old name of 'reduction division') then the whole process would be quite incomprehensible. Why go to all this trouble – formation of sex cells from special reproductive glands, elaborations of methods of fertilization, construction of a complicated mating chamber, and so on – if nothing more comes out of it than the bundling together of two sets of chromosomes more or less temporarily into one space and then their eventual reseparation? If this was everything that came out of the zygote, what would it be worth? Now the reduction effect of meiosis is obviously a necessary function, in the first place, to ensure that the haploid sex cells continue to be capable of being fertilized; but this is actually only a mechanical requirement. Two other effects are at least as important.

2. – rearrangement of the genomes
3. – reconstruction of the chromosomes.

How must meiosis and mitosis be alternated in order to achieve these three aims?

The secret lies in the fact that in meiosis it is not individual chromosomes that lie in the equatorial plane, but *pairs.* Everything that follows derives from this, and so if we want to understand correctly we must begin right at the beginning. We start with the fact that only diploid cells (never haploid cells) can divide by meiosis. They are called diploid cells just because they have two sets of chromosomes:

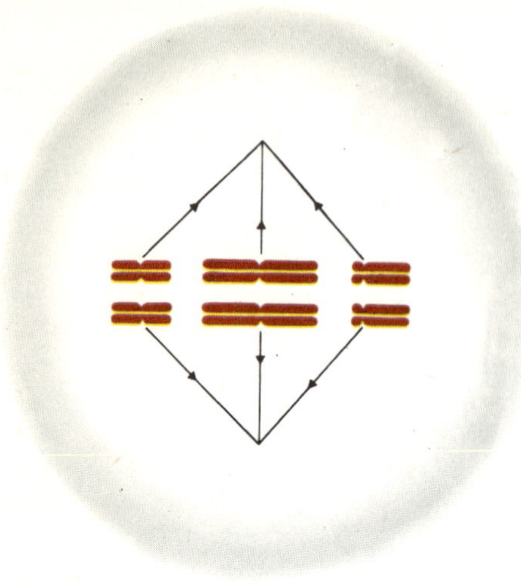

39. During meiosis the chromosomes do not lie singly in the equatorial plane, but in pairs.

for example a haploid number of three chromosomes becomes 2 × 3 or six in the diploid nucleus. In other words each chromosome is present in duplicate, one from the mother and the other from the father. Such chromosomes that belong together are called chromosome *homologues*.

The start of a meiosis is exactly similar to the prophase of a mitosis. The contents of the nucleus, which up to now have been uniformly coloured, begin to sort themselves out into chromosomes, which at first are very slender threads; for this reason this period of development is called the *leptotene stage* (from the Greek *leptos*, narrow). Immediately, however, they shorten and thicken, and at the same time become entangled together. In the stage that now follows, called the *zygotene* (from the Greek *zygos*, yoke), the first phenomenon that is specific to meiosis can be recognized, when the threads come together in pairs (conjugation). To begin with, only isolated pairs appear, but later, in the so-called *pachytene* (Greek *pachys*, thick) stage the threads are all in pairs. It will be less confusing if we refer to the threads as chromosomes – and it is quite correct to do so – though, as photomicrographs show, even in the pachytene stage of meiosis the chromosomes are never as short and thick as they are in mitosis; they do actually appear as 'threads.'

On detailed examination it will be seen that it is always only the homologous chromosomes that come together in a pair. Every time, the two are equally long, equally regularly or

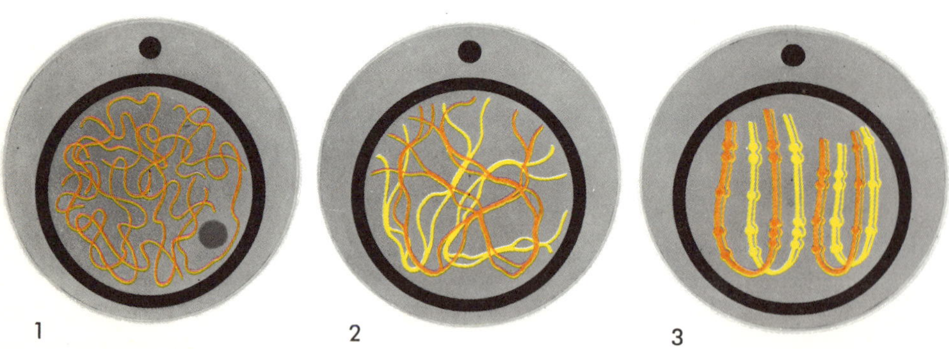

40. Diagrams of stages of meiosis: 1 and 2, leptotene; 3 and 4, zygotene; 5 and 6, diplotene; 7, anaphase; 8 and 9, the second cell division.

4

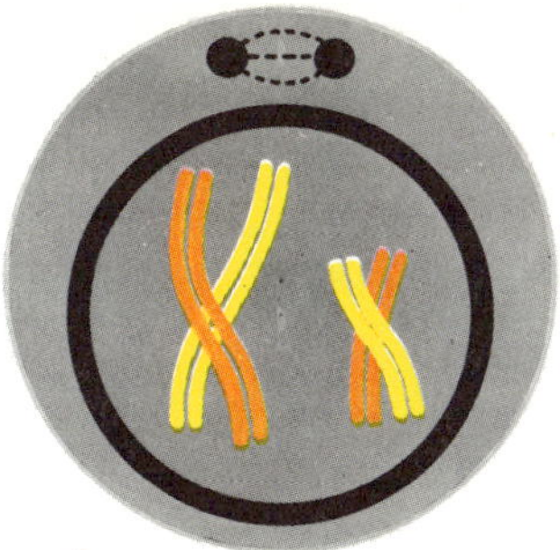

5

6

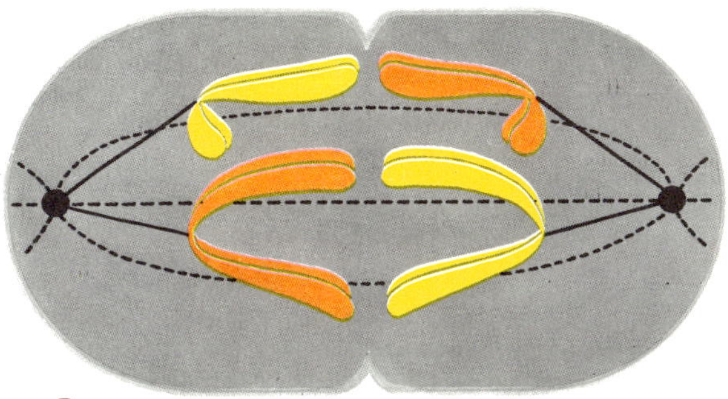

7

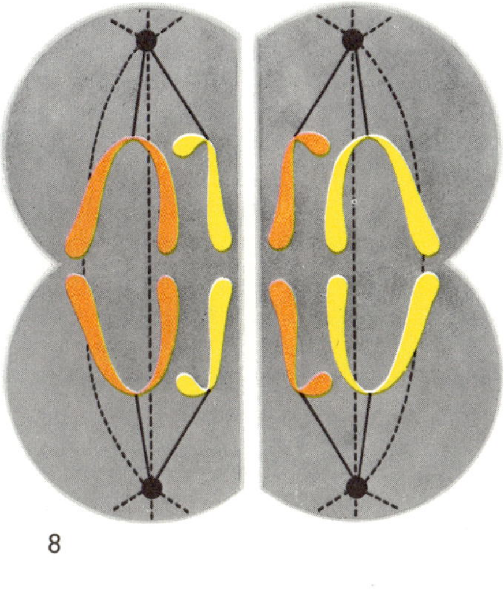

8

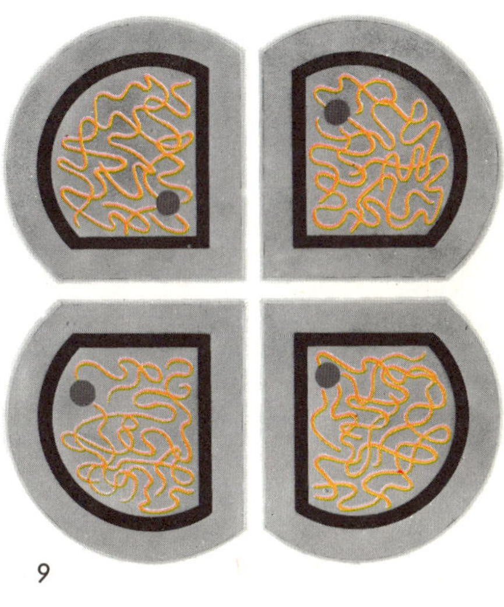

9

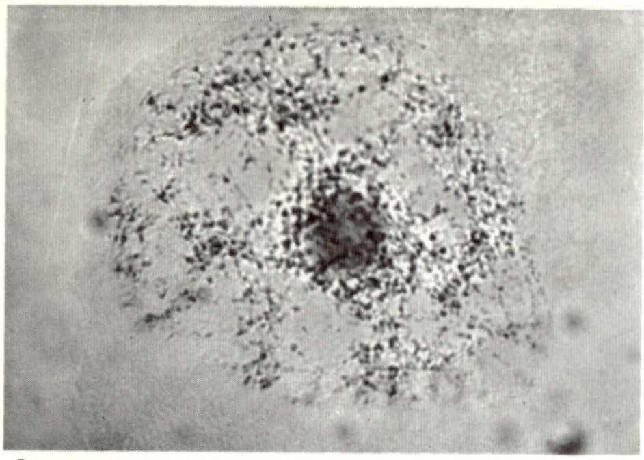

1

2

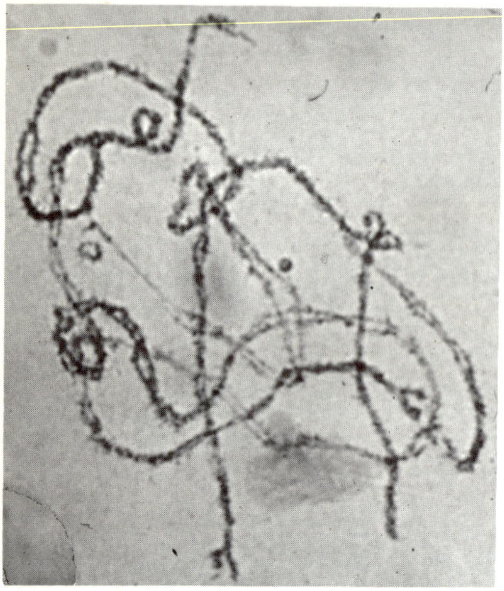

3

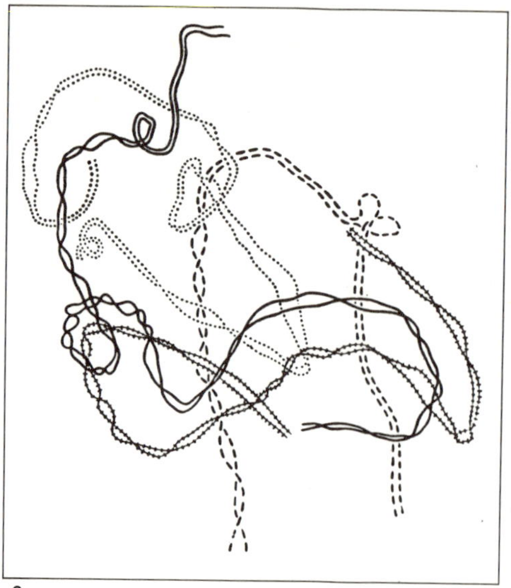

3 a

irregularly coloured. In short, every point that can be identified on one of the chromosomes can be matched by a similar point on its homologue. Thus in the pachytene stage instead of, say three chromosomes in the haploid stage, we have three chromosome pairs. In the *diplotene* stage (Greek *diploos*, doubled) which now follows, the conjugations are reversed, and the paired chromosomes move further apart again. At the same time they shorten still more. Now the nuclear membrane is dissolved, and the stage of *diakinesis* is reached (Greek *diakinein*, to pass through). The prophase of meiosis is ended, and is followed by metaphase and anaphase just as in mitosis. A telophase can now be easily distinguished, in which the chromosomes are dispersed, and immediately the chromosomes arrive at the poles a normal mitosis follows. The chromosomes split lengthwise into two chromatidia, and so we have four haploid nuclei, each with one set of

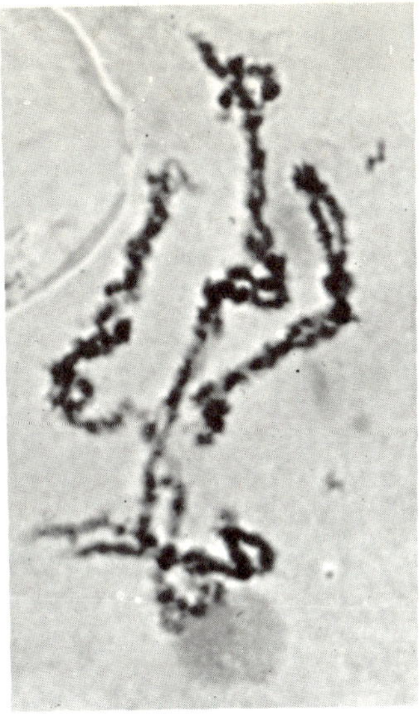

4

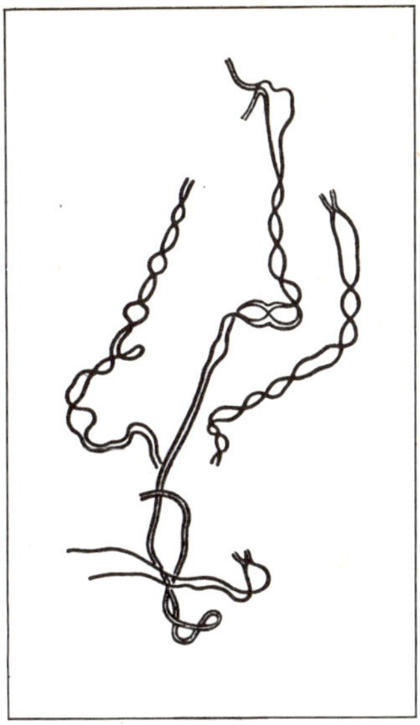

4a

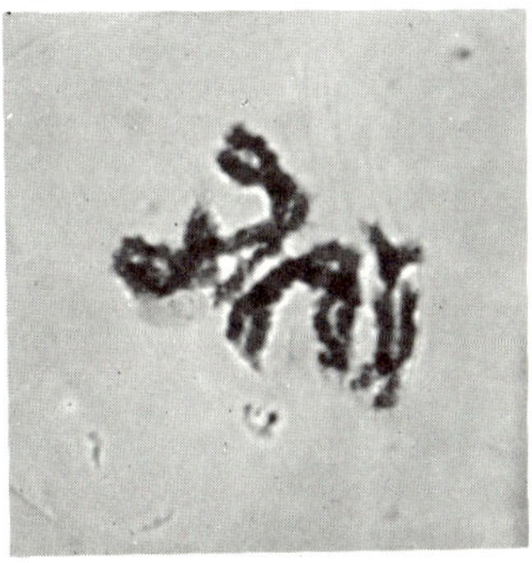

5

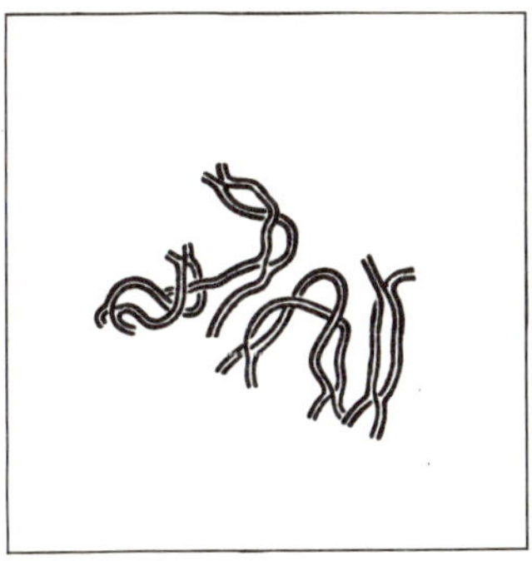

5a

41. Stages of meiosis: 1, the resting nucleus; 2, leptotene; 3, zygotene; 4, diplotene; 5, late diplotene; 6, diakinesis; 7, anaphase beginning; 8, telophase; 9, beginning of the second cell division (6–9 over page).

chromosomes. Then the nuclear membrane is re-formed, and the new cell walls, and meiosis ends with the existence of four cells. Individually these are called *gones,* and collectively they form a *tetrade* (Greek *tetra*, four); hence the expression *tetrade division* for meiosis.

This first, greatly simplified account of meiosis explains how a reduction in the number of chromosomes is brought about. What about the other two effects?

Meiosis – Effect Number Two
The Rearrangement of the Genomes

Let us first consider something that even today is still a complete mystery – that is, by what means the chromosome homologues always find their way to each other, and, we believe, bring about conjugation. In the almost hopeless confusion of chromosomes during the leptotene stage, the one thing that is most unlikely is that every pair of chromo-

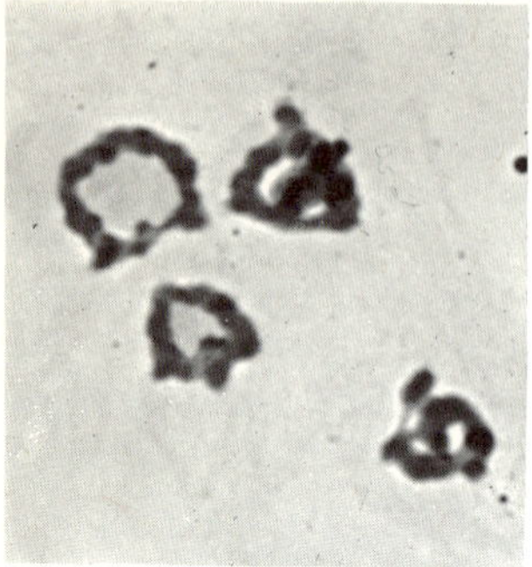
6

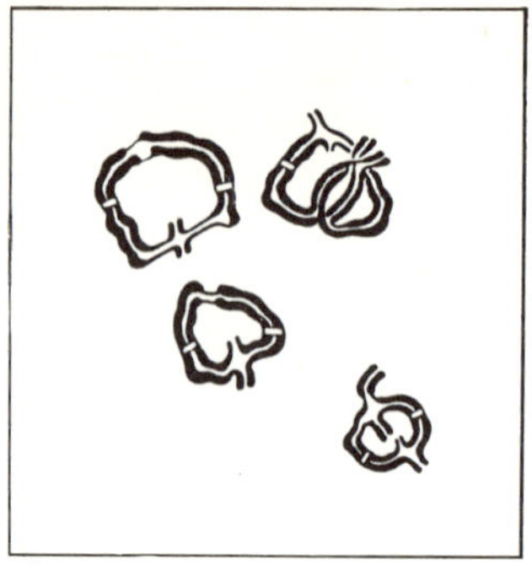
6 a

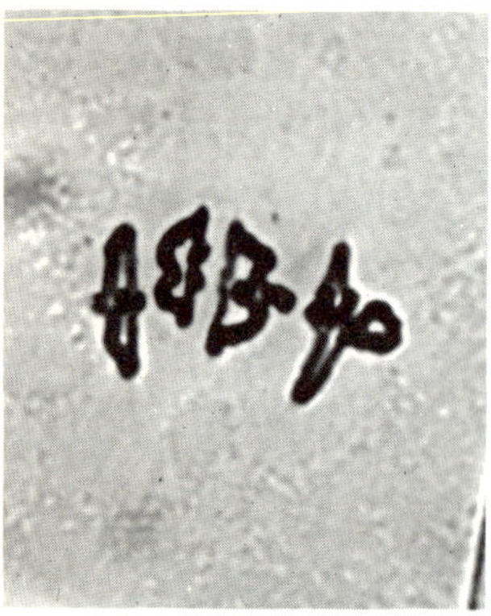
7

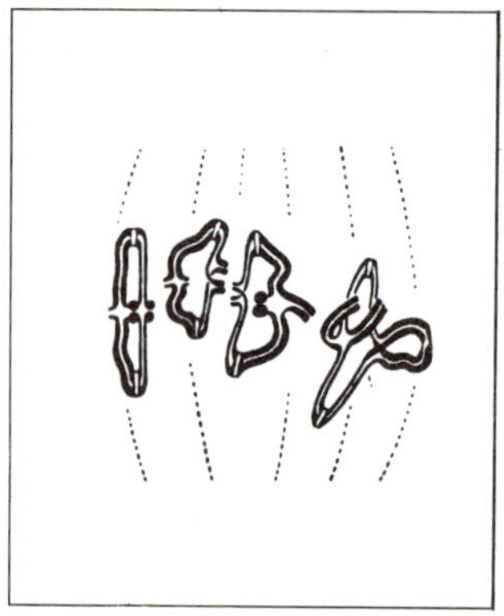
7 a

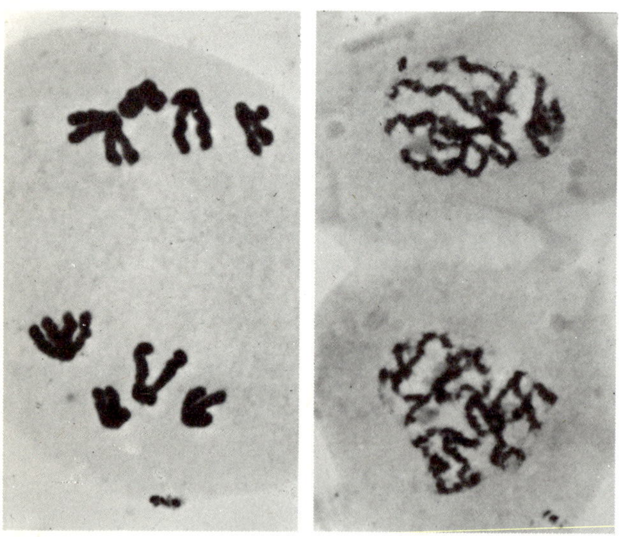
8

somes has arranged itself with all the paternal chromosomes on top and all the maternal chromosomes underneath; so that after diakinesis one daughter nucleus receives only maternal chromosomes and another daughter nucleus only paternal chromosomes. In practice the separation is a random one. Assume for the sake of simplicity that one daughter nucleus comes to lie above, and the other below, so that one nucleus receives the upper chromosome of each pair, the other the lower. *Which* chromosome, the maternal or the paternal, is uppermost, is determined by no regular rule; it is not ordained, but depends on the more or less random disposition of the partners at the start of meiosis. Consequently the chance that two adjacent pairs of chromosomes will be disposed in the same way is 50:50 with the likelihood that the two will be different.

The effect of this random division is that it is only on the rarest occasions that maternal and paternal genomes will be sorted out from each other exactly, so that the gones that result from meiosis will once again be identical with the sex cells from which the zygote was formed. It is much more likely that the gones will have some maternal chromosomes and some paternal ones.

Taking as example a cell with the haploid number of three chromosomes, two of the four gones (on the average) are likely to have two maternal chromosomes and one paternal one, while the other two are likely to have two paternal ones and one maternal.

Accordingly the chromosomes appear in a new combination, in which, however, the maternal and female chromosomes almost never have exactly the same gones – remember spot mutations, p. 83 – and now one nucleus may contain an assembly of genes which before fertilization were to be found in different nuclei (genomes). *Thus the information content of the cell has been altered.*

Of course it is not individual genes that are recombined in this way, but only complete chromosomes – that is, reproductive groups (see p. 94). Those genes that are distributed along a particular chromosome remain together. In spite of this, the genomes have been rearranged, and so requirement number two of meiosis has been achieved.

Meiosis – Effect Number Three
Reconstruction of the Chromosomes

It was pointed out above that the account of meiosis had been greatly simplified, but this is not good enough if we are to understand point number three. We must realize that the actual situation is more complicated. It will take us beyond the limits of vision, but this is not the first time; we have dealt with things at molecular level before.

Going back to a fact that we learned previously in connection with mitosis, during prophase the chromosomes are already split longitudinally into two chromatidia. This is also true of meiosis, and even already at leptotene. But this means that what form into pairs at zygotene are not two undivided chromosomes, one male and one female, but four chromatidia, two male and two female. A chromosome pair is really a tetrade of chromatidia, containing the sister chromatidia from two chromosomes.

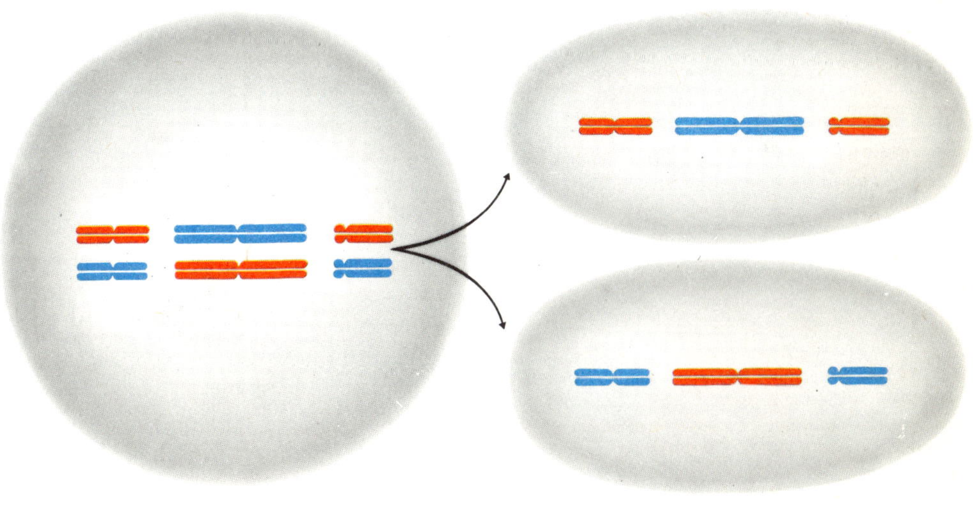

42. It is rare for a meiotic division to segregate the paternal genome into one cell and the maternal one into the other. Much more often the offspring contain some chromosomes from the mother and some from the father (shown here in different colours). The genomes become rearranged.

Now, this would be comparatively unimportant, and would not alter the final result of meiosis, if conjugation and separation (zygotene and diplotene) took place without any interaction between the two partners in conjugation. But this is not the case. Already at diplotene genuine *crossings over* can be observed. These are called *chiasmata* (singular *chiasma,* from the Greek word for *cross*), and actually it is the chromatidia that cross over. The diagram shows a chiasma schematically (fig. 43).

In this case, it looks as if breaks occurred in two places in two chromosomes that overlie each other, and the broken pieces were 'wrongly' joined together again. We shall see very shortly that it is not in the diplotene stage that the breaks first appear and then join together again, but that they were already present much earlier. Nevertheless the fact remains that bits of homologous chromosomes may be interchanged. After crossing over, chromatidia exist that are no longer entirely male or entirely female, but which combine sections which come from the male sex cell and parts from the female.

This has serious consequences for the offspring – that is, for the gones and the organisms which develop from the gones. The following account describes the case where just one single chiasma appears in a single pair of chromosomes. All the other chromosomes are assumed to remain unaffected, and the four gones are all quite different. Only the first (the letters in the diagram (fig. 43) may be ignored) bears the maternal chromosome unchanged, the fourth the unchanged paternal chromosome; but the second and third gones, in place of these, have chromosomes that have been cut up, and are now part maternal, part paternal. These second and third gones are not identical, but complementary – that is, the one carries so to speak the paternal bit that is missing from the other, and has given up a maternal bit in exchange.

43. Two chromosomes break and rejoin 'incorrectly,' with the formation of a chiasma or crossing over.

Let us take as an example of this type a fungus, because it is easy to rear and the generations follow each other quickly. The one that comes to mind is the mildew fungus *Neurospora crassa,* nearly related to *Neurospora sitophila,* the orange coloured bread mould that establishes itself in bakeries. This *Neurospora crassa* is abundant and is very successful when used for studies of heredity. The normal form, or wild type, may be thymine positive (Th+, i.e., it can by itself manufacture thymine) and streptomycin sensitive (S^s) – this is an artificial example, but one which has advantages since it allows us to make use of mutants that are already known to us. From this we have made, perhaps with the aid of nitric acid, a double mutant. In the first place this has become thymine negative (Th−). Suppose the gene for this lies at the left end of our chromosome. Secondly it has also become streptomycin resistant (S^r); let the gene for this lie at the right end of our chromosome.

Let us now cross the double mutant with the original form, and time and again the only forms without a chiasma are the original forms, namely the wild type and the double mutant. Since the mutated genes all lie on the same chromosome they always remain together. They are concentrated into a so-called reproductive group, or, if we want to express it more simply, they are tied together. (Normally a reproductive group is the same

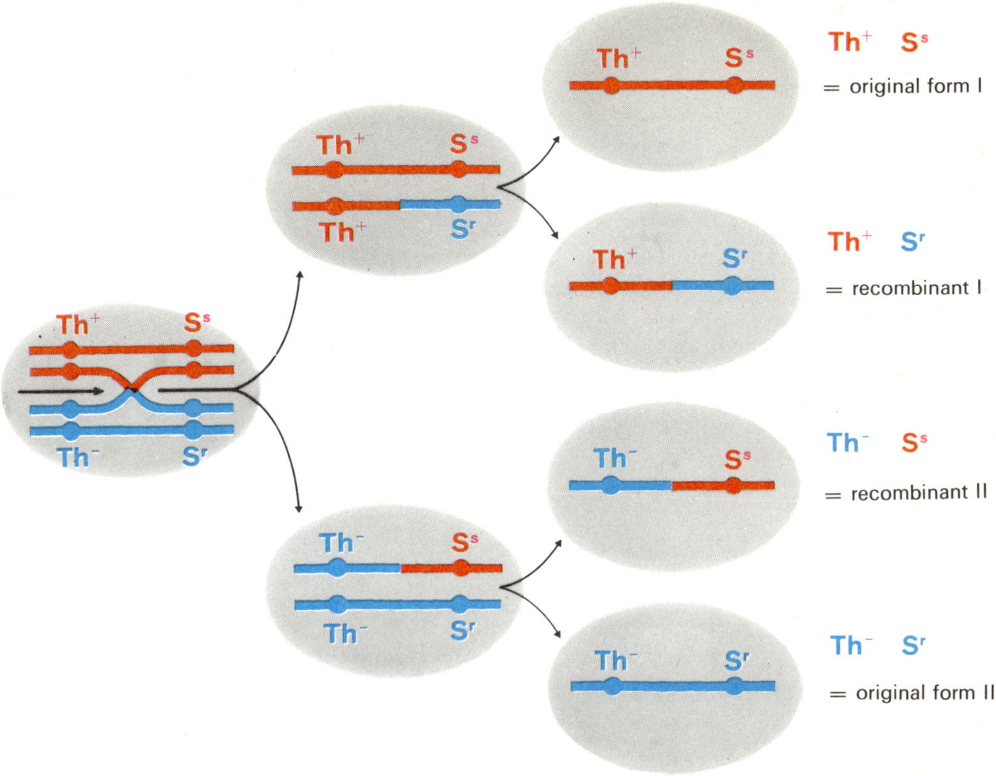

44. Even one crossing over is enough to cause reconstruction of chromosomes, and hence recombination of hereditary factors.

thing as a chromosome.) Naturally the same goes for the unmutated genes of the wild type.

In other words, if a chiasma appears between the sites of the genes, among the offspring will be found:

1. The wild type (Th+, S^s).
2. The double mutant (Th−, S^r).
3. A type (Th−, S^s) that is thymine-negative, and at the same time streptomycin sensitive.
4. A type (Th+, S^r) that is independent of thymine and at the same time streptomycin resistant, and as a result is distinctly better than the streptomycin-sensitive wild type.

As a result of the reconstruction of the chromosomes, the chiasma produces a new combination of characters, genes, or information units, known to the geneticists as a *recombination*. It may properly be said that one result of a chiasma is a recombination. If we note that a chiasma 'seldom comes alone,' that in one chromosome several or even many chiasmata may appear, and that every chromosome (with the exception of the special sex chromosomes) is likely to be affected in this way, then it will be clear that the meiotic system of reproduction greatly increases both variability and the number of possibilities for selection. Phylogenetic evolution can proceed much more quickly by this means than by mutation and mitosis alone.

Let us look at it from another point of view. A chiasma leads to a breakdown of the coupling of genes. For more than thirty years it has been known by experiment, which we do not need to discuss here, that the genes in a reproductive group lie in a line, rather like pearls in a necklace. The probability that a chiasma will appear between two particular genes is the greater the farther these are away from each other. The percentage chance of recombination is equal to the percentage chance of being involved in an interchange, and this is proportional to the distance apart of the genes. The smaller the separation, the less the chance of interchange. So these percentages can be used as an index of the (relative) separation of the genes, and since, moreover, these are additive, and are measured from definite locations, a regular chromosome map can be constructed, showing the genes in their relative positions.

Calculations must, of course, be made with extreme care. Thus if only two chiasmata arise close together, then only the section of chromatidia that lies between them will be exchanged; the free ends stay unchanged.

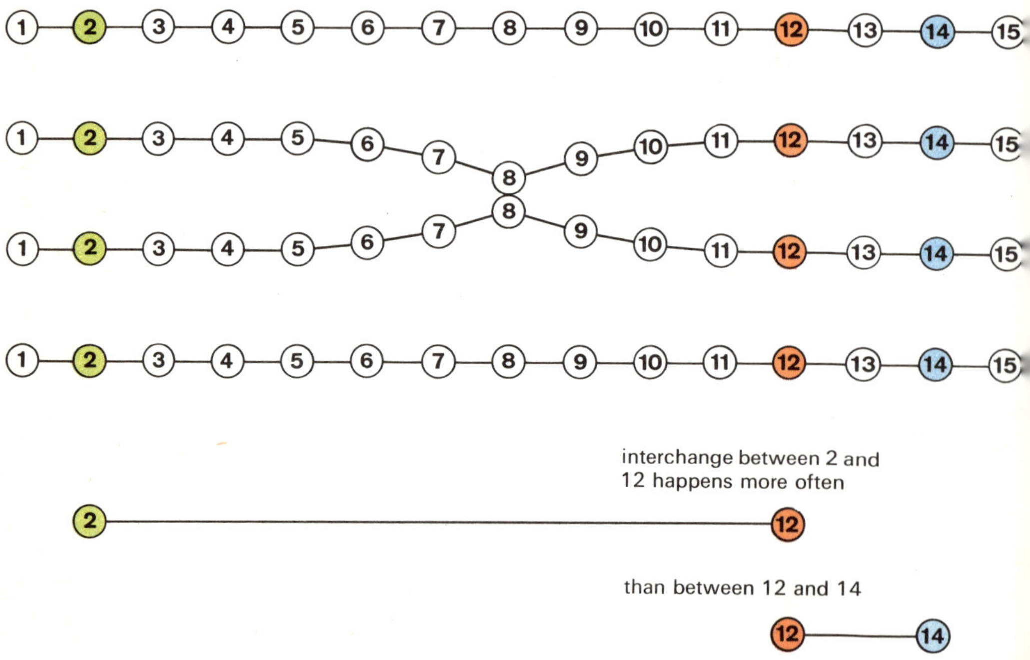

45. The relative probability of interchange of genes when they are close together and when they are far apart.

46. Chromosome maps for the fruit fly *Drosophila melanogaster*. The numbers indicate the relative distances apart of the genes. The nomenclature for the genes is internationally agreed upon.

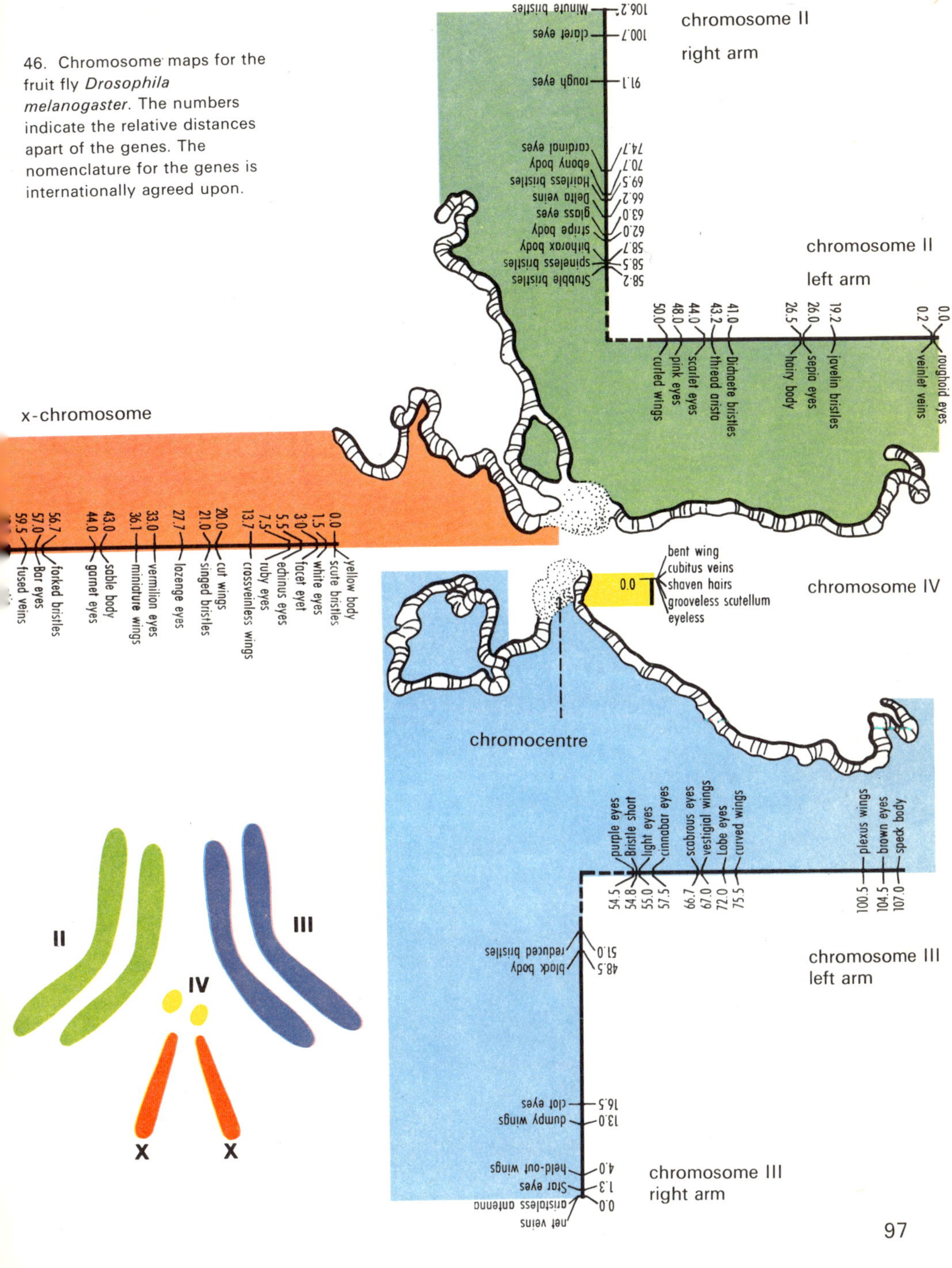

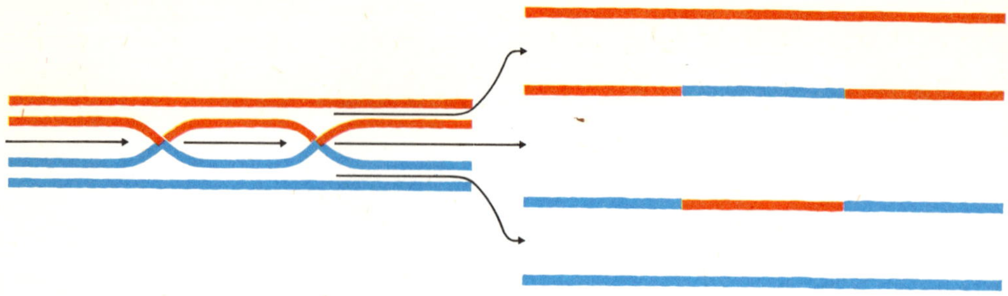

47. 'Back' crosses' occur when there is more than one crossing over.

2.07 How does a chiasma come about?
By Duplication, Breaking Off, or 'Copy-Choice'?

Uneasiness may arise at this point for another reason. Textbooks of heredity talk about the formation and breaking up of coupling groups, or combinations of genes, and explain the latter by the mechanism of 'crossing over.' They say nothing about chiasmata. What has a chiasma to do with crossing over?

Now, judging by the names, chiasma and crossing over should mean the same thing. In practice however, a chiasma designates the shape of a cross that can be seen under the microscope, whereas crossing over refers to the interchange of parts of the chromosomes, which needs to be demonstrated in the offspring as a recombination of genes or of hereditary characters. It is entirely possible, and indeed very likely, that crossing over has been finished for a long time before the mechanism of meiosis is set in motion – the latter has only to provide for an orderly separation of the chromatidia (segregation) and for passing them on to the daughter cells. But then a chiasma is the *result* of a crossing over that has happened previously. The crossing over itself is not visible under the microscope, because the four chromatidia lie too close to each other (though the phenomenon of recombination can be identified in the next generation). When they are pushed apart, however, the 'sister' chromatidia tend to cling together a little longer than those that are not 'sisters,' and this is what leads to visible chiasmata.

If that is so, how does crossing over come about, and what is its *molecular basis?* At the moment we have no conclusive answer to this, but rather, two possibilities to discuss.

1. It happens during zygotene, if 'reduplication' also occurs and the four-strand state (the chromatidium tetrade) is present. Then two breaks must occur simultaneously on two of the four chromatidia, and of course, in homologous positions; a double break, so to speak. This results in a 'cross over' adhesion. Now, it is very unlikely that two breaks would occur in homologous positions at the same moment – so it may be that this is the work of an enzyme, of a specific 'recombinase.' This is not quite such an airy suggestion, since we know, for example in the fruit fly *Drosophila*, that crossing over has been observed only in the female, during the formation of the egg cells, and never in the male. This suggests a sex-linked enzyme.

2. The difficulty described above, i.e., the improbability of things happening just so,

was first circumvented by the theory of 'copy-choice' (or template-choice). This is what it means: crossing-over does not come about *after* reduplication, in the completed tetrade of chromatidia, but during the process of reduplication. Let us go back to the zip-fastener mechanism, by means of which a double strand of DNA (*not* of chromatidia!) gives rise first to two single strands and then to two double strands. If then on one of the two single strands the complementary strand is built a little more quickly, and in addition this jumps across ('chooses the other template'), then the second complementary strand will find its template already occupied, and will be forced to switch over to the other template where it still remains perfectly free (fig. 49).

Illuminating though this model is – and it can be understood only on the molecular level, i.e., within the double helix of DNA, not in terms of chromosomes and chromatidia – it also has its faults. For example, it explains only the crossing over between two chromatidia, but not the observed fact that sometimes all four chromatidia recombine. In this case we must postulate a second double-break of the same kind. Further difficulties for the copy-choice model arise out of the semiconservative reduplication, but we may excuse ourselves from pursuing these. In these circumstances we must accept

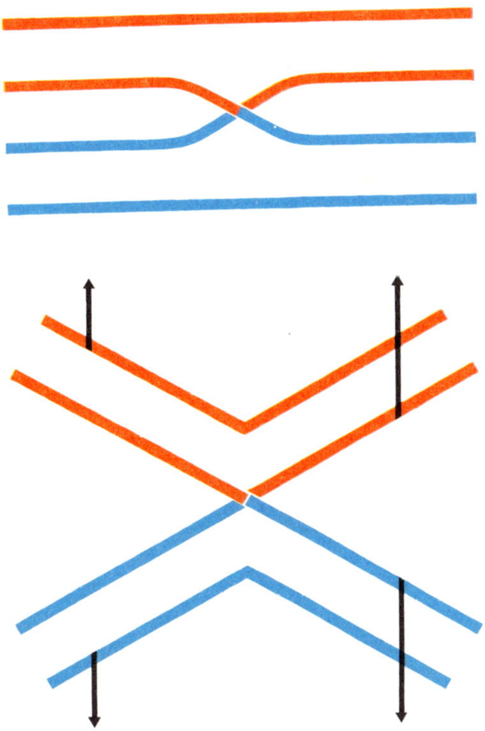

48. Above, exchange of pieces resulting from crossing over. Below, sister chromatidia stay close together longer than those that are not sisters, and this gives rise to the crossed figure, which can actually be seen under the microscope.

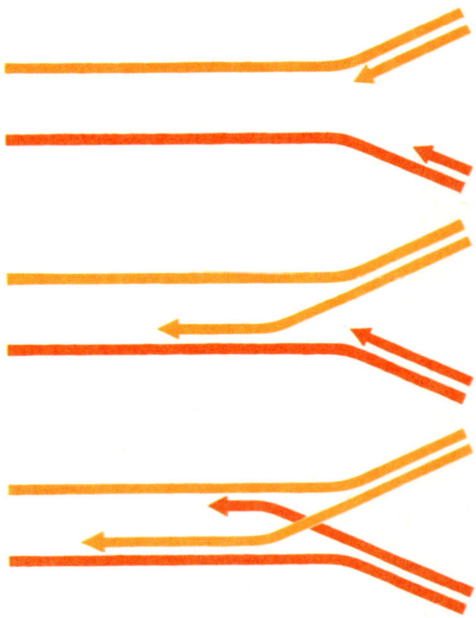

49. The copy-choice mechanism: the complementary strand of the yellow chromosome (also shown in yellow) grows more quickly than the complementary strand of the other (red) chromosome. If it also 'springs' across and forms up with the red chromosome instead of its own yellow one, then the red complementary will find that it has no place alongside its own partner, and will perforce have to pair with the yellow strand.

the fact that a satisfactory explanation of crossing over on a molecular plane has not yet been found.

Experiments in Recombination – Are They 'Better' than the Electron Microscope?

Since an approach at molecular level has not brought a satisfactory explanation of crossing over, we are likely to get better results if we take up once again the question of percentage crossing over (or percentage recombination). As a rule the recombination percentage between two complete genes is 50 or less. Fifty per cent is only another way of writing a Mendelian segregation of 1 : 1. It means that these two genes are free to combine with each other, and that is always the case if the two genes lie on different chromosomes. The probability that these two chromosomes will be together after meiosis is therefore 50 per cent (cf. figures 42–44). If, however, the genes lie on the *same* chromosome the recombination percentage is zero, as long as the coupling is not broken. All values between zero and 50 per cent are a measure of the frequency of breaks in coupling, of the frequency of recombination, of the relative separation of the two genes. But figures of less than 0·02 per cent have been recorded. If we use the average length of the chromosome as a basis for calculating the *absolute* distance apart of the genes (we need not follow this rather complicated calculation here), then we obtain distances apart of only a few ångstrom units, sometimes even only one Å. But this is another way of saying that the genetic method enables us to separate 'spots' on the chromosomes, i.e., on the DNA, which are only a few ångstrom units apart. So the analysis of recombinations leads straight back to molecular dimensions!

If a gene, to give an example, includes at least 150 codons = 450 nucleotides then it will be more than 1,500Å long. Thus recombinations are possible *within the gene itself*.

It is still too early to define what a gene is, but we can now say one thing: a gene is not a recombination unit, because recombinations can take place inside a gene. No more is the gene a unit of mutation, since even a part of a gene, for example a single base, may mutate.

We should do well, therefore, to go on, as long as we can, thinking of the gene as a functional unit, and we shall be in the best company if we do so! To conclude the chapter on recombination, let us make a chart of the genes. This is to anticipate our subject a little, since in this chart a gene is represented by a bacteriophage genome (and in fact one of the r_{ii} group), and the bacteriophages first come into the picture on p. 107. But this does not matter. We know that the gene is subdivided, and even know the location of the spontaneous mutations, which have arisen independently, and which may recombine. Each spontaneous mutation is symbolized on the chart by a small square, and there are only a few spots which cannot mutate at all. In contrast other spots mutate repeatedly.

One of these is especially surprising, because hundreds of independent mutations – so many that in the diagram they must be shown by means of a funnel-shaped indicator – have taken place here. This is obviously none other than one of the 'hot-spots' that we learned about (fig. 50).

2.08 Diplonts Are to Be Preferred

So far we have considered only the comparatively simple case in which fertilization and meiosis follow immediately after one another, and where the first division of the zygote is a reduction division. Since this results in four haploid gones, such organisms may justifiably be called *haplonts*. They pass

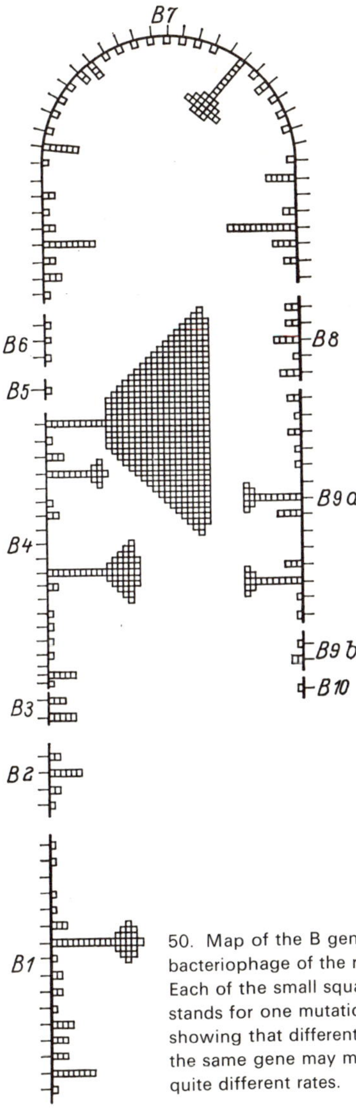

50. Map of the B gene of a bacteriophage of the rII group. Each of the small squares stands for one mutation, showing that different parts of the same gene may mutate at quite different rates.

almost their entire life with a single set of chromosomes, i.e., in a haploid condition. This is the case, for example, in many algae and fungi.

However this is by no means the general rule. Many plants, almost all animals, and naturally man too, are *diplonts*: the fertilized ovum divides mitotically, and so naturally delivers only diploid cells, with a double set of chromosomes. First of all the effect of meiosis on the structure of the sex cells is counteracted; only the gametes, i.e., egg cell and sperm, are haploid, and the product of fusion of the two is already diploid. So here we have reversed meiosis and fertilization alternating. It is no accident that it is just the most highly developed groups of organisms that are diplonts. Diploidy brings real advantages. It is assumed, of course, that organism is a heterozygote – that is, that the two genomes are not completely alike. Now, in the diploid cell each chromosome is double, and in the homologous chromosomes every gene is therefore doubled. If we call this assembly of genes an *allele,* then we have one maternal allele and one paternal one. Both alleles are involved in the construction of the complete character, and to different extents. Suppose we assume that in the fruit fly *Drosophila* one allele is specific for white eyes, and the other for red eyes. The complete fly, however has red eyes: the allele for red eyes has completely overwhelmed the allele for white eyes, and so is *dominant,* while the other is *recessive*. In the plant *Mirabilis jalopa* it is different. Here there is an allele for white blossom and an allele for red blossom. When crossed, these two alleles merge together, and give the plant rosy blossom. The character that is developed is intermediate between white and red; the two alleles restrain each other mutually.

As most readers learned at school, dominant and recessive genes are extreme cases, between which all intermediate stages are possible. Moreover, the *degree of dominance* of an allele is by no means constant, but can vary until the dominance is completely lost; in its place an intermediate state may appear, or the previously dominant allele may even becomes recessive. In this context additional factors come into play that are internal, but also environmental. Of these, it may be said that a diploid organism has at its disposal reserve information, which can be utilized in certain circumstances.

Moreover, a diplont may *accumulate* recessive alleles, up to a quantity that would be fatal to a haplont. Occasionally, recombination of these recessive alleles may give rise to advantageous new types – as we have already seen in the matter of streptomycin resistance.

Diploidy seems to be so 'desirable' to many haplonts that they have provided themselves with a substitute for it. This is particularly

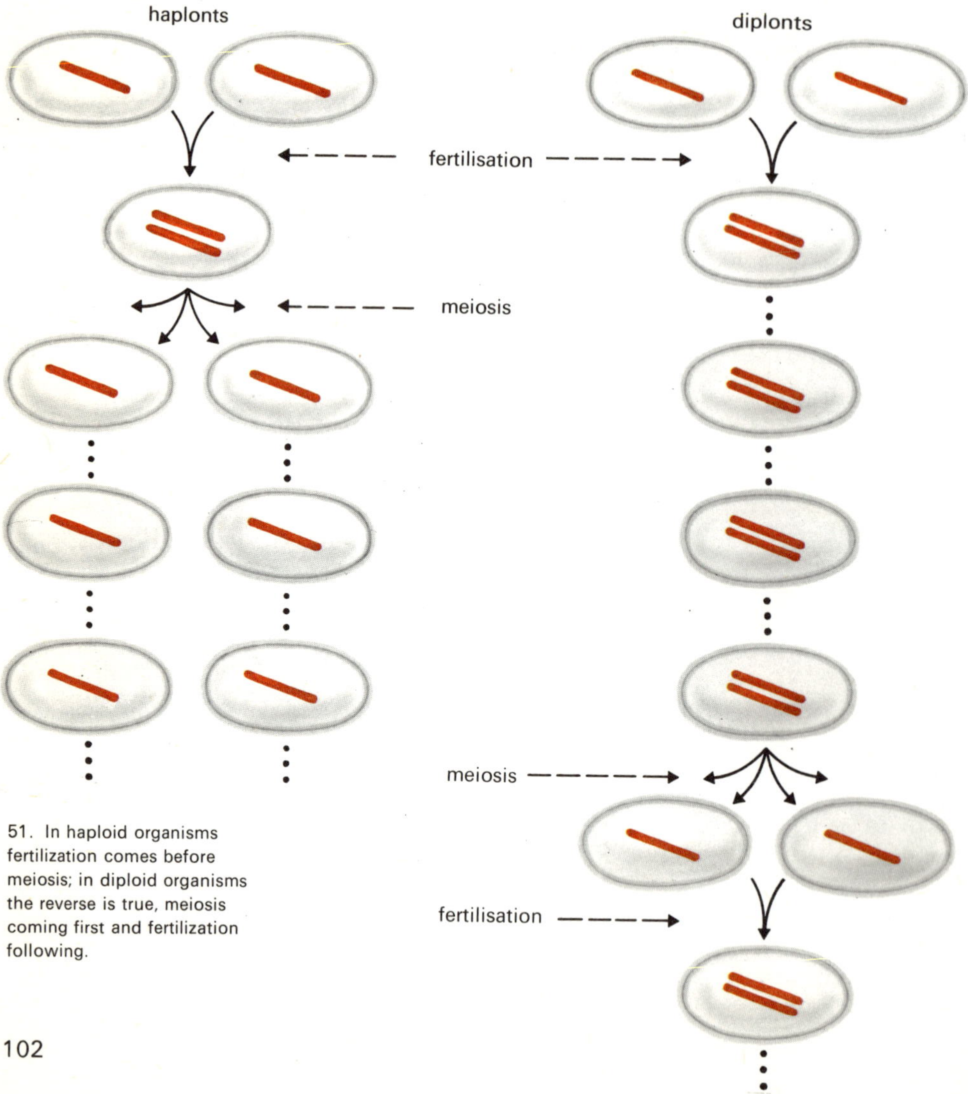

51. In haploid organisms fertilization comes before meiosis; in diploid organisms the reverse is true, meiosis coming first and fertilization following.

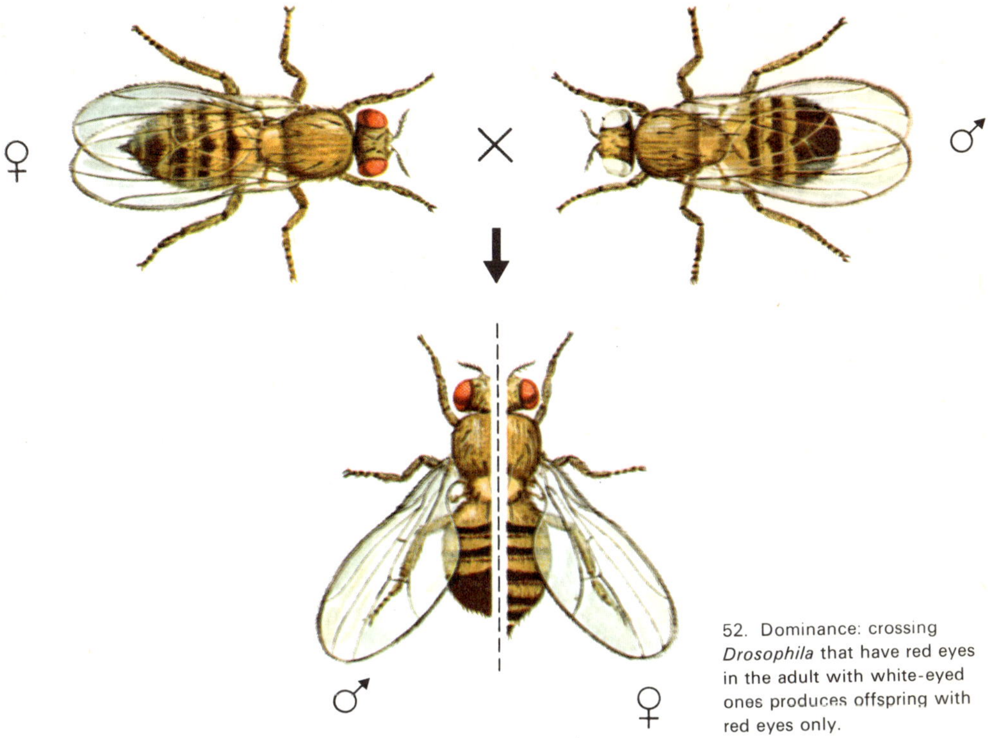

52. Dominance: crossing *Drosophila* that have red eyes in the adult with white-eyed ones produces offspring with red eyes only.

true of fungi. In *Neurospora crassa,* which has been mentioned before, two different cellular threads (which are called *hyphae* in fungi) can grow together, and in their cells two haploid nuclei of different ancestry lie close together. This sort of thing is called a *heterokaryon* (from *hetero,* different, and *karyon,* nucleus). Of course, the two haploid nuclei cannot fuse together, and recombination is not possible.

It is different in a few moulds, such as *Aspergillus* and *Penicillium* (the genus, several representatives of which produce penicillin). Here true nuclear fusion occurs, even if very rarely, in the ratio of about one to one million. It is possible to have a renewal of haploidy, now involving recombination – in a ratio of perhaps one in a thousand – not now through meiosis, but through a half-random division of the masses of the chromosomes.

Since recombination plays quite a subordinate role in heterokaryonts, it may be concluded that the presence of two different sets of information units must be advantageous. Otherwise – if this device were superfluous – it would long ago have disappeared.

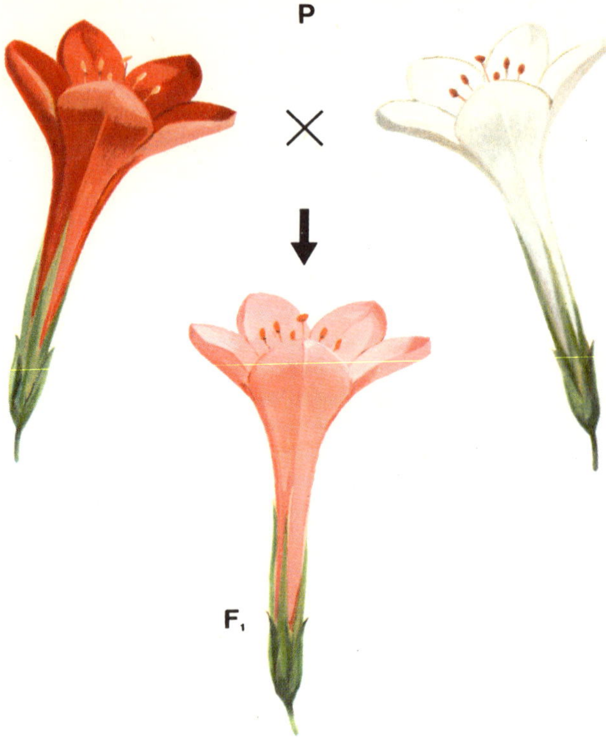

53. Merging of characters: in *Mirabilis jalapa* (the common four o'clock, or marvel-of-Peru) the offspring of a cross between a plant with red blooms and one with white have pink flowers.

54. A *heterokaryon* is what is produced when the ends of two threads of haploid cells merge together but the cell nuclei do not fuse; the new cells then contain two different nuclei.

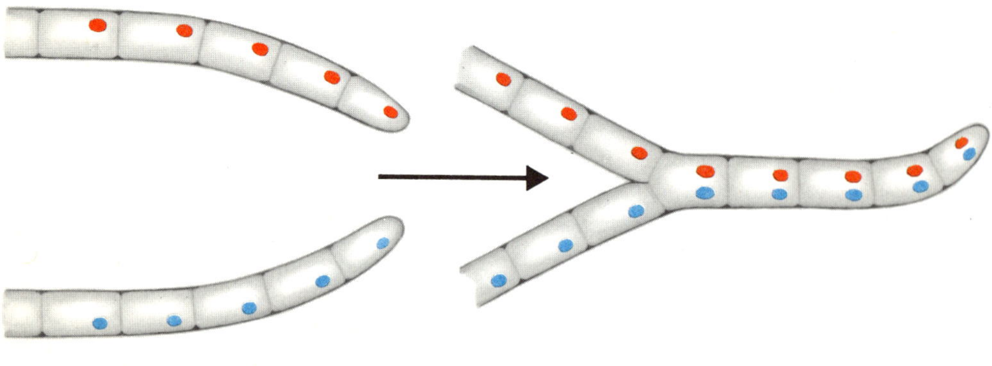

Still More Sets of Chromosomes?

If it is an advantage to have two sets of chromosomes instead of one, then it might be thought that the more sets the better. This is true, but only within certain limits. Organisms, mostly plants, exist which are tri-, tetra- and even polyploid, i.e., which have three, four, or more sets of chromosomes, and which are actually the better for it. This is not the general rule though. On the contrary diploidy is so very much the normal state of affairs that the occurrence of higher numbers of genomes or of chromosomes can be designated as *genome mutation*.

This is a third kind of mutation, following after gene mutations (see p. 83) and chromosome mutations. To take an example, if meiosis should fail to occur during the formation of the ovum – as can sometimes happen – then the ovum will remain diploid. Fertilization by a haploid spermatozoon or a pollen grain will then give rise to a *triploid* zygote, and if no adverse effects come from this, to a triploid organism. Often, though not always, the cells are then bigger, and the whole plant is tougher and more rank. It is interesting to realize that a not inconsiderable proportion of the plants we use, and among them sugar beet and the real peppermint, are triploid. Of course this also has a disadvantage. During meiosis the chromosomes of what is for the moment the third set (and which may not always be one complete set of genomes) have no partners to pair with, and so most of them are 'lost.' Such plants, as a result, are infertile. In sugar cane the diploid and tetraploid must be crossed before seed can be produced, while the real peppermint can only be propagated by cuttings (mitotic inheritance)!

Tetraploids with four sets of chromosomes are at least common enough to arise in more than one way: by the fertilization of a diploid ovum with a diploid pollen grain, or else by a malfunctioning of mitosis, in which the chromosomes are doubled, but they are not drawn off into two nuclei, remaining instead within the one cell. The second method can be reproduced experimentally by making use of the so-called poisoning of mitosis. The best known and most effective mitosis poison is colchicin, a very poisonous alkaloid from the meadow saffron, *Colchicum autumnale*. (The most famous poisoner of Greek antiquity, Medea, came from Colchis, hence the name.) Tetraploids do not usually suffer from distortion of meiosis, and so they can form diploid gametes, which are fertile.

Nevertheless fertility is often reduced, particularly among the higher levels of 'ploids.' This is a kind of disturbance of mitosis, in that the multiplication of the sets of chromosomes is limited. Hence the higher 'ploids' rely more heavily on 'reserve information.' For this reason such plants are more adaptable, and the view expressed above is supported by the fact that polyploid plants predominate in regions with a harsh climate.

Moreover, in polyploids by no means all the characters are equally strongly developed, as would be the case if all the hereditary material multiplied in the same proportion. Often particular quantities, for example, size of cell or nucleus, may be increased, while others, such as the rate of taking up of chemical substances, may be reduced; and these things do not remain constant. We must conclude from this that different genes do not operate independently, each by itself, but that between them they produce the most varied effects. More about this later, on p. 206.

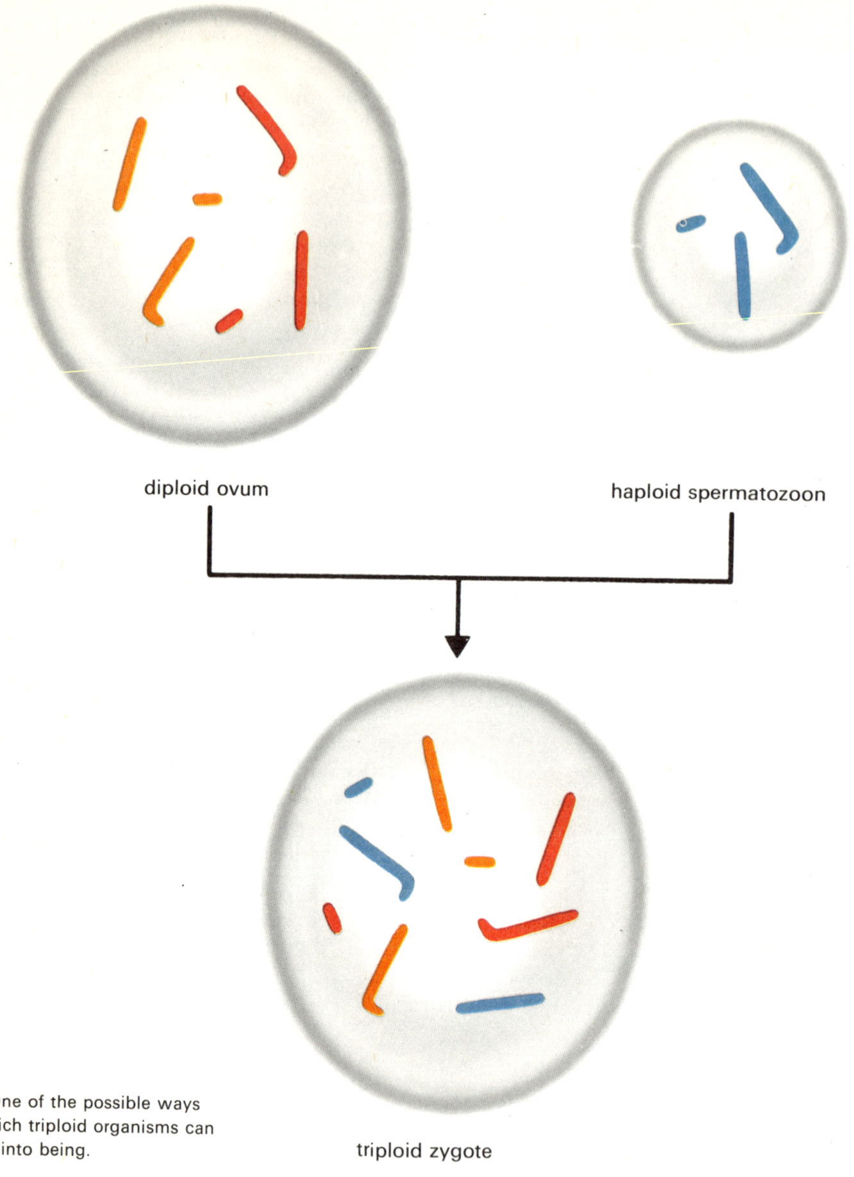

55. One of the possible ways in which triploid organisms can come into being.

2.09 Bacteria, Too, Undergo Recombination

Parameiotic Systems

We are always hearing it said that bacteria are particularly successful in adapting themselves to a new environment – in developing resistance against antibiotics such as penicillin, and so on. This is true only if the conservatism that is imposed upon the cell by mitosis can be overcome by mutation and selection, as we have already learned. But we also know that the recombination of hereditary units that have mutated is speeded up considerably if sexual processes cease – and in higher organisms this means fertilization and meiosis. We must inquire whether the same thing, or something similar, is also possible in the bacteria, which have no nuclei. Since the bacteria are especially suitable for this, not least because of their enormously high rate of reproduction, it should be comparatively easy to demonstrate. In fact, it is soon evident that bacteria – or at least some of them – have evolved methods which amount to being a complete analogue of a very successful recombination. However these methods of recombination work without meiosis, and so are known as *parasexual* processes or systems. There are at least three different methods like this, and it seems desirable to know about them, not least for medical reasons because of their connection with disease and infection. Many ideas about meiotic inheritance can be understood without going into details, but it will also be necessary to learn some new facts and describe some new phenomena. To understand these, it is best if we first of all deal with viruses and their genetics. What is a genuine virus, and what sort of inheritance does it have?

2.10 Bacteriophages

A Virus Is a Virus Is a Virus

This is not an echo of Gertrude Stein, but is adapted from a saying of André Woff (Nobel prize winner for medicine, 1965) made in 1959. What he actually said was that the idea of a virus still cannot be clearly defined even today, but that to say that a virus is just something that has an identity is almost tautological. In spite of this the ordinary person has a good idea what a virus actually is (singular virus, plural viruses). He knows from his own body that viruses exist. He knows that dangerous diseases are produced by viruses; polio, meningitis, smallpox, measles, and not least, various forms of the common cold are all virus infections. Among animals there are, for example, foot-and-mouth disease and fowl pest, while plants suffer from tobacco mosaic virus, potato-X virus, and many more. And finally there are – happy to say – viruses that even attack and eradicate bacteria, the so-called *bacteriophages* (from the Greek *phagein*, to devour), or devourers of bacteria.

We have already met one characteristic of a virus, namely its 'infectiousness,' its power to infect living organisms; a second characteristic is particularly puzzling: the virus has no individual metabolism. Away from the infectious cells it is inanimate, and often of a completely crystalline structure. Yet it possesses the property of heredity. What about the reflections with which we began this chapter: without metabolism there is no heredity, without heredity, no metabolism? Now, that is just as valid here as it was there. If viruses have offspring, and pass on to them their own characteristics, yet have no metabolism of their own, then they must 'borrow' one from somewhere. Viruses have been compared to pirates: they raid a merchant ship (i.e., host cell), and seize the entire cargo (metabolism), converting it to their own use. The virus behaves like a marauding band of genes. Inside the host cell, the virus takes over the metabolism and uses it to produce its own offspring, yet away from the cell it is an insignificant, completely powerless chemical substance.

So are viruses living creatures? This is once again a matter of definition. If metabolism is taken as the criterion of life, then viruses are lifeless. On the other hand, if heredity is the sole criterion of life, then viruses are living. But are we prepared to go so far as to accept a mechanism of inheritance that produces offspring only with the help of a borrowed metabolism?

A third and equally surprising characteristic of viruses is that they contain only one kind of amino acid, either DNA or RNA at any one time. There even exists one kind of bacteriophage that is composed of *single-stranded* DNA. This is called ϕX 174, and stands out like a unicorn among the other ruminants in a zoo.

With the help of these three characteristics – no metabolism, a hereditary mechanism, only one kind of nucleic acid – viruses can be clearly distinguished from cells or cellular tissues; but what a virus is, we still do not know. In the following pages we shall confine ourselves to viruses that devour bacteria, the bacteriophages, or *phages* for short. These will give us enough opportunities to become familiar with the characteristics of viruses, and especially with their genetics. A phage is so small that it is not visible under the light microscope. Many are so small that one cubic millimeter may contain 10^{14} – 100,000,000,000,000! – 'individuals.' If we want to identify them, the electron microscope must be used, or else the virus must be identified by its effects on bacteria.

What Does a Phage Look Like?

There are hundreds, if not thousands, of different phages. In order to avoid diversion of effort by wasteful research it has already been proposed in the United States, and agreed, to confine investigations to a small number of phages, but to study these as intensively as possible. This comprises almost all the phages that can multiply only in the b-strain of *Escherichia coli*. (This shows immediately that phages can by no means attack any kind of bacterium, that on the contrary they are highly specialized to a particular kind of host cell.) Seven different phages of this group exist, which are most conveniently labelled T_1 to T_7.

Fig. 57 shows a structure that cannot always be clearly distinguished in electron microscope pictures. The phage T_2 consists of a head and a tail, and at the tail end, a plate with six spikes (like those in running shoes or in snow tires) and also six 'tail feathers.' The DNA is placed in the head, spread over a framework of protein.

T_2 resembles T_4 and T_6, the other even numbered phages. In T_1 and T_5 the tail is longer and thinner, but in T_3 and T_7 it is very short and hardly visible. The 'unicorn' phage that we mentioned above phage ϕX 174 is quite different; this is a polygon, which consists of twelve subunits, and has no tail.

2.11 Life Story of a Phage

A Tale of Piracy

This is it in brief: when a phage meets an infective bacterium it attaches itself to the

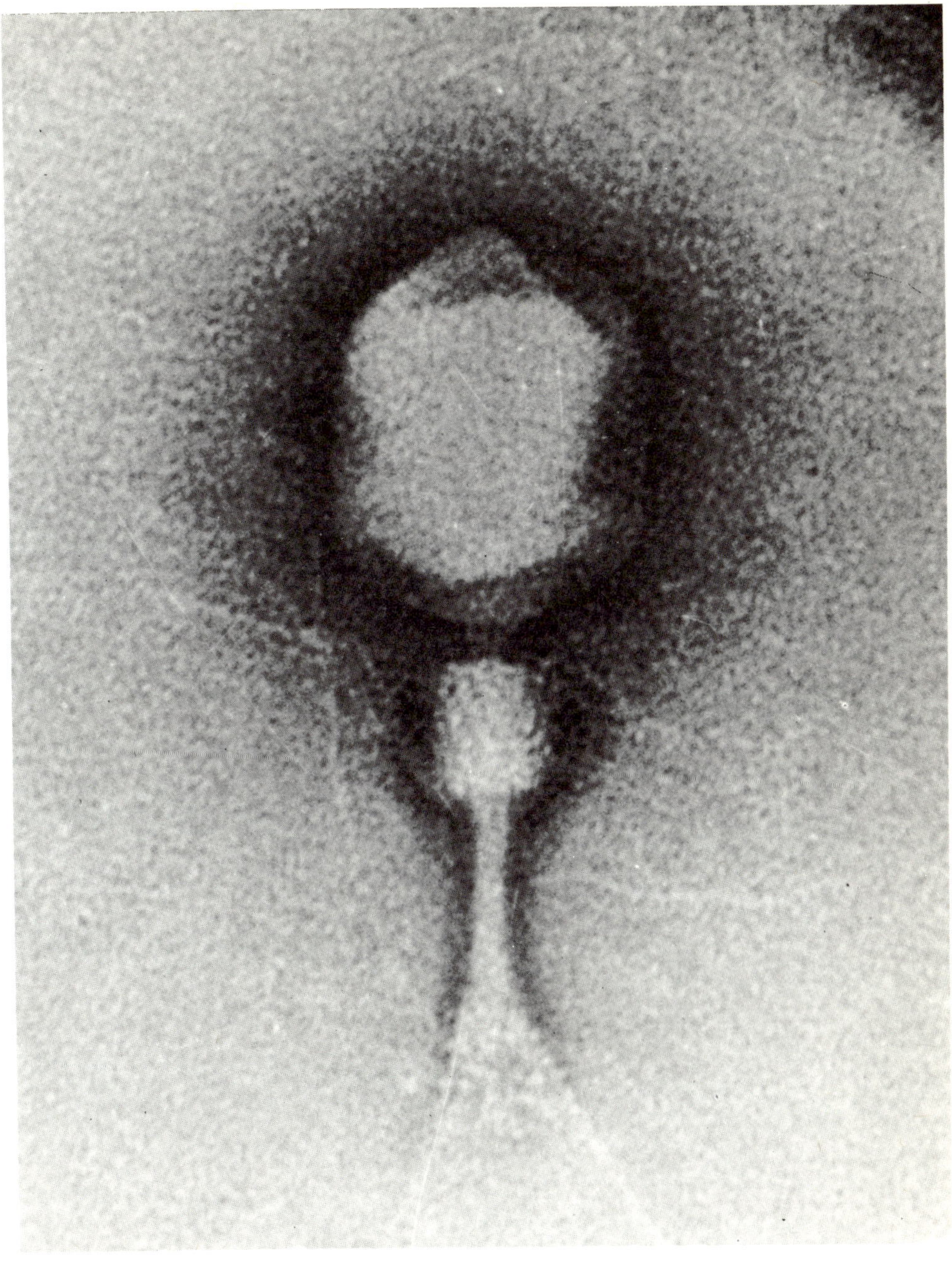

56. An electron-microscope picture of the bacteriophage T_2

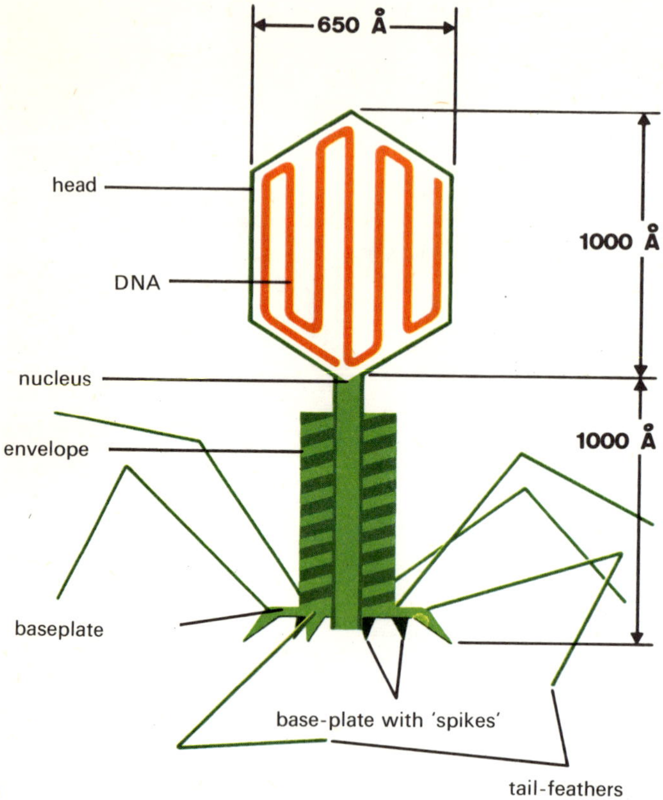

57. Diagram of the construction of the phage T_2.

bacterium with its tail outwards, burrows through the cell wall into the interior, and for some time is no longer visible. After ten minutes the cell of the bacterium is seen to contain phage particles, which in twenty minutes have increased in number to one or two hundred. At this point the bacterial cell breaks up, a process called lysis, and the one or two hundred new phages are released to wait for new victims.

The phage and the bacterium meet by chance, and since the phage specializes in one particular kind of bacterium, it must first test whether the bacterium is one that it can infect or not. To all appearances the 'tail feathers' act as 'eyes' and 'examine' the outer surface of the bacterium to see if it is suitable. What actually takes place between the 'tail feathers' and the cell wall is an immunity reaction, which we shall discuss in Chapter Six.

The next step is for the phage to produce an enzyme, which originates from its tail, and with this to bore a hole in the solid wall of the bacterium, by dissolving its constituents over a small area.

The way is now clear for the DNA. The outer covering of the tail of the phage contracts, and acting like a hypodermic syringe, squirts the DNA of the head, and only this, into the bacterium. The external covering of the phage, its tail, spikes, 'tail feathers' and all, remain outside. If the bacterium is broken up and examined at this stage, it is found not to contain any infective phage as such: the phage – or more precisely, its DNA content – has passed into its so-called vegetative stage. But the DNA is by no means idle. For one thing it produces (or rather orders the cell to produce) proteins, and even enzymes. In this way it makes possible the synthesis of phage-DNA; this DNA is actually foreign to the bacterium, and contains several special components. As a result the cell under attack begins to build phage-protein and phage-DNA independently of each other. When enough of each of these is present, new, complete particles of the phage begin to appear in the host cell. This process may surely be labelled *morphogenesis* (specialists call it morphopoiesis), and compared with morphogenesis in multicellular organisms.

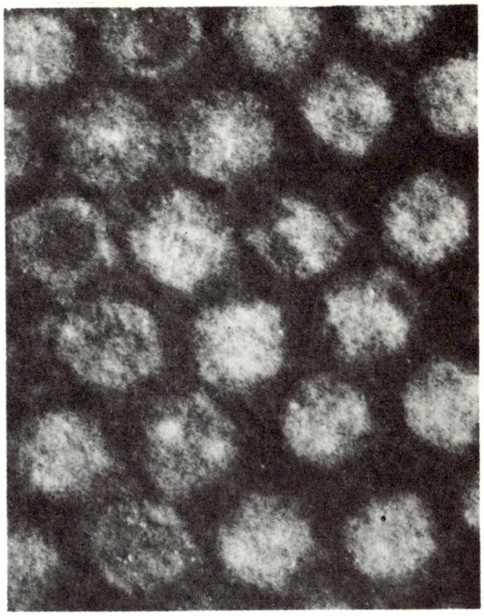

58. Electron-microscope picture of phages of type øX174

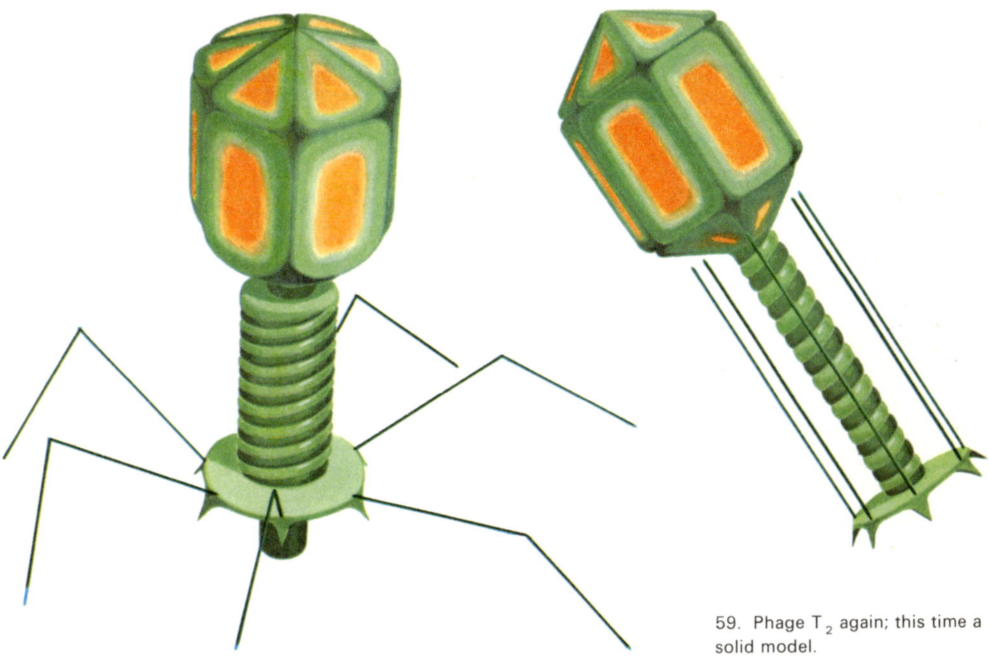

59. Phage T_2 again; this time a solid model.

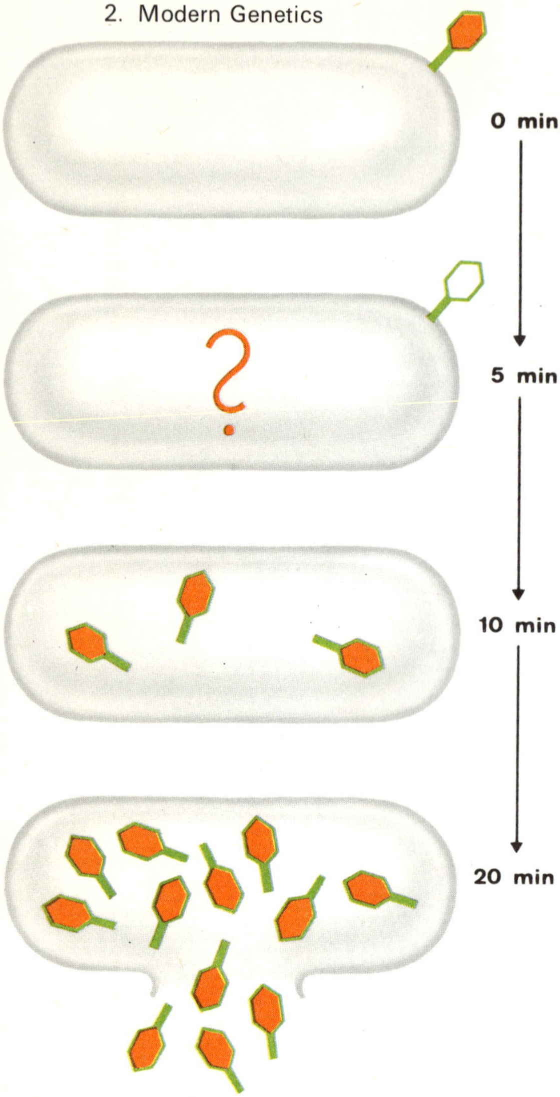

60. Life story of a phage from infection through the vegetative stage to the release of the daughter phages.

In both cases the process is still comparatively incomprehensible but it certainly seems that phages offer the first possibility of tracing morphogenesis back to molecular phenomena.

A particularly fascinating phenomenon takes place among phages when the DNA, which is contained in the head, and surrounded by a membrane of protein, is liberated by means of an 'osmotic shock' (this is a grand term for the simple fact that the phage is taken out of the culture medium and placed in distilled water). This more or less moribund DNA is then displayed as a tiny droplet of liquid protein, spread over a large water surface; the protein has the tendency to spread itself all over the water surface, and so the protein and the phage-DNA are spread out in contact with one another. Using very subtle techniques it is possible to lift off the layer of DNA as a sort of film, and to photograph it under an electron microscope, where it can be seen quite distinctly that the loops belong to one double strand of DNA, which has both a beginning and an end.

Both by measurement and by calculation the total length of the strand appears to be 34 μ or 34,000 $\mu\mu$. But the overall length of the head of the phage is only 100 $\mu\mu$, and therefore the double helix of DNA must be folded several hundred times. This convolution takes place in the host cell during the time that the phage particles are being built up – i.e., within a few minutes, since, as we have seen, the complete life cycle takes only twenty minutes. A comparable exercise would be to try to get about eight yards of thread into a thimble in the shortest possible time. Perhaps this is something similar to the contraction of the chromosomes during mitosis and meiosis? We do not know yet.

The end of this story, at least as far as it concerns the reproduction of phages, can

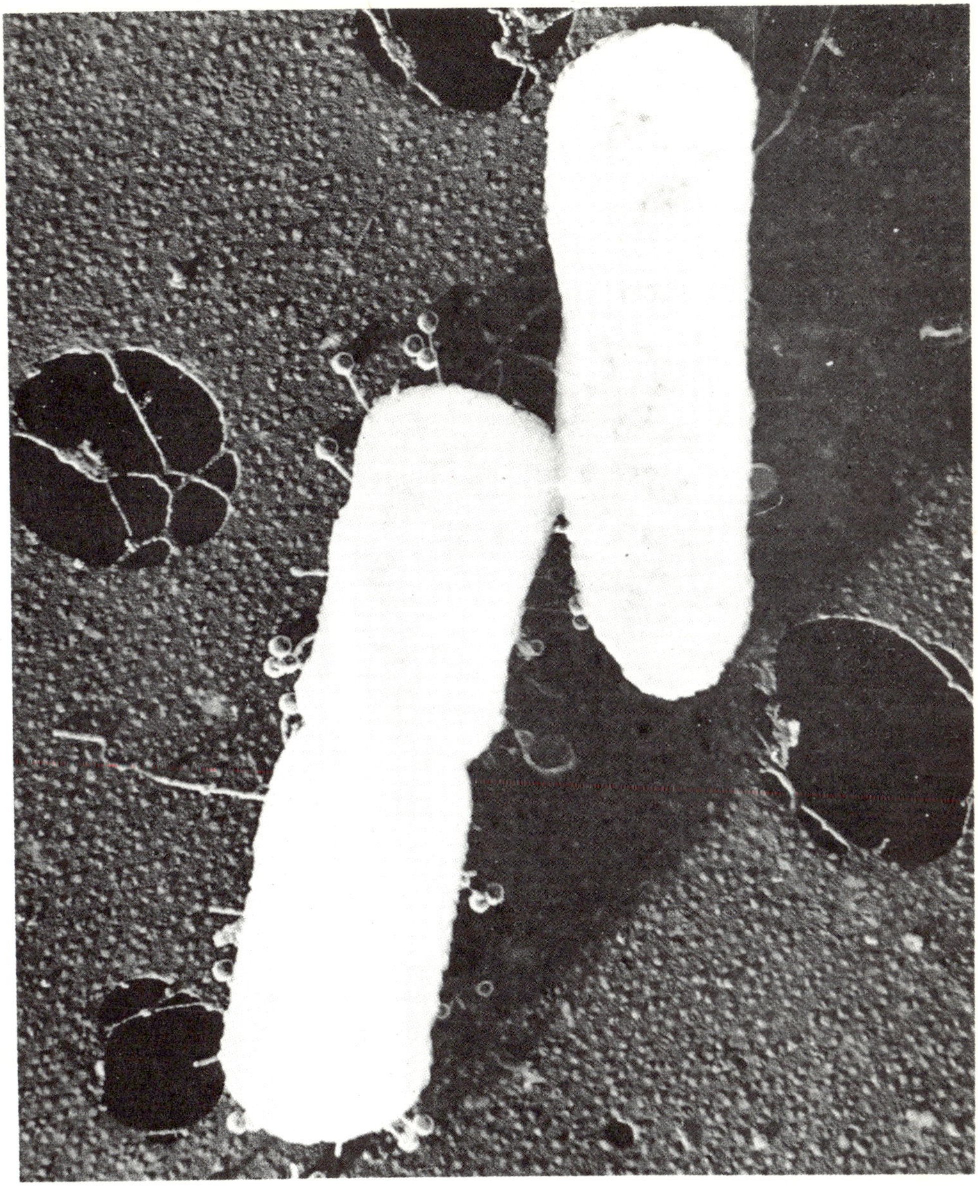

61. A bacterial cell attacked by a number of phages, in conjugation with a second cell that is free from phages.

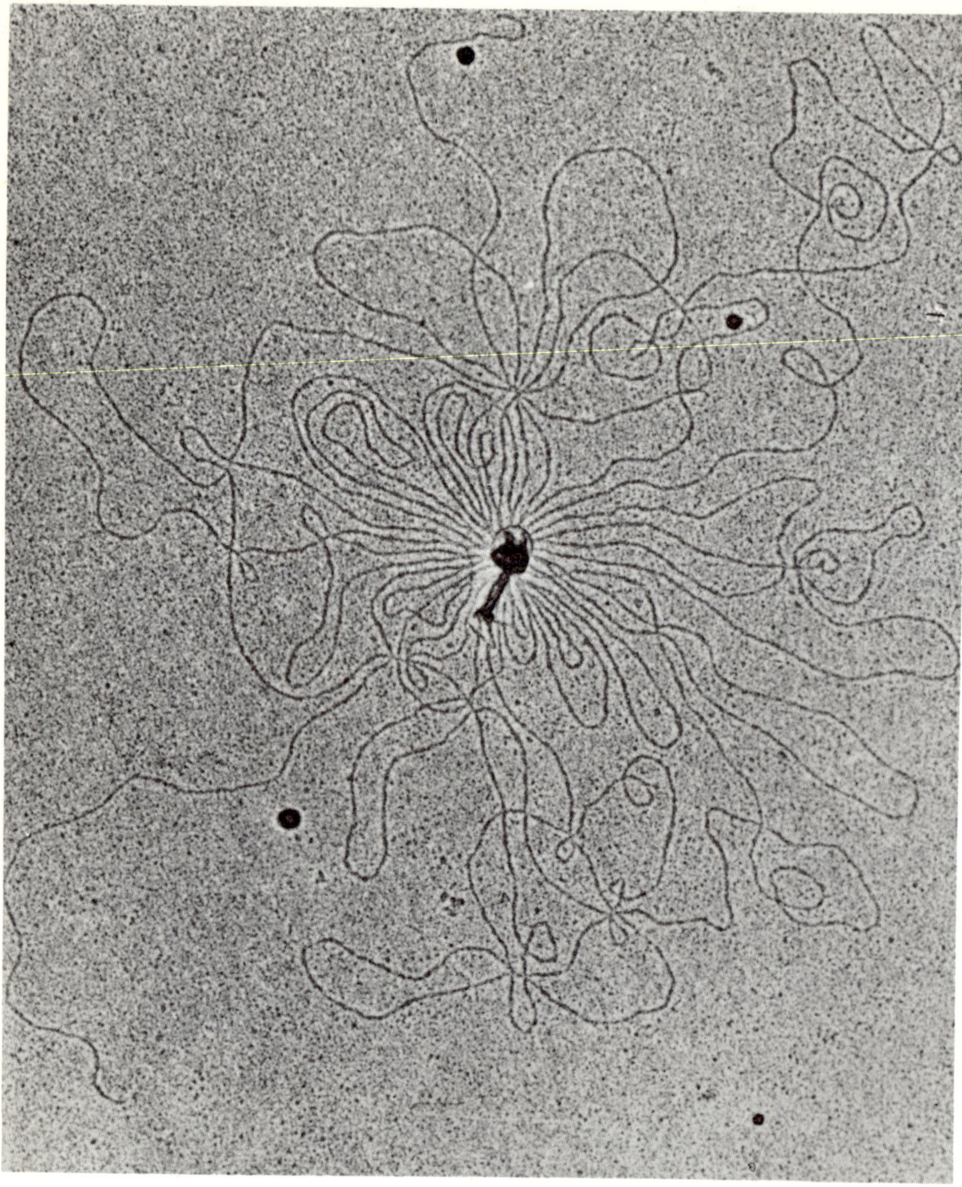

62. A genome from a phage, spread out to show a beginning (top) and an end (lower left). Originally all of this was housed in the head of the phage.

quickly be told. Almost all the available material of the host cell is consumed in building up the phage particles; what remains goes to make lytic enzymes, similar to those which break through the cell wall when the bacterium is being invaded. *Of the bacterium practically nothing remains.* This is journey's end; the pirates have won (in the case of the phage, they have also multiplied a hundred-fold); the merchant ship is sunk.

Phages, Too, Mutate

Phages multiply more rapidly than bacteria normally do. This can be clearly demonstrated if bacteria are mixed with as many phages as possible. The suspension, which was formerly turbid as a result of the presence of the bacterial cells, soon becomes completely clear, because all the bacterial cells have been destroyed. More information can readily be obtained by reversing the numbers, i.e., mixing a few phages with a lot of bacteria, but this time not in solution in a test-tube, but as a layer on a substratum of nutritive material, on an agar plate. After two or three days this is placed in contact with a culture of bacteria of the same size. Then, but only in those places where an phage is present, the result is infection, breakdown of the infected cells, release of one or two hundred daughter phages, infection of another one or two hundred bacterial cells, etc., etc., etc. In these places a 'hole,' or *plaque,* in the bacterial culture may result from even a single phage. Moreover, the size and shape of the plaque is characteristic for each breeding strain of phages.

Occasionally plaques appear in which the margin is fringed, or which are especially big. These are where mutants have arisen, in the latter case the so-called r-mutants (r = rapid), which multiply at an unusually high rate – and this brings us back to genetics. Besides, we know about r-mutants already, and there are three kinds of these, r_i, r_{ii} and r_{iii}. The gene map (fig. 50) is derived from phages T_2r_{ii}. Now let us bring together, again on an agar plate, a lot of phages and bacteria. The whole plate is cleared, and the bacteria are broken down, but with a few exceptions.

From these is built up a population of *resistant* bacteria, cells which have become resistant to the phage through mutation. The strain *Escherichia coli b* mutates to *coli* b_2 – i.e., it is resistant to phage T_2. And now, try the reverse again. Mix these T_2-resistant bacteria with a large number of normal T_2 phages, so many, in fact, that there is likely to be a mutant phage among them, and in fact isolated plaques will be observed here and there. Obviously these must have arisen from T_2 phages, but these are now able to break down resistant B_2 bacteria. They have mutated to a *wider host range* (= h, hence T_2 mutates to T_2h). So now we have two mutants, r and h. What more is needed to make experiments in recombination? Recombination must take place, because one bacterium can be infected not just with one, but with two different phages equally well. To put it briefly, such an experiment can be made. Among the offspring we actually find, not only the original type, but also recombinations, which are equally r (rapid) and h (with

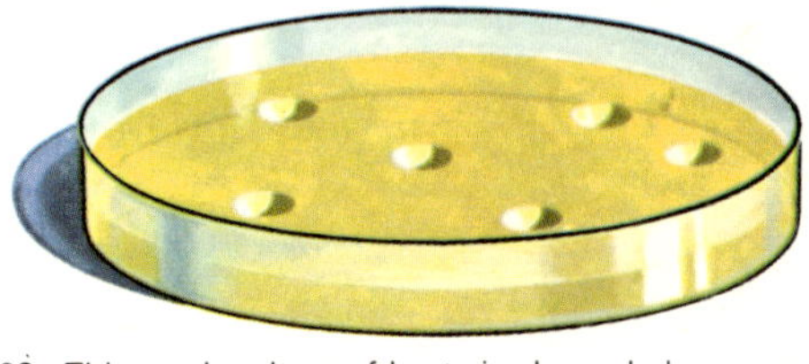

63. This weak culture of bacteria shows holes ('plaques') where lytic phages have been at work.

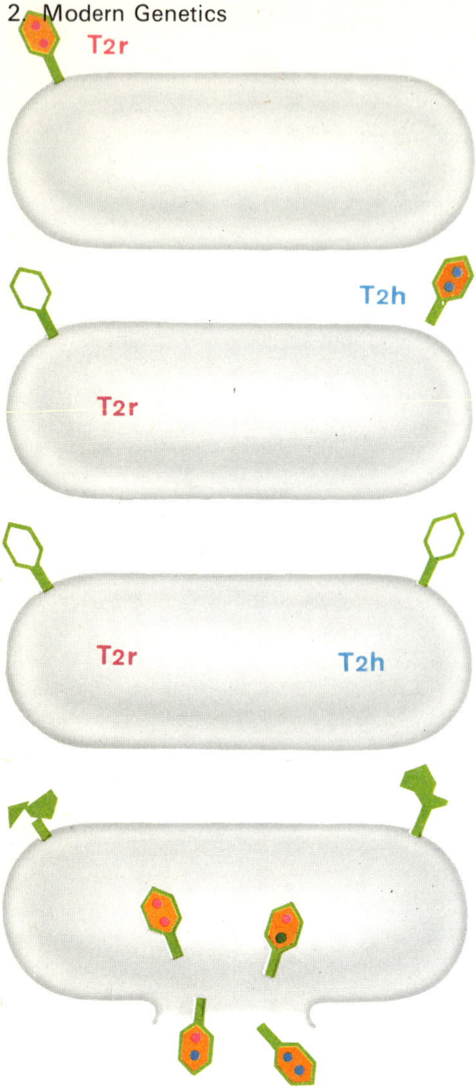

64. A double infection of a bacterium with two phage mutants; T_{2r} and T_{2h} produces daughter phages in which both characteristics are present, in recombination.

a wider host range). It is not a big step from this to the breeding of as many mutations as are possible. The cross is made, the recombination percentage is worked out, genome maps are drawn up (the r_{ii} groups have already been introduced in fig. 50). The DNA from a phage – and not only that from the T_2 phages – behaves during recombination like a genome, and it is therefore permissible to speak of a *phage-genome*. It bears all the information that the phage needs – yet it cannot replicate itself without assistance, but only when it is inside a bacterial cell. It is unfortunate for the infected bacterium that it cannot recognize the harmful genome as such, but confuses it with a genuine one, and even prefers it. It wears itself to death doing its dangerous duty.

Phages Tamed

A few years ago the known behaviour of phages as it had so far been described was so typical that it did not seem to require a special name, but then another kind of phage was discovered which behaved very peaceably. Nowadays such phages are called 'benign,' or 'temperate,' in contrast to the first group, of 'virulent' phages, which are always malignant in action.

If a temperate phage enters a bacterial cell it *may* behave like a virulent phage and destroy the host cell, but it *may* make peace with its host. In that case it remains in the cell, entering into an association with genomes of the bacterium, rather as if a pirate signed on as a crew member in the ship he intended to plunder. The bacterial cell, on its part, treats the phage-genome like a bit of itself, replicates it, and passes it on to the cell's own offspring. Although the host cell, and its offspring, have thus absorbed the phage in its entirety, they do nothing with it. Both are changed: the phage is 'tamed' – and is called a '*prophage*'; and the bacterial cell is left intact, not broken down, and is called a *lysogen*.

This is already a hint of trouble to come. The word *prophage* means 'taking the place of a phage,' just as *lysogen* means procuring lysis, or dissolution. Ought we not to say, then, that the struggle is only postponed, that the apparently so benevolent pirate will later on start a mutiny?

Before we answer this question two things must be clearly stated:

1. The decision whether the infective phage shall become virulent or shall change into a prophage in the genome of the bacterium lies *not with the phage, but with the host cell,* or more precisely, with the external and internal environment of the cell. In a cell in the lysogen condition the phage has no option but to be benign.

2. A lysogen cell which contains a phage in the prophage condition is protected against every other infection, it has become *immune* – though this does not exclude the possibility that occasionally it may incorporate into its own genome a second phage, either of the same kind or a different one. A strain of bacteria has been found – from pus produced by a staphylococcus – which contains no fewer than five different prophages.

Will the truce last? No, it will not. Sometime or other, for reasons unknown to us, the cell which has hitherto been lysogenous becomes virulent. Many benign phages can be made virulent by artificial means, for example by radiating them with ultra-violet light, or by using chemical reagents. But the opposite is also possible, namely that a lysogenous bacterial cell will get rid of its prophages and thereby make a full recovery. This can be termed 'artificial' healing.

All this is somewhat confusing, particularly the multiplicity of possible modifications. Unfortunately we cannot dismiss the temperate, or benign, phages as curiosities, of no particular importance. Quite the contrary – and for two reasons. The first reason, is, so to speak, statistical: nowadays it seems fairly certain the lysogenous condition is very widely distributed, that it is to be regarded as the normal condition, in spite of certain exceptions.

To make the second reason clear we must inquire how and from what source the prophage comes to be present in the lysogenous cell. It is certain that it lies in the immediate neighbourhood of the genome of the bacterium, but it is very probable that it is not incorporated into the genome, but is attached to it in a very special place – so to speak, in conjunction, like the homologous chromosomes in meiosis, in a kind of persistent synapsis. In this situation it is *handed down like a bacterial gene.* Moreover it is assumed that it is not paired for its entire length, but only with over a middle section (the c_1 region), while the lateral sections remain free. This type of pairing apparently makes it possible for the prophages, during the changeover into the virulent state, occasionally to seize a small bit of the genome of the bacterium – but we shall come to that soon.

If the phage can exist like a bacterial gene, then it is not entirely clear why it does not go on to reproduce itself and then produce lysis of the cell, but instead, gallantly imitates the behaviour of the surrounding bacterial genes.

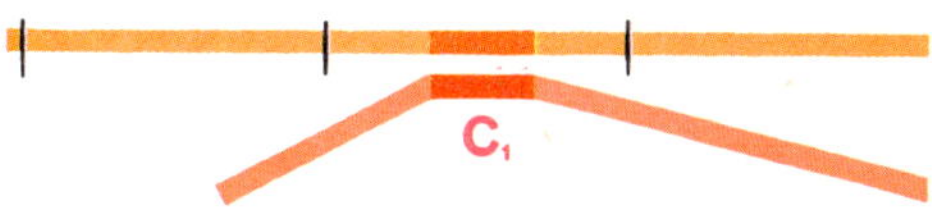

65. The phage genome (red) is attached to the bacterial genome (orange) only over the short section C_1.

There seems to be a factor that prevents the unrestrained multiplication of the phage, and this factor is not part of the genome, but is located in the surrounding cytoplasm. It damps down the activity of the prophages, and for this reason is called a *repressor*. We shall be going into repressors in detail in Chapter 4. The repressor operates as an *immunity factor*. If the repressor is destroyed, for example, with ultra-violet rays, then the phage is activated, overflows into the cytoplasm, and proceeds to multiply: it has been transformed into a virulent phage.

Evidently we have found a new kind of genetic element here which can operate either in the genome or free in the cytoplasm. Several of these exist, such as in 'sex factors' of bacteria (see p. 127) and in RTF, the **R**esistance **T**ransfer **F**actor, the terror of all clinics. The term *episome* is used for this kind of unit of inheritance, which is only mentioned briefly here, but which will be discussed later on.

2.12 Phages Carry Harmful Genes

Transduction

The statement made earlier that a prophage can take along with it a small piece of a bacterial genome needs to be discussed in more detail. It has been known since 1952 that this was so, not in *Escherichia coli* (which seemed to be an exception, though later on this was found to be true of this bacterium, too) but in *Salmonella typhimurium*, which causes mouse typhus.

The wild form of the bacterium can undergo a whole range of mutations, among which there is one which cannot manufacture arginine unassisted (arg −) and one that is resistant to streptomycin (S^r). Now the following combinations of mutations can be produced (how they are produced will be told in the next section): arg − S^s and arg + S^r. The phage P_{22} is allowed to multiply intensively on the latter of these; the infective phages that result are used to infect the first strain, arg − S^s, and this is spread on agar plates which contain either no arginine or no streptomycin. In both cases isolated colonies multiply, perhaps one out of 100,000. The cells have become arg + or S^r respectively, and so there must surely be a phage which has brought over this gene from the previous host.

That both characters should be transmitted equally is at least rare, and sometimes even quite out of the question – as can be well understood. It has been calculated that a phage genome is about one hundredth of the length of a bacterial genome; at any rate genes that are separated by no more than this distance can be transmitted together. Beyond this distance joint transmission is much less likely. We might make a marginal note here that this provides an exquisite little ruler by which a gene map might be made for bacteria.

The factual evidence that a recombination may come about with the co-operation of phages is so well established that the term *transduction* has been applied to it. Yet how does it take place? Clearly the simplest case would be if the phage P_{22}, acting on a strain of arg + S^r, which it broke down, shattered the bacterial genome and absorbed a fragment into itself during maturation. Then this fragment would be simply attached to the second host cell (arg − S^s). But this is not so. On the contrary, it must be accepted that the phage attaches itself to the bacterial genome (cf. Fig. 65), and that then an exchange of pieces (a crossing over?) occurs between a free length of the phage genome and the bacterial

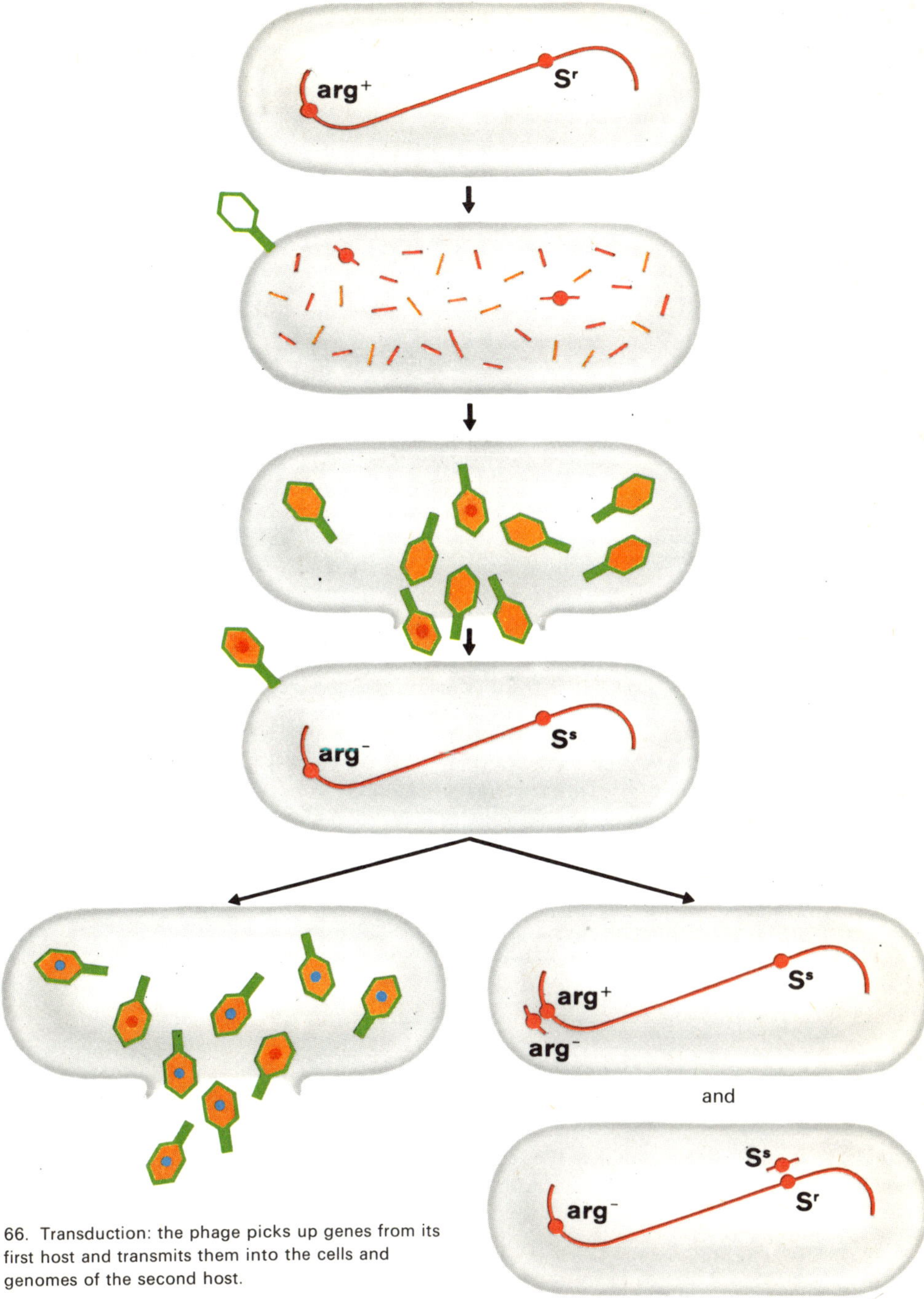

66. Transduction: the phage picks up genes from its first host and transmits them into the cells and genomes of the second host.

genome. The phage is now carrying a piece of the bacterial genome, e.g., arg+, *as an integral part of itself* (and in exchange has given up a piece of its own original genome). A parallel process can obviously take place in the newly infected host cell, the carried-over bit of the 'old' bacterial genome, with arg+, being traded against a bit of the 'new' bacterial genome. In this way the infected (lysogenized) bacterium has become arg+.

In the so-called *unrestricted transduction*, every position in a bacterial genome can evidently be filled, and interchanged. In *restricted transduction* it is different. This involves only the so-called *gal-region* – in which occur genes that are able to ferment galactose, gal+ for short; and those which through mutation have become inoperative, and are gal–. The operator is the phage λ (lambda), which can evidently only establish itself in the neighbourhood of the gal-region as a prophage. If it breaks down an infected cell immediately, and thus does not lysogenize it, then it is ineffective in transduction. It must first have gone through the prophage stage, the attachment stage, and then been released by ultra-violet light, before it can bring about transduction! It is still not known precisely why the phage lambda is not free in its choice of its place of attachment. Possibly the prophage and the gal-region may have to come together, as the homologous chromosomes have to be brought together in higher organisms. This view is supported by a further peculiarity of the phage lambda; it does *not* exchange its fragment of bacterial genome, and so has to replicate itself complete with its bit of gal-genome, exactly, and simultaneously with the replication of the bacterial genome.

By analogy with the term *heterozygote*, there is something called a *heterogenot*, and in this instance a 'heterogenot for the gal-section.' Finally a special feature of transduction must be mentioned, which we shall be interested in again later, in connection with episomes. This is the so-called *abortive transduction*. During this process the transferred section of the bacterial genome is not incorporated (exchanged), but remains lying in the cell, *without reproducing itself*. In spite of this it is effective; but at the next cell division only one of the two daughter cells can obtain this fragment, the other daughter cell going without. After many cell divisions have taken place, all the clones are without this fragment, with one solitary exception. This can be made into a very pretty experiment, if it is possible to find a fragment that carries the gene for the formation of a flagellum. If this can be passed by transduction into a *non-flagellated cell*, then this cell will form a flagellum and become mobile. Every

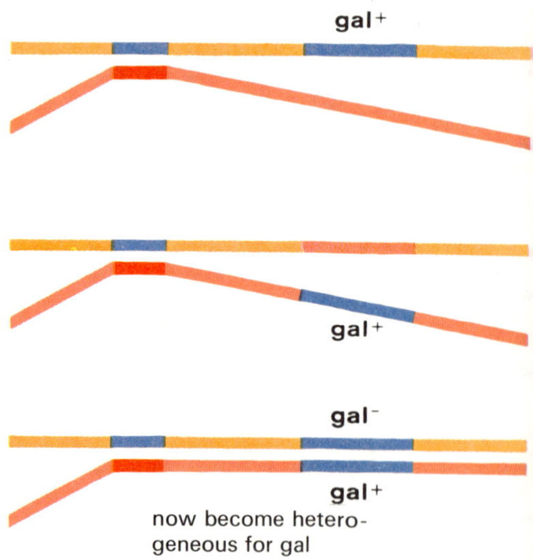

67. Heterogeneity: top, a phage genome (red) lies alongside the gene gal$^+$ the bacterium, middle, the phage picks up the gene gal$^+$ by exchange and transmits it to the phage's next host cell, bottom, if the new host is gal–, it becomes 'heterogeneous for gal,' since it now has equal potentialities for being either gal$^+$ or gal$^-$.

cell division, however, produces one immobile cell, and from each such cell a colony of immobile cells develops. It is therefore possible to follow the progress of the single flagellated cell by means of the mobile daughter colonies that it produces.

2.13 Autonomous Genes?

Transformation

Whereas during the process of *transduction* the genome fragment is transported by a phage, in the so-called *transformation* process the phage is superfluous: this time the genome fragment finds its own way into the cell, and exists there, so to speak, autonomously.

The starting point was a discovery made in 1928 by Griffith, and which today is a landmark in the history of genetics, though at the time it was hardly noticed. Species of *Pneumococcus,* the causative agent of pneumonia, when in the virulent phase, have a firm cell wall, or capsule. If they are cultured on an agar plate, they form smooth, shining colonies, with an even margin (indicated by the letter S). Colonies that have mutated have often lost the capacity to form a capsule, and so their colonies are dull and rough (indicated by the letter R). At the same time they have become avirulent; if mice are infected with the avirulent form, they suffer no harm. It is different if at the same time S-cells that have been killed are injected; then some of the mice go down with blood poisoning. From the blood of these individuals pneumococci can be isolated and cultured on a plate, and they prove to be a pure strain of S-cells. So obviously something out of the *dead* S-cells has passed into the living R-cells, and has *transformed* them into S-cells. Later it was proved that the transformation can continue by itself outside the host cell, in a test-tube (1931), and then (1933) that it will work with cell-free extract, i.e., with cell-fragments. But it was not until 1944 that it became self-evident that the transformation principle was DNA. The one thing that can inhibit transformation is the DNA-destroying enzyme DN-ase. Up to that time the legendary 'hereditary principle' had been sought among proteins, but here for the first time it had been demonstrated that this consisted of DNA.

In the interim a whole range of other examples of transformation have been found. One such occurs in *Haemophilus influenzae,* which was responsible for the great epidemic of influenza after the First World War. What happens in this case? This question needs to be divided into two:

1. In what form is the genetic material passed from the 'donor' cell to the 'receiver' cell? And,

2. How does it become incorporated into the genome of the receiver cell – as cannot otherwise but happen, if the offspring of the transformed cell are to breed true?

The next point is the dimensions of the DNA that is transferred. During the preparation, which consists of smashing up the cells in a homogenizor, the original bacterial genome is shattered into fragments, which have a fairly uniform molecular weight of about five million – that is about one two-hundredth to one five-hundredth of the total amount of genome. Since the S-gene in the above experiment is obviously the only gene, and is shared between the donor and receiver cells, this gene ought to be present only in every two hundredth (or five hundredth) fragment, and be able to transform these.

Small fragments, those of a molecular weight about 500,000 are apparently not picked up, or at least they are not effective. Similarly single-stranded DNA, which can

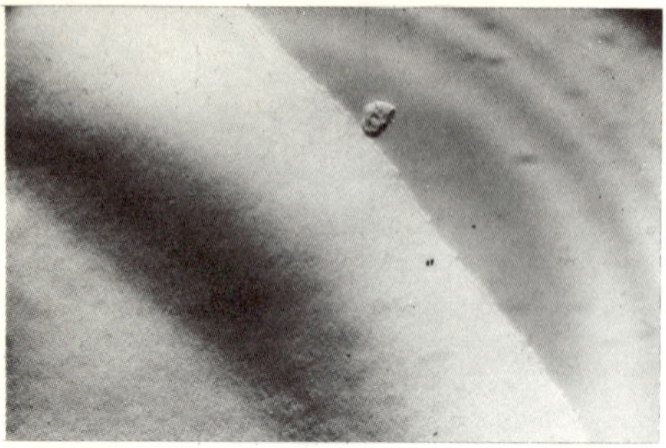

68. R cells and S cells can be distinguished from each other even with the naked eye. Colonies of S bacteria are shiny, with a smooth surface (above); R cells are matt, with a rough surface (below).

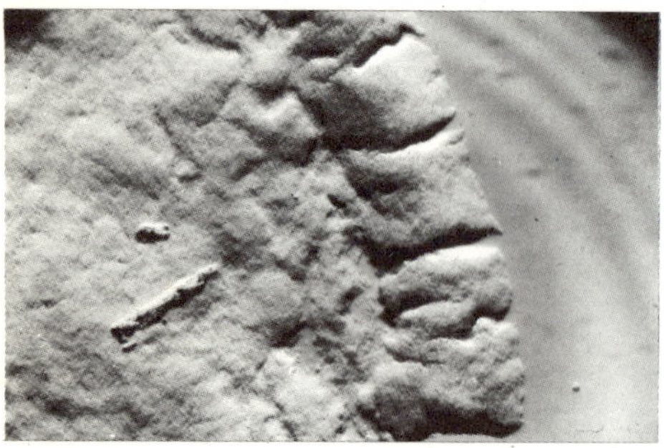

be made artificially, is not taken up. So we have the remarkable phenomenon, which we still do not fully understand, that only double helixes are taken up, but that *only one of the two strands* is necessary for the transformation (see below). Otherwise the take-up goes forward smoothly and very quickly; after ten seconds the DNA has already penetrated the cell, and is no longer vulnerable to outside attack by DNA-ase.

And now the second part of the question, about the incorporation of the genome into the receiver cell. The first step is obviously a pairing, a conjugation with the homologous position on the bacterial genome – this would correspond to the zygotene of meiosis. This is followed by an exchange of pieces, like a kind of crossing over, breaking and joining together again, just as we saw it in meiosis; but we must not forget that in this case we are dealing not with the chromosomes of higher organisms, but with double helixes of DNA. What happens in detail is just what we did not know previously, and we still know no more about how the fragment finds its way to the homologous position on the

69. Transformation: either living R cells or killed S cells are harmless by themselves, but if they are both injected together, they are fatal.

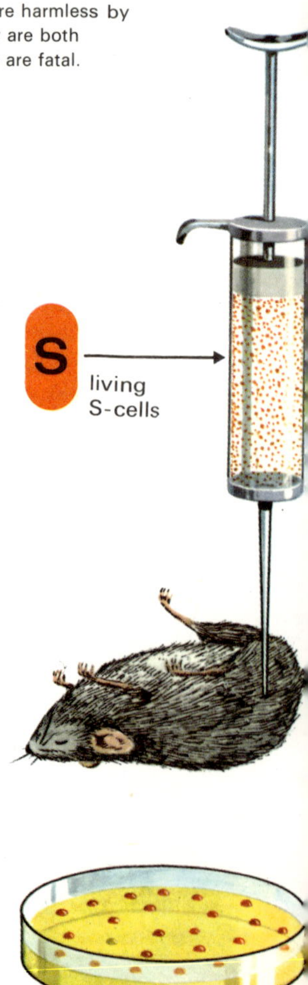

genome. One thing has been discovered, however: the incorporation of the foreign fragment of genome goes on just the same, even if no DNA synthesis takes place. Obviously in this case the exchange does not take the form of a copy-choice mechanism (see p. 99).

Important as it may be for the R-cell to get back its capsule factor, and with it, its virulence, the consequences for the patient may be fateful. All the same, at the present time there is no more than polite interest in this phenomenon.

The experiment can proceed in the reverse direction (S → R), that is R-cells that have been transformed into S-cells can be transformed back into R-cells. Furthermore, as many different genes as possible, whether

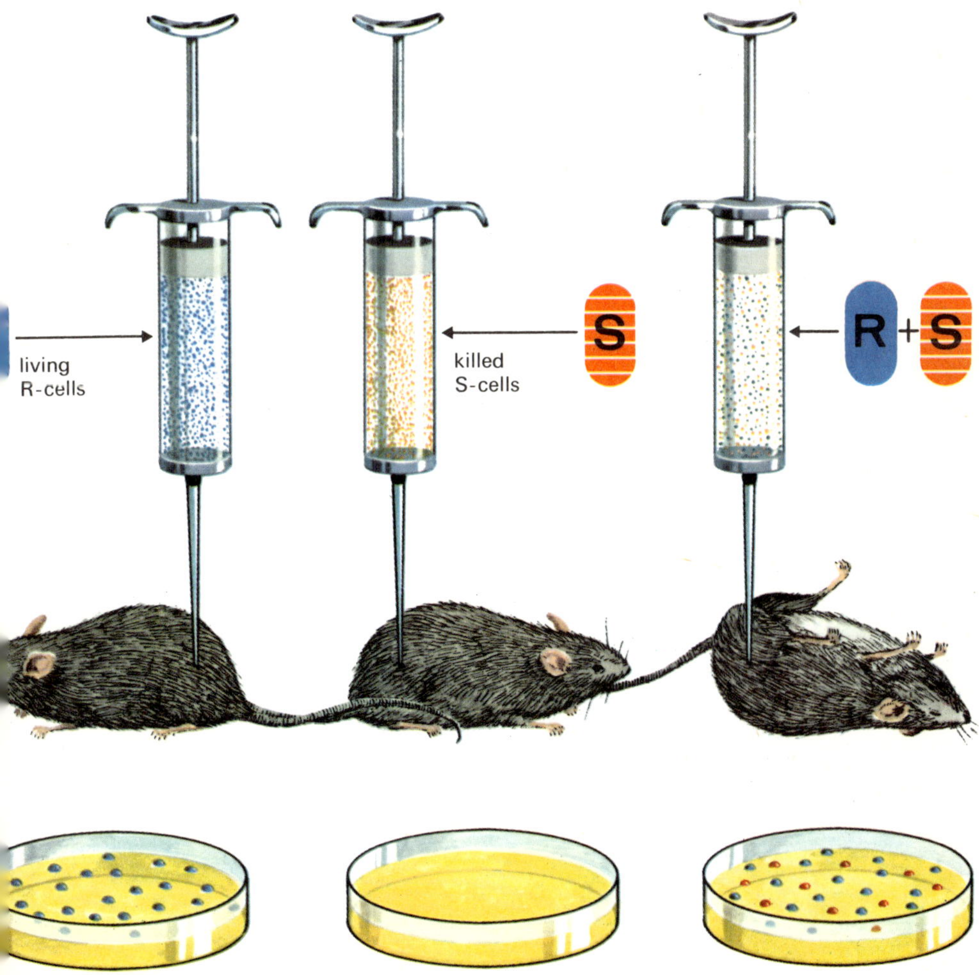

mutated or not, have been successfully – and without any great difficulty – tricked into entering a receiver cell; indeed a receiver cell can be 'planted' with particular genes deliberately. That is of the greatest clinical importance, and it should not be forgotten that transformation is an invaluable genetic tool, in particular in matters to do with recombination. This process may be greatly complicated by the action either of phages (transduction) or of complete bacterial genomes (conjugation, which is discussed in the next section).

Finally, it should be noted that transformation, along with transduction, is a parasexual system, which makes recombination possible.

2.14 Conjugation in Bacteria

Biochemical Mutants Can Be 'Stamped Out'

The third, and provisionally the last, parasexual system is, or depends upon, conjugation. It was discovered twenty years ago and since then has been clarified a good deal. This success was mainly due to a method known as selection technique, or marker technique, for which reason this method will be briefly described.

If we want to make experiments in recombination, or crossing over, it is necessary to have as many different, precisely defined mutants as possible. The so-called biochemical mutants are well-characterized, and hence easy to recognize. Bacteria such as the fungus *Neurospora crassa,* from which the first biochemical mutants were isolated, can normally make for themselves almost all the organic substances that are needed for protein building: all twenty amino acids, all the vitamins, and so on. These wild strains mutate to what are called auxotrophic forms (Greek *auxein,* to grow; *trophein,* to nourish), which have lost the power to synthesize either one particular amino acid or one particular vitamin. In order to produce growth, they have to be supplied not only with the minimal amount of food, but also with the amino acid or vitamin that is lacking. For instance, if the missing constituent is histidine (his for short), or threonine (thr), or lysine (lys), then the bacterium is his – (histidine minus), or thr –, or lys – ; if the form can no longer operate on galactose, then it is gal –. For this reason it is easy to isolate and easy to recognize again.

To begin with, as many mutants as possible are induced in a large population of the wild strain of the bacterium, either by ultra-violet light or by nitric acid, giving a variegated mixture of different kinds. Next, the cells that remain unchanged are separated from the mutants, and for this purpose, use is made of a peculiarity of penicillin, which is to kill only those cells that are in process of division. The population concerned is put into a nutritive liquid; penicillin is added, but no amino acids are supplied. Only the unchanged cells are able to grow and divide in this medium, and as soon as they start to divide they are killed by the penicillin. The mutants that lack amino acid are not killed, because in the absence of amino acid, they are unable to grow and divide. After the dead cells and the penicillin have been thoroughly washed out, only mutated cells are left behind.

These are then placed on a solid nutritive medium, with agar, in a Petri dish. This time the medium is a 'complete' medium, i.e., it contains all the twenty amino acids, which ensures that the cells will be able to grow. In a day or two, spots and disks can already be seen with the naked eye. An ordinary stamp, covered with a sterilized pad, is pressed onto the Petri dish, and a few cells from the colony adhere to the pad. This is then pressed onto another, sterile agar plate; the cells adhering are taken up by the new plate, and so reproduce exactly the positions of the colonies on the original plate. The decisive trick about this is that the new agar plates never contain all twenty amino acids, but each time there are only nineteen; naturally the missing one is different in each plate. The same technique is used that was described in Chapter 1 (see p. 60) for deciphering the code on the amino acids when nineteen of the amino acids were normal and the twentieth was made radioactive. In the example shown in Figure 70 the amino acids are histidine, arginine, trytophane, and proline. The arrows indicate where a colony is now missing; its cells need the very amino acid that happens to be missing from the plate, and so the colony is not able to develop. It is only necessary to find the corresponding place on the original culture in order to be able to say, This colony is his –, or try – (in this experiment arg – and pro – are not involved). Such colonies can be isolated and cultured, and from them can be obtained dozens of 'marked' strains, which can be used for crossings over of all kinds.

'Asymmetrical Sexuality'

Crossings over in bacteria can be achieved in practice. If cells from two *different* auxotrophic strains of *Escherichia coli* K_{12} are brought together, then among the offspring can be found recombinations which are now doubly auxotrophic (prototrophic, or anauxotrophic), so that they resemble the original wild strain. In this conjugation it is necessary for *two bacterial cells to make contact,* unlike transduction, where a bacteriophage transports a bit of the bacterial genome from cell to cell, and unlike transformation, where a free fragment of genome migrates to the 'borrower' cell. If the two strains of bacteria are placed in a U-tube and separated from each other by glass wool, which allows phages and genome fragments to pass through but stops complete bacterial cells, then recombination does not take place.

The fact that such a contact actually exists can be demonstrated under the electron microsçope, where a fine bridge, presumably made from the cytoplasm, and probably containing DNA, links the two cells. This does not lead on, as it would in a nucleated organism, to a fusion of nuclei, or even of whole cells, to a fertilization between ovum and spermatozoon. Instead, after a time the two partners separate again. Hence the use of the term *conjugation,* which implies no more than the formation of a bond; but during this conjugation an exchange of genetic material must take place, otherwise a recombination could not be explained.

Here already we can see some similarities, as well as clear differences, between this system and the sexual system based on meiosis, but we are immediately confronted with other phenomena that are new, very surprising, and not really simple, and which *apply only to bacteria.*

The first of these is a disappointing fact to learn: that only one out of a million, or even ten million, cells is a genuine recombinant. This is why the bacterium really breeds true. Moreover, this ensures that prototrophy does not come about through back-mutation.

Further experiments with the different mutants proved that all the genes in a single coupling-group are bound together, and that the genome is thus a *linear* fragment, or, as it may be expressed, but not entirely correctly, that bacteria have only a single chromosome. However, when this was followed by an attempt to calculate the conjugation percentage, with the object of constructing a

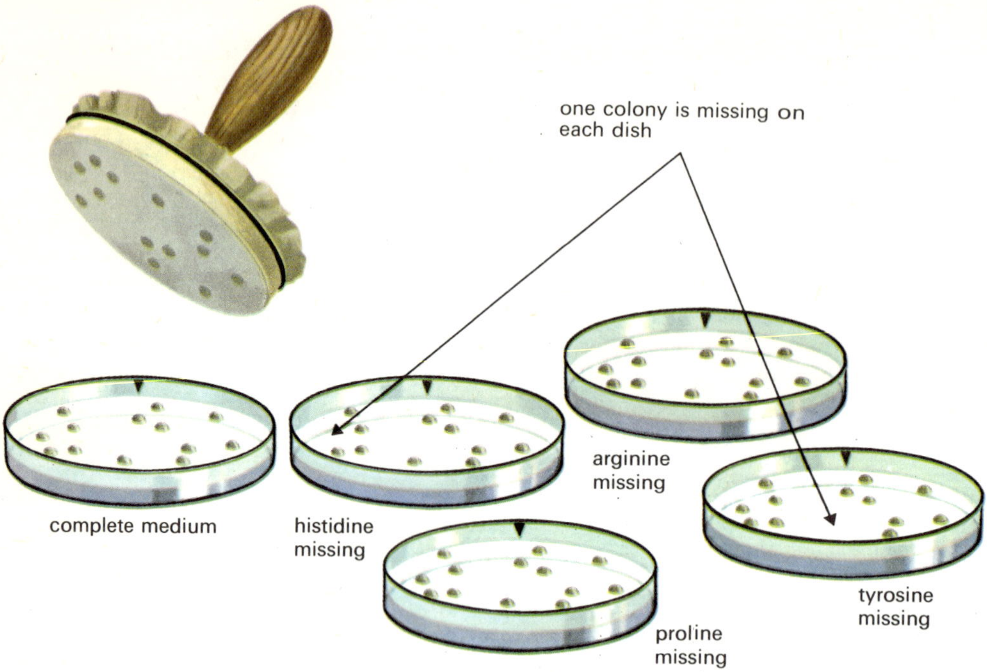

70. Use of the 'stamping' method. The dishes contain nutritive media from each of which one, different amino acid has been omitted. The stamp impresses the same pattern of bacteria into each dish, and the failure of a colony to grow indicates that the amino acid missing from that dish is essential to the growth of the bacterium ('biochemical mutants').

genome map, this did not confirm the hypothesis of a single linear structure. What does this signify? The first important discovery was, that of the two cells in a conjugation, *only one* carried out the recombination, and that for the recombinants to be preserved as offspring, it was only necessary for this one cell to survive. So once again we have the system of *donor* and *recipient*. The donor hands over genetic material to the recipient, but the reverse operation is not possible. In this instance conjugation is a one-way process – hence 'asymmetrical' sexuality. Obviously it is possible to speak of the donor as being 'male' and the recipient as being 'female.'

A second new phenomenon is that *Escherichia coli* K_{12} exists in two substrains. One of these will not cross over with itself, and so it cannot produce any recombinations; it is called F− (for 'fertility negative,' or sterile). The other substrain will form conjugations within itself, and especially *with all F− strains.*. This belongs to the so-called F+ ('fertility positive,' or fertile) group.

Thus:

F− × F− → sterile
F+ × F− → fertile
F+ × F+ → fertile, but usually much less so than F+ × F−.

It follows from this (as well as from other experiments, which we need not go into) that *the F+ cells are donors and the F− cells are recipients.* Who can distinguish, however, whether a cell is, or will become, either a donor or a recipient?

The answer lies in a result which was obtained independently in several laboratories. Contact between an F+ and an F− cell resulted in the F− cell becoming a donor itself. Thus a *factor F* had been transmitted, and that very easily. After one hour almost all the F− cells had already become F+, even though the numbers of recombinants were in the proportion of only one to one million. Thus the factor F could be transmitted independently of the normal bacterial genome ('chromosome'). (Fig. 74.)

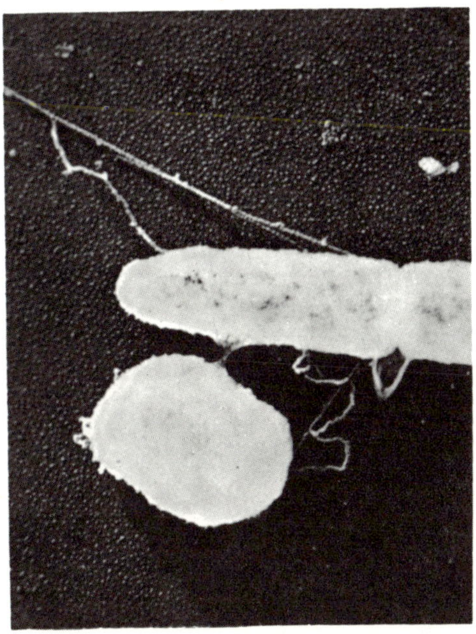

71. Conjugation shown in an electron-microscope picture. A bridge of plasma between two bacterial cells enables genetic material to pass from one to the other.

Without any further discussion, the factor F can be tagged the 'sex factor.' Cells with the sex factor are F+; cells without it are F−. An interesting and surprising thing is that a small number of F+ cells is sufficient to convert all the available F− cells into donors in the shortest practical time. The process goes much more quickly if the bacteria are able to multiply. The sex factor multiplies itself independently of, and faster than, the bacteria, and spreads like an epidemic through the recipient population (we shall come back to the epidemic idea later). On the other hand the F factor may be irretrievably lost, either spontaneously, or by radiation – or by treatment with dyestuffs containing acridine!

It must be concluded from all this that the F factor is a *genetic* factor; that it exists independently of any actual bacterial genome; it reduplicates itself and multiplies without assistance. When an F+ cell has passed on its factor and thus ceased to be a donor cell it automatically becomes a recipient cell (F−) and in this guise may come into contact with a new donor cell.

If the reader thinks that the phenomenon of conjugation is now clear, he is making a big mistake. We are about to encounter the F factor in a new context, but before doing so, we must mention a phenomenon which otherwise would be completely unexpected. This is . . .

Incomplete Genome Transmission

The successful crossing over between a strain A as donor and a strain B as recipient may be correctly written thus:

$$\underrightarrow{A \cdot F+ \times B \cdot F-} \qquad \text{(i)}$$

This process can always operate only in the direction of the arrow, and never in reverse – hence the term 'asymmetrical,' or 'polar,' sexuality. However, a variety of strain A may exist which has lost its sex factor; indeed, we have heard of such a strain before. This strain is F −, and therefore must be written as A · F −.

Now we make use of the fact that a donor strain can convert a recipient strain in the shortest possible time from F − to F + by transferring its F factor – and without passing over any genome material.

$$A \cdot F+ \times B \cdot F- \rightarrow B \cdot F+$$

In this way B · F can be converted into B · F +. But this time the reciprocal crossing in the reverse direction is possible

$$\underleftarrow{A \cdot F- \times B \cdot F+} \qquad \text{(ii)}$$

The direction of transfer of the F factor is now obviously reversed. This is confirmed in practice. If genome A is introduced into cell B (i), then the two genomes A and B are both present; if genome B is introduced into cell A (ii), then again the two genomes are present together, but this time in cell A. Nevertheless the two situations differ considerably from each other. In the recipient cell in each case there are always only a few genes from the donor cell, and the great bulk of the genes are those from the recipient cell.

After some controversy at the beginning, the right explanation was soon reached. As a rule the donor cell transmits *only one piece* of its genome to the recipient cell, and never, or only in the rarest cases, the complete genome. The transfer is incomplete, and the resulting zygote is equally incomplete, since it contains only one complete genome, that from the recipient cell; from the other cell it has received only a fragment. It is then a 'merozygote' (from the Greek *meros*, a part).

The next stage in the speculation was soon verified by a famous experiment with 'interrupted conjugation.' To repeat this it is necessary to 'make ourselves a strain' (in the manner described earlier, on p. 124) which has on its genome the sequence of genes A, B, C, D, E. The genome is given an F factor, and so becomes F + and will serve as a donor. The recipient is marked with the equivalent alleles a, b, c, d, e. Undisturbed crossing over leads to recombinations in the following proportions (taking A as 100 per cent): B, 90; C, 75; D, 45; E, 25. The *absolute* frequencies of these recombinations are, of course, very very much smaller!

But now let us interrupt the progress of conjugation by tearing the pairs apart in a high-speed centrifuge, and that for varying times. Then let us spread them out on plates and count them. The result is more or less nil for the first eight minutes; no recombinations can be detected. The earliest appearance of recombinations is after eight minutes, when they become visible on the agar plate, but only those with the following formulae:

after 8 minutes A b c d e
after 9 minutes A B c d e
after 10 minutes A B C d e
after 17 minutes A B C D e
after 25 minutes finally A B C D E

The only way to explain this is that during conjugation, the tip A of the genome (thinking of the genome as linear) enters into the recipient cell and is *gradually reinforced*, so that after 25 minutes it has all been fed in, right to the tip E. Indeed, this 'timetable' can be used as a genome map.

Superman Hfr

Actually, the term *Hfr* should have been introduced earlier, since the experiments described above can be carried out only with an Hfr strain. What does *Hfr* mean?

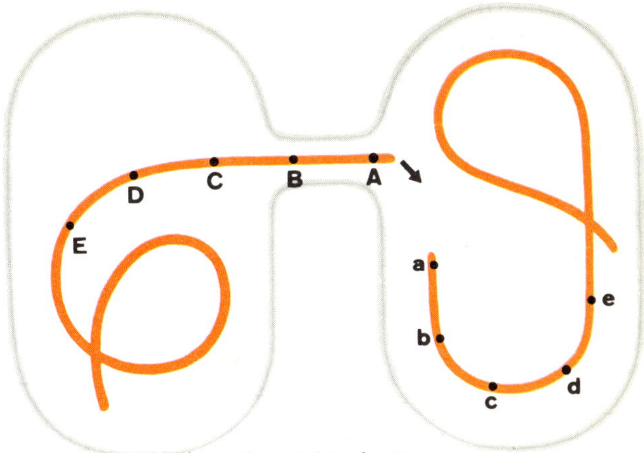

after eight minutes

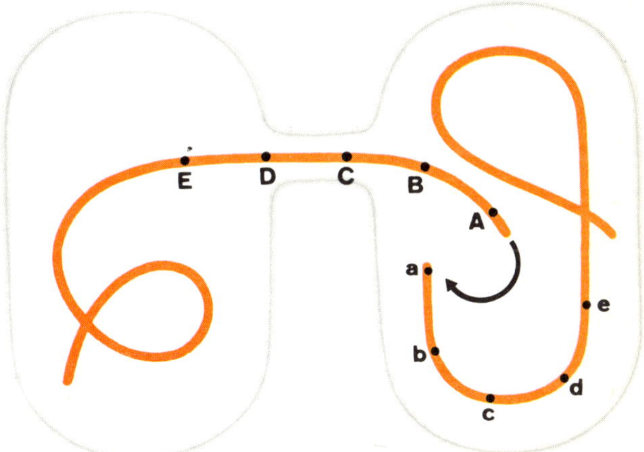

after nine minutes

72. The longer the two cells remain in conjugation, the longer is the piece of genome that passes from one cell (the donor) to the other (the recipient).

All Hfr strains that are known arise from F+ populations. If one of these is brought into contact with an F− strain, it produces about *a thousand times as many recombinations* as in a normal F+ × F− conjugation.

Hence the name *Hfr*, for **H**igh **f**requency of **r**ecombination. Hfr cells behave like supermen – but only in respect to the transmission of genomes. For this reason, they can only very rarely achieve the result that is child's play for an F+ cell – that is, changing an F− cell into an F+. Often they seem to tire of their intensive masculinity, and then they behave just like a 'normal' F+ strain (and they can even be isolated from such a strain). Thus Hfr must also have the sex factor F, but in a different form from that in the F+ strain. It was a characteristic of the F+ strain that the sex factor could multiply itself independently of the genome; it is a characteristic of Hfr strains that the F factor is *built into the genome* and can only multiply in association with this. An obvious parallel will at once be drawn between this and the system of prophage/virulent phage.

This explains why the offspring of a crossing-over Hfr F− always remain recipients and never become F+. Indeed they very rarely become Hfr cells, or even zygotes, but all remain undeveloped. This can be explained if the F factor is placed at the end of the genome and pushes the genome in front of it into the recipient cell, like a shunting engine. If this is so, then the F factor would only reach the recipient cell in the very rare instances where the entire genome penetrates, and only then could an Hfr cell be created. This has not yet been proved.

The Genome Is Located in the Tail of the Bacterium

The sexual system of *Escherichia coli* and other bacteria therefore differs considerably from a meiotic system, and it is not surprising that at first the recombination percentages turned out to be extremely contradictory, and so a genome map could not be made. Later on, too, there were still difficulties, even taking into consideration the incomplete transfer of the genomes, and the complications arising from the sex factor, which had been discovered in the meantime. If the genes set out along the genome were labelled a, b, c . . . x, y, z, it would be expected that a recombination (i.e., an exchange of bits) would occur especially frequently between a and z, since these were farthest apart. This does happen, for example, in the Hfr strain H, but in other strains, such as strain B 2, or strain AB 312, the recombination percentage is extremely low, so low that one would conclude from it that genes a and z were situated immediately next to each other. Table 4 sets out seven different Hfr strains, with the sequence of transfers, indicating the order of genes along the genome.

Table 4

Hfr Strain	Sequence of Transfers
H	a b c d e f g . . . .
1	b a z y x w v u. . . . . .
2	e d c b a z y x . . . . . .
3	g f e d c b a z. . . . . . .
AB 311	m l k j i h g f e . . . . .
AB 312	s t u v w x y z a. . . . .
AB 313	t s r q p o n m . . . . .

A *formal* explanation of this is possible only – and this is the last surprise for the present – if the bacterial genome is not stretched out straight (i.e., linear), but takes the form of a ring, so that the genes a and z are joined together. Some scientists still doubt whether any *real* ring exists (but see Fig. 31).

The peculiarity of the Hfr strain lies, therefore, in the fact that the F factor turns the 'ring genome' into different positions (naturally the position is the same for any particular strain), places itself at one end, and pushes the other end of the bacterial genome into the recipient cell. Fig. 73 shows the ring genome schematically, with genes a, b, c, x, y, z, and the starting positions of the F factors for all the strains from H to AB 313 as given in Table 4; and finally the *direction* in which the severed genome is pushed.

A transfer does not become complete, therefore, until the Hfr factor has pushed the entire genome into the recipient cell, and *itself travelled with it.* This happens com-

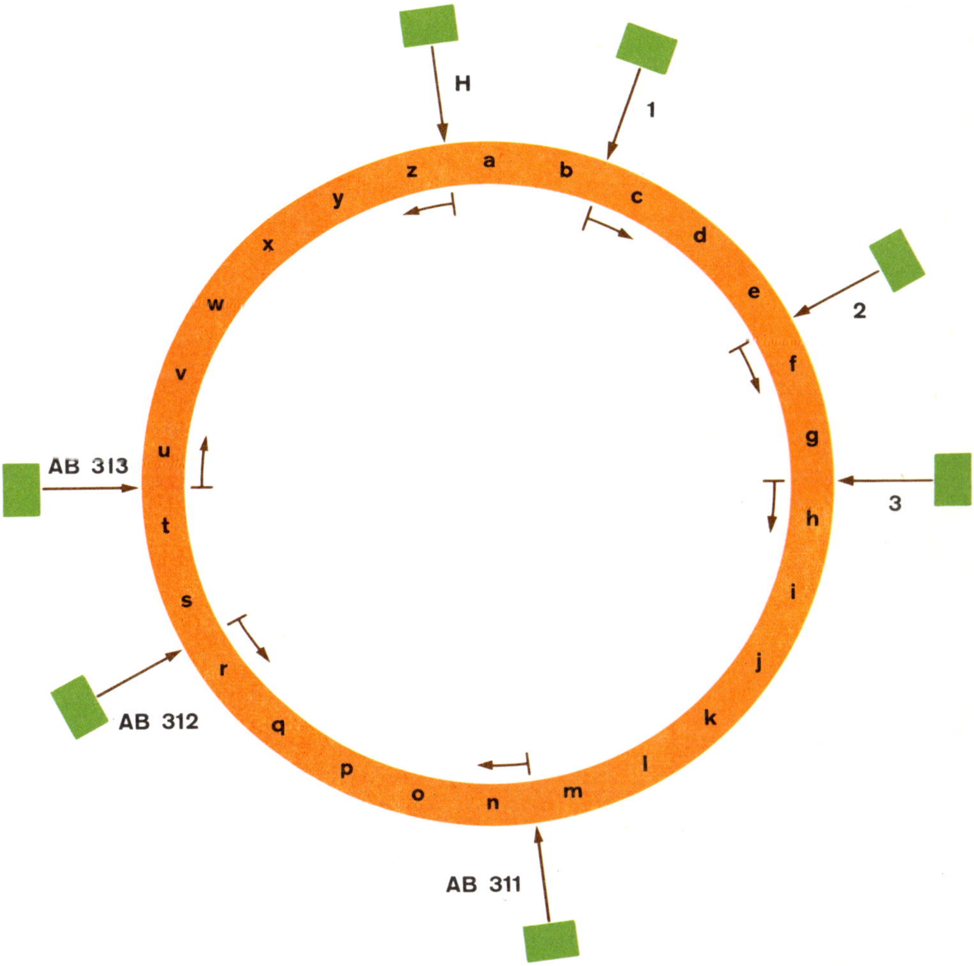

73. A 'ring genome' from a bacterium. The arrows show the points at which the F factor (green) penetrates the ring genome, and also in which direction the genome slides into the recipient cell.

paratively often in some strains, which are called **V**ery **h**igh **f**requency, or Vhf strains. Much more often, however, the pairing is interrupted before the entire genome has been transferred; usually the fragment comprises only about 10 or 20 per cent of the whole.

How the break comes about is not yet known. Also unexplained is how the fragment that is inserted becomes incorporated into the original genome of that cell. Is it a crossing over, as in higher organisms? So far we do not know.

Finally, what happens to those fragments of genome which are exchanged for the inserted piece, and which come to lie alongside the altered genome? Their fate is obscure. Probably they are not replicated at the next division, and 'get lost.'

Summarizing the whole picture, in *Escherichia coli* there are three types of sexual process:

1. *F factor absent.*

2. *F factor present, though not in the genome,* but free in the cell, where it multiplies itself independently of the genome. One or more factors can become altered into F –, but these may be reconverted into F +. Genome fragments are not transferred.

3. *F factor located in the genome,* and consequently the plastic F factors have disappeared and can no longer be altered into F –. This type does not permit an 'epidemic' outbreak of F factors. The ring-type genome is broken and 'injected' into an F – cell. Hfr cells transfer genome fragments, and *these only*. Occasionally the F factor can separate from the genome and move into the plasm; such a cell reverts to an F + cell, and is again 'infectious.'

Sexduction

The last-mentioned phenomenon, that the F factor can return to the cytoplasm, is very reminiscent of a prophage which leaves its position on the bacterial genome and, free in the cytoplasm, multiplies like an infection. We had seen that in this way the prophage could now and again pick up an adjacent piece of bacterial genome and 'transduct' it to another cell. Might this also be true of the sex factor F?

This has been observed in practice. Occasionally during the transformation from Hfr to F + a fragment of genome remains attached to the F factor, and is dragged along with it – exactly like a prophage (for example, prophage lambda, see Fig. p. 120). Because the F factor is highly 'infectious,' this genome fragment is transported into a newly infected bacterial cell. This gene obviously determines into what position in the genome the F factor will be incorporated – namely, that in contact with the homologous allele of the gene. The whole process is called 'transduction with the assistance of the sex factors,' or 'sexduction' for short.

At first sight this seems to be just a specialization, a curiosity without much meaning. We had better find out at once more about the question of episomes, and of the dreaded Resistance Transfer Factor (RTF) – or experience something many times worse. we may be in for serious trouble. However theoretical and abstract this may sound, at some time or other every one of us may find it a matter of life or death.

2.15 Bacteria Become Resistant through Infection

Hospitalization and RTF

It is already serious enough from a medical point of view that bacteria can become resistant through mutation and can then spread quickly through a population as a result of selection. All the same, mutations

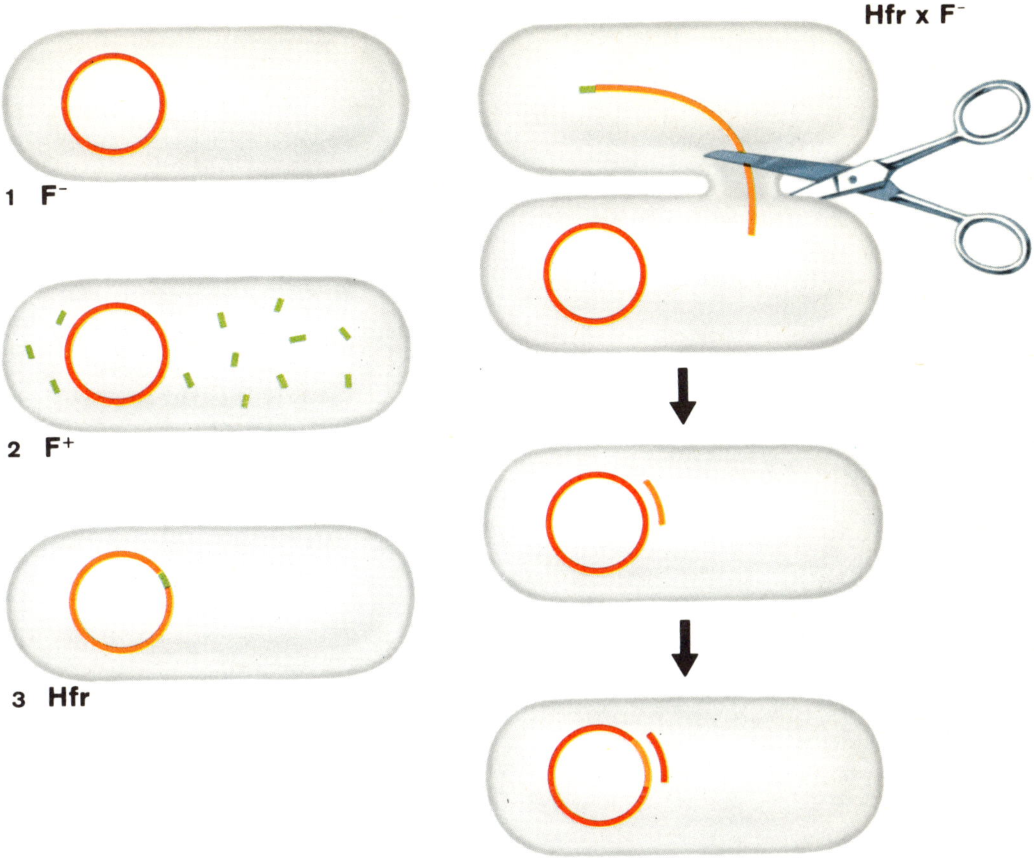

74. The F factor. On the left: 1, F factor absent (F^-). 2, F factor present (F^+), not yet in the genome, but free in the cell, where it reproduces itself independently of any genome. One or more F factors may be transmitted to cells that are F^-, thus making these F^+. Genome fragments are *not* transmitted. 3, Hfr. The F factor has penetrated into the genome, and consequently the F factors remaining in the cytoplasm wither away, and are no longer capable of being transmitted to F^+ cells. There is now no 'epidemic' spread of F factors. On the right: the effects of crossing Hfr and F^-. The ring genome is broken and passed into an F^- cell. Note that only Hfr cells, and no others, can transfer a genome fragment. Occasionally the F factor can detach itself from the genome and pass into the cytoplasm, whereupon the cell will become F^+, and again be infectious.

are rare, and spontaneous, double mutations ever rarer, while three-fold mutations must be regarded as altogether improbable. First, by mutation, a bacterium becomes resistant against sulphonamide, so its offspring are attacked with streptomycin; mutation makes it resistant to this also, so a switch has to be made to chloramphenicol (= chloromycetin), and so on.

Now, however, the bacteria have discovered something that is a thousand times more malevolent, the R factor. The possession of this factor makes the bacterial cell *simultaneously* resistant against four groups of antibiotics, namely against streptomycin, chloromycetin, the tetracyclines, and the sulphonamides. But there is worse to come.

A. This factor is transmitted during conjugation and spreads rapidly throughout the whole population, much more quickly than would be possible by the normal multiplication of bacteria.

B. A newly infected position on the genome is usually especially active in the business of conjugation.

C. The factor is not species specific; for example, it can be transferred at will between members of the so-called enterobacteriacae. There are various different enterobacteriacae, as follows:

1. The harmless colibacteria *(Escherichia coli)*, which are present in the human intestine, and to some extent are even necessary for normal existence.

2. The *Proteus* group, as well as a *Pseudomonas* group. These are equally harmless so long as they are in the intestine, but beware if they get into the urinary system! Obstinate nephritis and kidney inflammation are the result.

3. The *Salmonella* group, well known as the cause of typhus and paratyphus. One shudders to think of outbreaks of *Salmonella* poisoning that have arisen through unwise importation of foods infected with these bacteria.

4. The *Shigella* group, cause of bacterial dysentery. This disease broke out as an epidemic in Japan in 1957, and led to the discovery of the R factor.

5. The *Vibrio* group; its most dangerous member is *Vibrio comma (cholerae)*, the cause of cholera.

It must be admitted at once that a harmless colibacterium can sometimes acquire the R factor, and that this can spread like lightning in any of the enterobacteria. This happens not only in the normal environment of the bacterium, i.e., the intestine, but also in other places. It happens in foodstuffs and in dishwashers, for example, and even on towels that are used by more than one person – a practice that still goes on, even today. Indeed, hospitals are the places most seriously affected by this. The term 'hospitalization' sounds harmless enough, but if a patient should introduce into an isolation hospital a strain of bacterium that is immune to the antibiotics in use, the catastrophic result can hardly be imagined: no less than an outbreak of typhus or cholera.

These epidemics have not been mastered even today, but we now know a little about how they happen – and this is a prerequisite to being able to take effective measures against them.

The R factor is carried and transmitted by means of the RTF, the **R**esistance **T**ransfer **F**actor, and this RTF conducts itself in exactly the same way as a temperate phage and as the sex factor F. All three, as well as at least two other factors, are *episomes*.

2.16 Resident and Wandering Genes

Episomes – a New Kind of Genetic Units

An episome can be characterized as being an autonomous genetic structure, which can alternate between two forms.

Either it is free in the cytoplasm, where at any time it may multiply more rapidly than the genome. In this case the episome becomes 'infectious' in the same way as a lytic phage, as F+, and as RTF. The infection follows, independently of any transference of genome.

Or it is attached to the genome of the cell,

and under control. In this form an episome can multiply only at the same time as the genome of the bacterium, and moreover is transferred only in association with this – as a prophage, as Hfr, and as RTF (no distinctive name has yet been given to this type).

An important point here is that the two forms are obviously mutually exclusive. If the episome lies in or on the genome, then it is absent from the cytoplasm, so that the infection also disappears. On the other hand, by the use of ultra-violet light or acridine dyestuffs the phage (Hfr, RTF) can be 'induced' to leave the genome and move into the cytoplasm, where it begins to multiply rapidly.

The outstanding importance of the concept of episomes for medicine is now clear, but it also has wide-ranging consequences for genetics. If there exist 'vagrant' genes, which can wander between genome and cytoplasm, and in the process change their potentialities, then the idea of a gene must be thoroughly revised. The gene is now no longer the unit of mutation, of recombination, and of function, and it is also no longer bound to the genome or to the chromosome. Scientific opinion is still not unanimous on the subject of how this should be formulated, and with good reason. Before we can do this we need at least to know whether the change of location of the episome is random, or whether there is some functional unit that regulates it (if possible, a regulator gene?).

We will let this lie, temporarily, since the aim of the present chapter was to answer the question (p. 66), By what process is the genetic information of the older generation passed on to the individuals of the next?

We have investigated this process, and we have also seen that several mechanisms exist – and that the bacteria, more than any other group, are distinguished for a considerable array of phenomena that have been investigated with their aid: transduction, transformation, conjunction, sexduction, and so on. In only one thing have they not been successful: in evolving a genuine system of meiosis. However, their parasexual system allows them a very effective means of multiplication, and an extraordinarily rapid means – without which they would probably have died out long ago – of adapting their hereditary material to suit new conditions.

The answer to the second question that we set out on p. 66 – namely, How is the genetic information that has been transmitted used to build up the final characteristics? – will be studied in the next chapter. For this we need a rather more precise knowledge of the fine structure of the cell, and of its 'substructure.'

3. A GLANCE INTO THE CELL: NEW INFORMATION FROM THE ELECTRON MICROSCOPE

3.01 What Are Electrons Doing in a Microscope?

The processes described above will have made it clear that the phenomena associated with living organisms take place not only with the *assistance* of molecules – this is a truism – but also under molecular *direction.* If we want to learn a little more about living organisms, we must resign ourselves, for better or worse, to penetrating further into the unfamiliar world of molecules, and looking around there, all the more so because it has recently become certain that molecular concepts actually have a basis in reality. The strange logic of thinking in molecular terms is impressive, even if one is not very fluent in it, but it can also be irritating. Sometimes a small doubt creeps in when we see the certainty with which the scientists draw their conclusions, with what apparently careless abandon they leap from one end of a chain of chemical reactions to the other. They observe that a suspension of millions and millions of bacterial cells is suddenly in a state to ferment galactose, and they have an instant explanation: This can only be a phage at work; it has transducted the gal-region!

In the present instance the abstraction is certainly carried as far as the reader can be expected to follow it, and since most people are happier with the concrete than with the abstract, we have tried to help by giving diagrams. This cannot take us very much further, however, since these diagrams themselves are really only models, and so involve a degree of abstraction. The reader would no doubt be happier if the molecules that are doing the 'directing,' and the various phenomena of molecular biology, could also be *seen,* and if possible, photographed and made into a film. We have all heard and read about the electron microscope, and wonder whether this creation of modern electro-technique is not capable of making the molecular dimension visible.

This is a very good question. In Fig. 62 we saw the double helix of DNA from the head of the phage T_2. That is indeed a giant among double molecules, but its giant size is confined to *length;* the cross-section of the thread, scarcely exceeds 10Å. Why are there no corresponding examples among molecules of m-RNA, from protein chains, and from enzymes?

To understand this, and to assess it, to know what is to be seen in examples of this kind, we must have a little to do with the electron microscope. We want to confine this to essentials, because this is a book about biology, not a textbook of optics and electricity.

On p. 100 we set the genetic method and the workings of the electron microscope alongside each other and compared them in their 'resolving power' – so we can begin with that. The human eye, like the microscope has also a resolving power. At a distance of 25 centimeters, the normal viewing distance, the human eye can separate two points, or punctiform objects, if they are about one

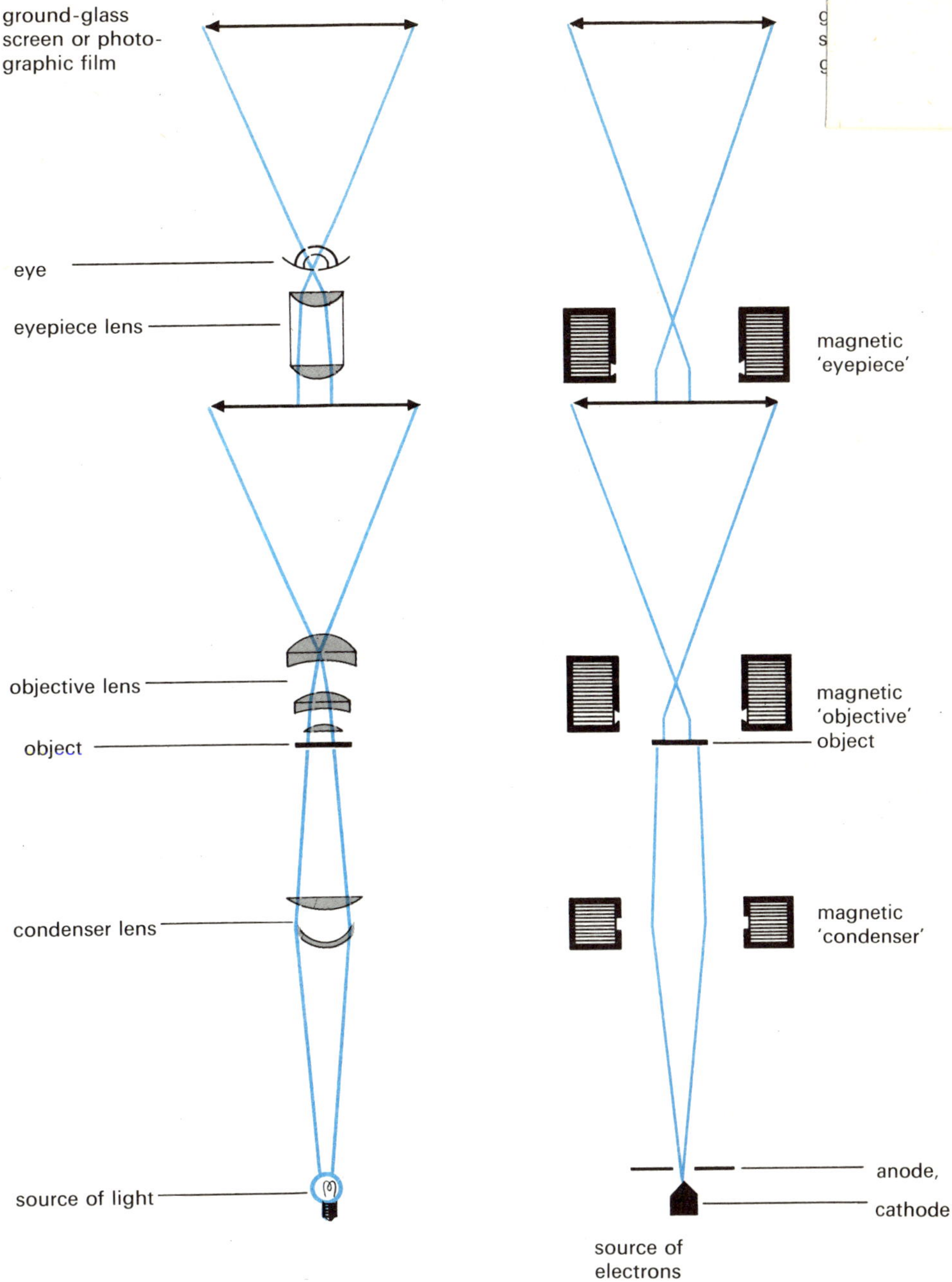

75. Comparison of the path of the beam in a light microscope and in an electron microscope. In practice the electron microscope works the other way up, see fig. 78.

tenth of a millimeter apart (the exact value for the distance apart is 0·073 mm.). The angle subtended at the eye by these two objects, called the *angle of view*, has therefore a limiting value of about one minute of arc (a complete circle is divided into 360 degrees, and each degree into 60 minutes of arc).

The light microscope increases the apparent separation of the two points, and therefore the angle of view, very considerably. This is done by using glass lenses, in a lens system. One part of this system, the so-called *objective* (because this is the lens that is held nearest to the object), projects a real (i.e., an actual) image, which is also called the intermediate image. This is then magnified again by a second system, the ocular (because it is placed close to the eye), acting like a magnifying glass. Finally, to complete the microscope, there is a *condenser* (Latin *condensus*, very thick), which collects the light. Occasionally, in place of the eye, there is a photographic film.

Objects that subtend an angle of only one minute to the naked eye are now spread out; they are much farther apart, subtending a much greater angle at the eye, and so allowing much more detail to be seen. Unfortunately the magnification of a light microscope cannot be increased indefinitely; there are definite limits, which are still a long way above molecular dimensions. For physical reasons, which need not be discussed here, the resolving power depends upon the wave length of the light that is falling on the object and passing through the microscope, and the limit is approximately one-half wave-length. So as to get the names right, let us recall the measurements of length insofar as they apply to microscopy:

One millimeter (mm.) = 1,000 micrometers (μm); this unit is abbreviated to the Greek letter μ and pronounced *mu*.

One micrometer (μm) = 1,000 nanometers (nm.), from the Greek *nanos*, a dwarf. Thus there are one million nanometers in one millimeter.

The next lower unit is the ångstrom unit Å. 1A = 1/10 nm. = 1/10,000 μm = 1/10,000,000 mm.

The wave length of light varies a little according to the colour: as a rough figure it lies around 0·5 μ = 500 nm., so that the resolving power (half a wave length) is about 0·2 μ = 200 nm. This is a considerable improvement over the human eye: from

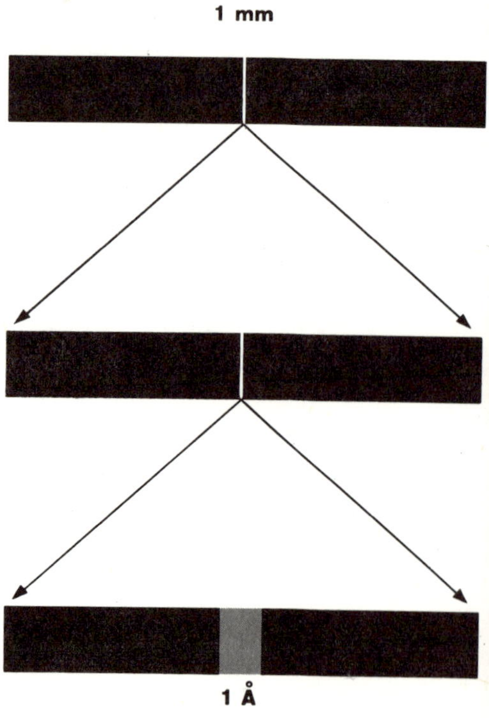

76. Comparative scales of lengths applicable to the needs of microscopy. If the top band is taken to represent one millimeter (1 mm.), then the thin white line may be taken (diagrammatically) to indicate one thousandth of this, one micron (1 μ). If this is now enlarged up to the full length, and a white line represents one thousandth of this, the white line is one nanometer (1 nm.). Repeating the process, one *tenth* (not one thousandth) of a nanometer is one angstrom unit (1A).

0·1 mm. = 100 μ to 0·2 μ!

On the other hand, the distance apart of two atoms in the same molecule is about one ångstrom unit; one molecule of grape sugar (glucose) has a cross section of about 5Å = 0·5 nm., and even that is one four-hundredth of the resolving power of the light microscope. Dimensions such as these are thus far outside the range of the light microscope. Is the electron microscope equally wide of the mark?

The electron microscope makes use of the following facts:

1. That so-called elementary particles, such as electrons – which are elementary particles with negative charges – when travelling at very high speeds behave like electric waves, and if they are given the appropriate treatment, can be handled like light rays.

2. That a relationship exists between speed and wave length; the greater the speed of the electron, the shorter the wave length of the electronic radiation (according to an equation formulated by Louis de Broglie in 1924).

The resolving power can be raised above that of the light microscope if the electrons can be made to move very quickly. Nowadays it is technically possible, with our present resources, to accelerate the electrons that emerge from a source of radiation to a very high speed; with voltages of 40,000 to 100,000 they reach speeds of up to 200,000 kilometers per second. From Broglie's equation – which takes account of other factors, e.g., the mass of the particle – the wave length then works out at about 0·05Å, which is one twentieth of the average distance apart of the atoms!

If the electron microscope could be operated in *exactly* the same way as the light microscope, focussed with lenses, and so on, then it should have a resolving power of a half wave-length = 0·025Å.

Actually this is not true even of the light microscope itself. The electrons have only a very minute mass; in spite of their high speed, they are absorbed by the glass of the objective and ocular, i.e., they 'stick fast.' Not even in air, which is present in the tube of the light microscope, can the electron beam travel far. As soon as the electron strikes a molecule of oxygen, nitrogen, or carbon dioxide it is retarded, or at the very least, deflected from its path, 'scattered.' So we cannot work either with glass lenses or in air, and this imposes a condition that is very troublesome for biological purposes: the electron microscope must be *evacuated of air*.

Fortunately ways exist of deflecting beams of electrons without lenses; the direction of a beam of electrons can be changed by using an electric (i.e., high-tension) field or a magnetic field, whichever is convenient.

Its operation is so similar to that of a glass lens that we even speak of electric or magnetic 'lenses.' Thanks to this fact, the electron microscope can be constructed to exactly the same plan as a light microscope, with the same sequence of lenses: condenser, objective, projector lens, and screen or photographic film. The term 'projector lens' mentioned here is apparently new, but it is also called a 'projection ocular' and exactly matches the ocular of the light microscope. It projects the electron beam onto a photographic plate or a fluorescent screen, for the human eye does not react to a beam of electrons. The fluorescent screen, like the screen of an X-ray apparatus, contains substances that glow when they are struck by an electron.

As Figure 78 shows, the direction of the rays in an electron microscope is the opposite of that usual in a light microscope; this is merely for practical reasons. The human eye looks down more easily than up, and so it is arranged that one looks down into a light

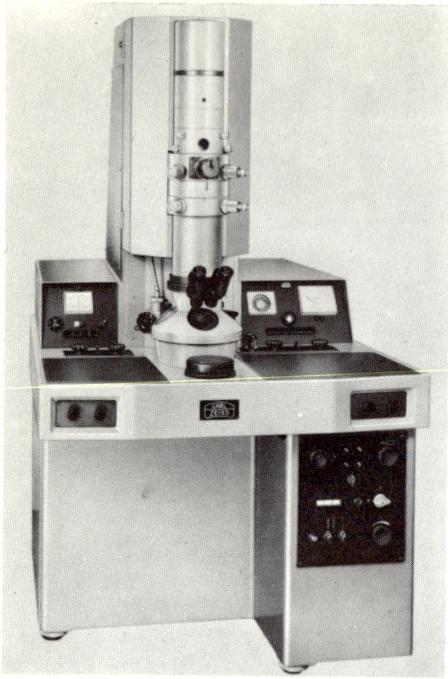

77. A modern electron microscrope. Its resolving power may be in the region of 10–12A.

microscope from above. The electron microscope, however, is usually so big that the observer would have to stand on a ladder; to avoid this, the microscope is reversed, and the eye allowed to look at a fluorescent screen at the bottom.

There we have the working principle of the electron microscope, in broad outline, and omitting all the physical details – but unfortunately the resolving power is not as high as we had hoped. The electric and magnetic lenses are not nearly precise enough; they are not able to control the beam of electrons as strictly as the glass lens controls the light beam. As a result, both the angle of view and the resolving power are smaller than expected. For magnetic lenses the resolving power, *at its maximum*, is about 3Å, not 0·015Å as we had hoped. In spite of this, we can compare this figure with the 0·2 μ = 200 nm. = 2,000Å of the light microscope, so it appears that the electron microscope magnifies about 700 times as much as the light microscope. Since the length of a molecule of sugar is about 5Å, we can actually see it under the electron microscope.

But what do we actually see on the fluorescent screen? Nothing more than the points of impact of the electrons – that is to say, a fluorescent spot on the screen corresponds to a spot on the preparation which has allowed an electron to pass through without either absorbing or deflecting (scattering) it, and by the same token, a dark spot on the screen indicates a spot on the preparation which has not allowed the electron to penetrate, but has either absorbed it or scattered it. Absorption

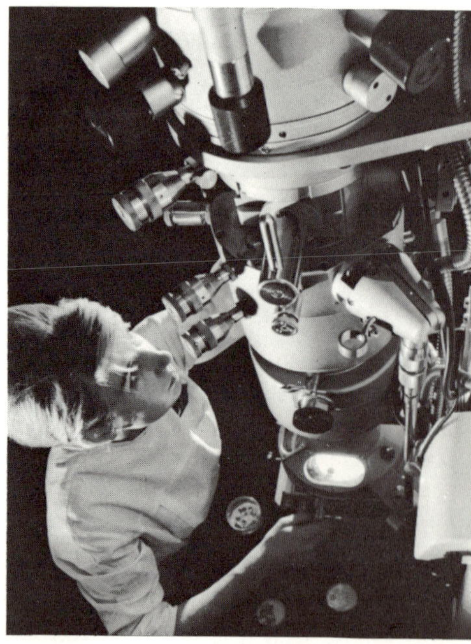

78. Detail of an electron microscrope. The source of the electron beam is at the top, the illuminated screen at the bottom; this is the exact reverse of arrangements in a light microscope.

and scattering, however, are dependent on the thickness of the preparation, or more precisely, on the product of density and thickness. The greater this is, the less penetrable is the preparation.

1. Variations in brightness on the screen give us information about the variations of density in the preparation – and nothing more. They do not, as in the light microscope, tell us anything about refraction, diffraction, or differences in colour.

2. If a preparation is uniformly thick, then we see only uniform brightness on the screen – there is nothing to break it up.

3. And finally, if a preparation is too thick, then we see nothing at all.

This has one disagreeable result. Even the smallest cells, bacterial cells for example, have a cross-section of more than 1 μ = 1,000 nm. = 10,000Å, and that is much too much! If we put a bacterial cell under the electron microscope, it will therefore not allow any electrons through, and will appear on the fluorescent screen as a solid black spot. If we want to find out about the details of the cell, we must cut it into sections, and very thin sections at that. Experience has shown that the thickness of a section must be less than 100 nm., and preferably, it should be between 30 and 50 nm., that is 300 and 500Å. Thus bacteria must be cut into at least twenty sections – the task is equivalent to shaving a sheet of typing paper into over a thousand sheets! Technically this is no longer a problem: in the last twenty years *ultramicrotomes* (Latin *ultra*, extreme; Greek *mikros*, small;

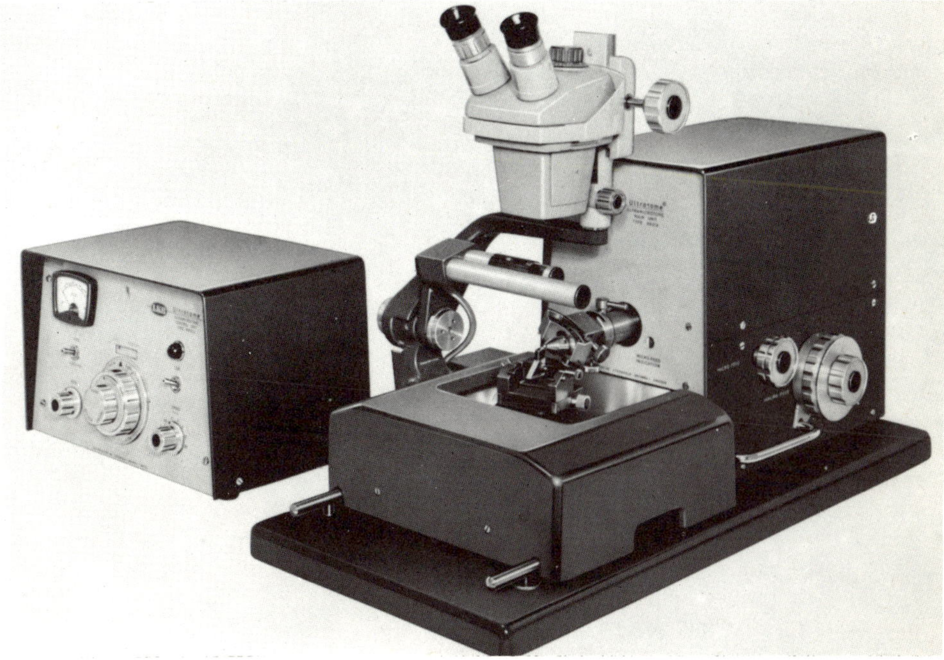

79. An ultramicrotome, for cutting particularly thin sections. On the right is the cutting part of the apparatus, equipped with a microscope for controlling the cutting visually; on the left is the electronic equipment that directs the cutting.

Greek *tomein*, cut) have reached a high degree of perfection.

Of course a bacterium, a yeast cell, or a piece of animal or plant tissue cannot be sectioned just as it stands; the material must be 'embedded' beforehand in some sort of artificial medium, which is liquid at first, and then becomes hard, a plastic. This medium must naturally have no microstructure of its own which might become visible under the electron microscope; it must be uniformly transparent to radiation.

This disposes of the last hope that it will be possible, in the foreseeable future, to study living cells under the electron microscope, or to film what is going on inside them. The cells that we can observe in this way are dead, embedded in an artificial medium, sectioned; above all, they have to be brought into a vacuum before they can be photographed. Moreover, a section only 500Å thick is so delicate that it can be shattered by the slightest vibration. Before it can be handled, and put into the electron microscope, it must be attached to a relatively firm base, that is, deposited on a slide. This is almost always a plastic sheet about 100–300A thick, which, just like the embedding medium, must be uniformly transparent. This plastic foil is reinforced by a tiny network of metal threads. The metal threads are, of course, opaque to the electrons, and the microscope picture is seen through the mesh of the net (see Fig. 80).

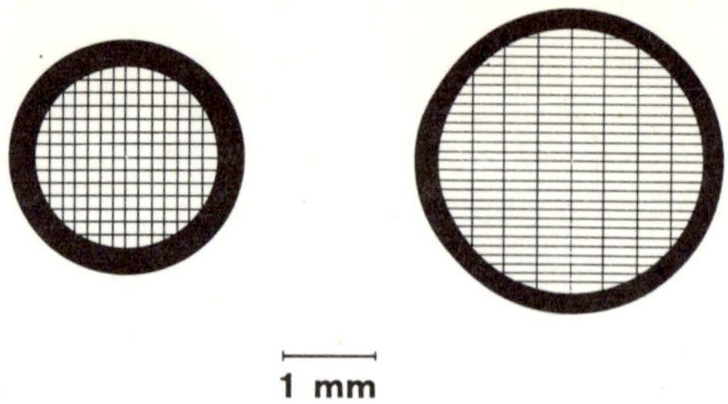

80. Specimens under the electron microscope must be supported on a mesh of the finest metal wires, to prevent collapse of the ultra-thin section and the foil on which it is spread.

And what sort of a view do we get of a molecule of sugar? What is theoretically possible turns out not to be feasible in practice. Such a molecule, 5Å thick, part of a section that has five hundred times the thickness, and surrounded by a large number of other molecules, similar and different, cannot be picked out – or at least, it cannot be picked out without using a trick, which will be discussed later. For this to be possible, the molecule in question must have a density that is very significantly different from that of *all* the surrounding molecules, and unfortunately that hardly ever happens. The component parts of most molecules are all very similar in density, and this density is very close to that of the surrounding water, or of the embedding medium; only very rarely is the difference sufficient to give a contrast under the electron microscope.

The relative density of any particular molecule is naturally greater when compared with a vacuum than when compared with water. As an experiment, therefore, a few drops of a very dilute solution of sugar can be spread over one of the plastic foils and allowed to evaporate, so that the molecules of sugar will lie separately on the preparation. But nothing comes of this either. The individual molecules are left lying on a plastic material that is 100–200Å thick, and this, however pure and structureless it may be, still gives an uneven background (cf. Fig. 61) against which the sugar molecule disappears. And even if we can distinguish the cell against its background, what good is that? For biological purposes we want to see *what is inside the cell*.

3.02 Inside the Cell with the Electron Microscope

What do electron microscope pictures show us of the details of the cell? Have they produced anything new, or do they merely repeat what is already known from the light microscope? Textbooks on this subject begin as a rule with the same picture of the cell, which is that given by the light microscope. Undoubtedly this is the correct thing to do for students; because of the tremendous difference in resolving power between the two microscopes, the students would probably be lost if given the details shown by the electron microscope, and be unable to see the wood for the trees. The details shown by the electron microscope can only be *correctly* interpreted by comparing them over and over again with photographs taken through the light microscope, which makes it all the more unfortunate that the electron microscope shows only differences in density, and the two sets of photographs are not strictly comparable.

The textbook usually begins its chapter on the cell with the words 'The typical cell consists of . . .'; then a fairly comprehensive catalogue of the components of the cell follows. If it is a botanical textbook, it confines itself to 'The typical plant cell . . .'; if a zoological textbook, to animal cells. Very likely, however, there is nothing general to say about 'the cell' pure and simple, for 'the cell' is little more than an abstraction, a statistical unit.

Cells, the smallest living units, are nevertheless very varied, and there are thousands of different kinds, classified according to their structure, function, condition, age, and so on.

The differences are often more important to the scientist than to the general reader; he can draw conclusions from them about the stage of development, differentiation, functional readiness, and many other matters. He will often find it difficult, in face of the multiplicity of detail, not to lose sight of the general picture which is his ultimate aim. The general reader, on the other hand, only wants, with the help of the electron microscope, to take a glance into the cell in order to find out what goes on there, what is the significance of the various parts of the cell, and what each can contribute to an understanding of the mechanism of life.

Accomplishing this task will certainly be made easier if we can make use of well-chosen photographs to show the structures which can be found in all or at any rate most kinds of cells. We can then try to put together the units to make a picture of an entire cell. This expresses the basic aim of all science: the breaking down of the cell into its smallest structures, right down to molecular level – the analysis, in fact – is not the final goal; some time or other it will be possible to reconstruct the complete living cell out of these details.

3.03 The Protein Structure of Ribosomes – This Time within the Cell Itself

It will certainly be simplest if we begin with the ribosomes. We know these already, from Chapter 1, as a kind of protein synthesis; we saw that they could be extracted from the homogenized cell-contents by the use of the ultracentrifuge, in such a pure state that they give a clear picture on a slide in the electron microscope (see Fig. 20). Can the ribosomes also be seen *inside the cell?*

Figure 81 shows us a small section of a young cell; this is a so-called epidermal cell (from the outer skin), from the root of a four-day-old radish seedling. We can recognize a

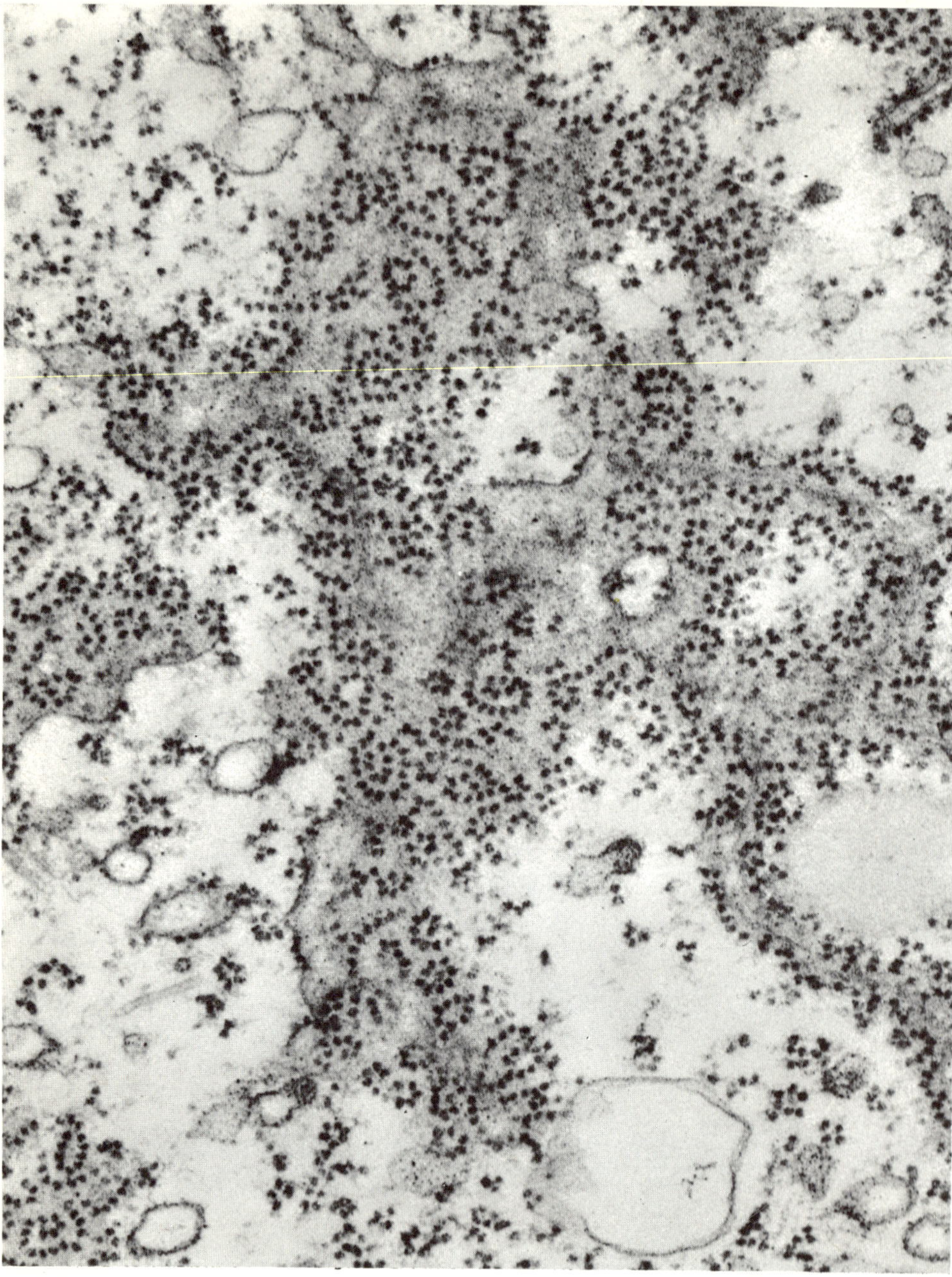

81. A preparation from the surface of a radish root, showing ribosomes, some of which are twisted into spiral polysomes.

more or less grey background, sprinkled with a few small bubbles, and then innumerable smaller, darker (because they scatter electrons) bodies, which are seen here *in situ* (Latin *situs*, place), or on the spot. Nowadays it is absolutely certain that these small corpuscles are actually ribosomes, the same ribosomes that can carry out the synthesis of proteins in the test tube, i.e., *in vitro*. Actually it took several years, and needed a great deal of comparative work, using different methods, before it could be proved that the *in situ* and *in vitro* particles were identical. We shall not try to check this, but put ourselves in the hands of the electron microscopists and the biochemists.

The ribosomes are very uniform in size. They are spherical in shape, with a cross-section of about 15 nm. = 150Å. Nevertheless, they are not quite uniformly distributed. Moreover, there are regions which are free of ribosomes, and, contrasting with these, other regions where ribosomes are aggregated. Furthermore, when they are reproducing they usually lie close to each other in chains, which are sometimes twisted into a spiral. They are probably held together by a molecule of m-RNA (see p. 54). And something else. Under higher magnification those in one region appear as if they made up a sub-unit, which might be one of the 30-S, 50-S, or 70-S units that we met earlier (p. 46).

Actually the irregular, apparently completely random distribution of the ribosomes is surprising. Certainly, several or even many of them may be combined together into polysomes – but the polysomes themselves are then scattered in quite a disorderly way over the background. May the background substance of the cell, i.e., the cytoplasm, be no more than a structureless mass of variable density, in which the ribosomes are simply scattered? This would mean that our concept of the cell as a system with an extraordinarily high degree of organization would have to be thrown on the scrap-heap.

There is something more still to be said about the section that is shown in Figure 81. It was cut almost exactly parallel to the side wall of the epidermis cell, and very close to this wall. If we now cut a second section (in a different cell), at right angles to the previous one, we get a completely different picture (Fig. 82). Even here there is also a background with variable particles, and the small bubbles are again visible. The new feature is several long 'tubules.' Some of them end blindly in the background material; others run out of the picture; others again seem to be tangled together. The last are shown clearly if two pictures are taken one behind the other, i.e., showing two successive sections. The 'tubules' are obviously empty, in the sense that their contents are structureless. This can be seen most distinctly where they widen out (lower right). The tubular walls, on the other hand, give a clear contrast, and constitute a membrane which is dark (because it scatters electrons) and runs from one end to the other. (Details later, p. 150.)

But let us return to the ribosomes, which we find as in Figure 81, irregularly scattered over the background, often in the form of polysomes. Furthermore – and this, too, is new – they are present in greater numbers, and very closely packed, on the outer surface of the 'tubules'; never are they found inside the tubes.

For the time being, we must just accept this fact. We do not know whether ribosomes settle for preference on the outer surface of the 'tubules' because they are 'active' in protein synthesis, in contrast to the 'inactive' ribosomes that are to be found in the surrounding medium, or whether they differ in some

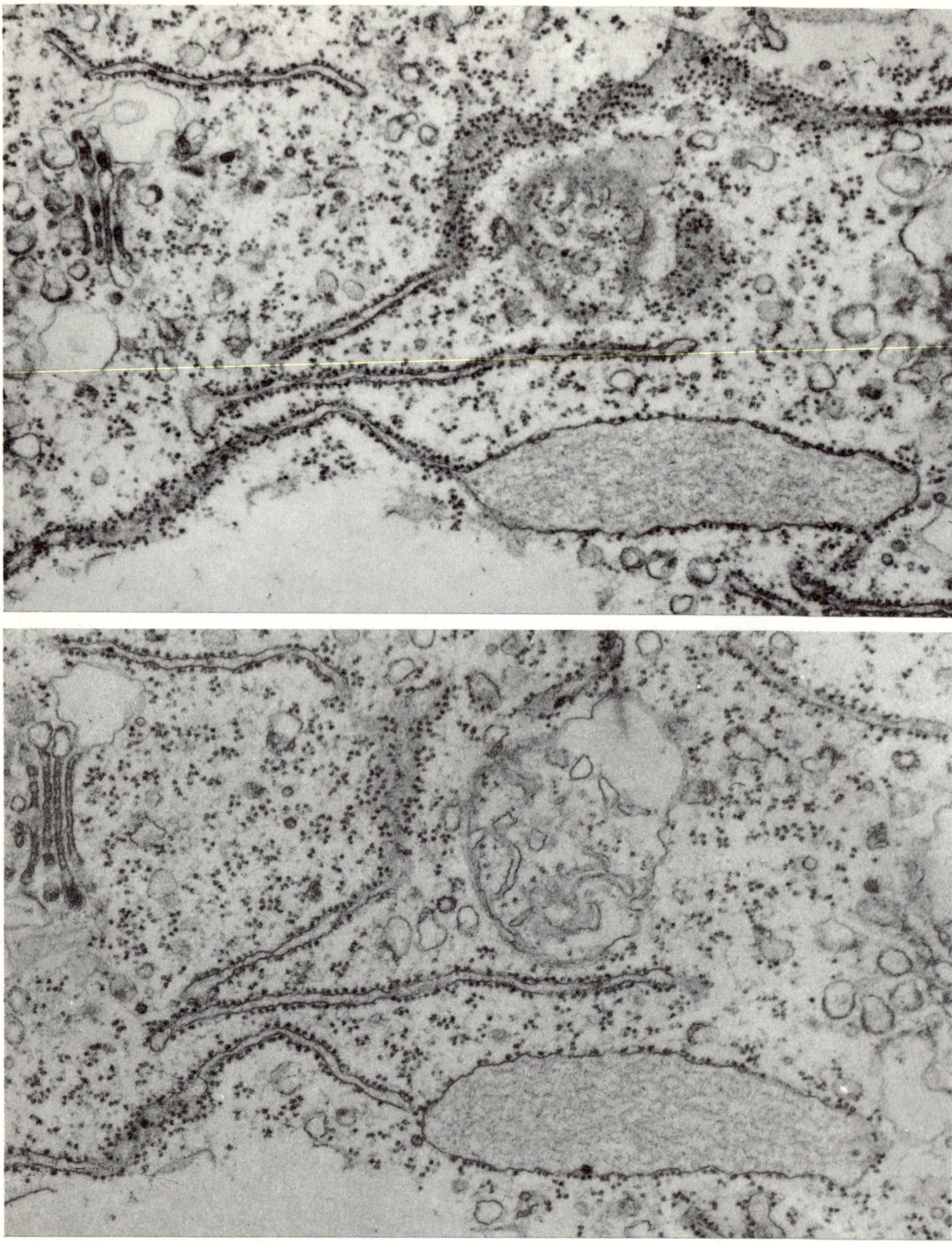

82. The same cell as in Figure 81, but sectioned in a different direction. Again, ribosomes are to be seen in great numbers, but there are also several cisterns belonging to the endoplasmic reticulum, and the ribosomes can now be seen to be predominantly lying on the membranes of the reticulum. The two sections, above and below, are immediately consecutive.

other way, if at all. On the other hand, it is certain that ribosomes are to be found in cells which are actively synthesizing proteins – thus, for example, in young cells or in those which are producing antibodies (of this more on p. 254); old, worn-out cells are poor in ribosomes or even ribosome free. The preference of the ribosomes for settling on the outer surface of the 'tubules' is remarkable all the same, and should make us look at this situation a little more closely. Up to now, as a precaution, we have always put the word 'tubules' in inverted commas, to emphasize the provisional nature of the name. In reality these 'tubules' are only rarely as round in cross-section as a true tube. Much more often they are flattened in cross-section, with the appearance of flattened blisters. This is further confirmed by the fact that the tubules in the upper and lower parts of Figure 82 appear in the same position, and since they are shown on successive sections, they must extend in depth to at least twice the thickness of a section; this would be about 600 to 1,000Å; on the other hand the 'overall length' *in the plane of sight* may sometimes be no more than 200 to 300Å.

For these flattened, hollow spaces – which naturally are not really empty, but whose contents do not seem to have any particular structure, and in which ribosomes are *never* found – the expression 'cisterns' has come into use (from the Latin *cisterna*, actually a pit for catching rain water). That they do not always have to be so narrow has already been established from Figure 82; occasionally, too, they may swell.

The cisterns communicate with each other, and the whole assembly is called the 'endoplasmic reticulum,' because it constitutes a net-like (Latin *reticulum*, small net), ramifying system of canals, cisterns, etc., embedded in the plasmic ground-substance. Figures 81 and 82 by themselves do not show this relationship very well, but it will become clear as soon as the next few diagrams are studied. Figure 83 shows a cell from a leaflet of the water hyacinth, which contains numerous cisterns, in a tangled array, with many interconnections, and in Figure 84 three cells from the pancreas can be seen in contact with one another. Here the cisterns are strongly narrowed, and so orientated that they run parallel to the margin of the cell.

However, before we devote our attention to this remarkable arrangement let us take a look back at the ribosomes of Figure 81. There is no trace there of an endoplasmic reticulum, in general; the scattered bubbles probably have nothing to do with this, since they have no ribosomes attached to them. This does not mean, however, that a reticulum is completely absent in this case. The section through the cell was cut very nearly parallel to the cell wall, and very close to it. Comparison with Figure 82 tells us that we often see a cistern from the edge, i.e., we look from the outside of the cell to the outer margin of the cistern, and so the ribosomes that are attached to it fill the field of view; in a cross-section of the cistern (as in Fig. 82) these ribosomes would be seen to lie around the outside of its walls.

After our pessimistic comments on p. 142 about the different constituents of the cell having only insignificant differences in density, it is really surprising that the ribosomes stand out so clearly from their background, with such a strong contrast. Are they actually so much denser than the matter that surrounds them – denser, for example, than the walls of the endoplasmic reticulum?

In actual fact a little bit of help has been given here, to make the contrast stronger. It appears that nucleic acids – and thus the constituents of ribosomes – have a tendency

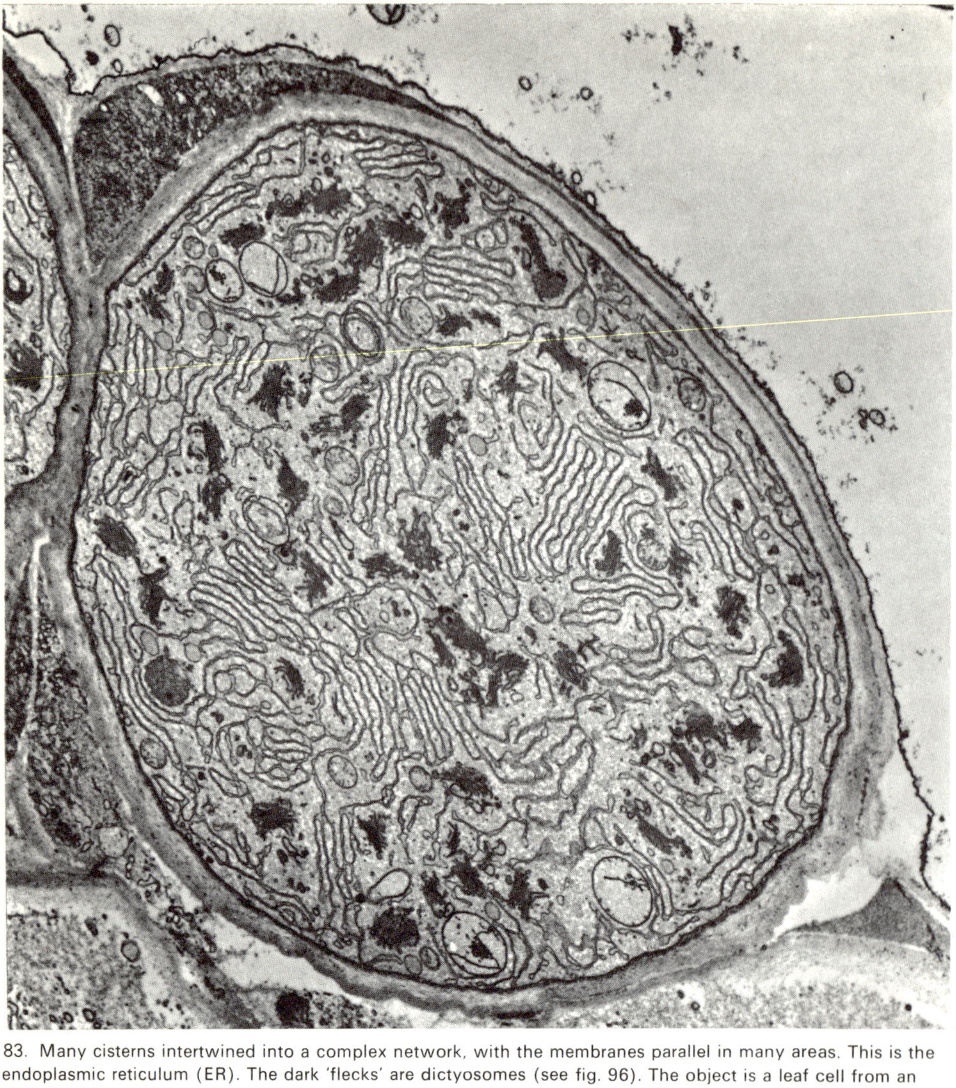

83. Many cisterns intertwined into a complex network, with the membranes parallel in many areas. This is the endoplasmic reticulum (ER). The dark 'flecks' are dictyosomes (see fig. 96). The object is a leaf cell from an aquatic plant.

to combine with certain salts of heavy metals, more intensively perhaps than proteins and fat-like substances. If we treat a section – or better still, the whole piece of tissue before it is embedded and sectioned – with uranyl acetate or lead citrate, then these 'preferred' substances attach themselves to the structures that contain nucleic acids, and so 'mark' them. Uranyl acetate is a compound of uranium (or uranium plus an acid) with acetic acid, and lead citrate is a similar compound of lead with citric acid. Uranium has an atomic weight of 238, lead of 207 – that is, they are 238 times and 207 times respectively as heavy as one atom of hydrogen; thus they are also very much denser. Consequently, they absorb or scatter electrons very much more than the molecules of water, albumin, or nucleic acids. The marking thus becomes visible under the electron microscope, since the structures to which uranium or lead is attached appear darker than their surroundings.

In many cases the increase of contrast is obtained by working the other way round. This is a very practical method for those substances or those cell constituents (and also for bacteriophages) – that for one reason or another are repellent to heavy metals. Then the *surroundings* of the objects that are to be picked out are, so to speak, impregnated with the contrast-producing substance. So the background appears darker, and the other objects remain clear, giving a picture that is the exact opposite of that produced by uranyl acetate. It is completely immaterial whether the object we wish to study is pale against a dark background or dark against a light background – under the electron microscope we are only interested in the degree of contrast, whether it is positive or negative. American scientists call this process, somewhat inaccurately, negative staining. Just as for the light microscope we can stain certain parts with a particular colour and so make them easier to interpret, so for the electron microscope we can stain parts with heavy metals; but naturally these will not appear in colour, but only increase the black-and-white contrast. The phage T_2 of Fig. 56 can be made visible by the negative-staining technique.

3.04 Now, at Last, for the Endoplasmic Reticulum

Transport Routes within the Cytoplasm

We have already met two manifestations of the endoplasmic reticulum: the irregular arrangement in the cell from a root hair, and the peculiarly uniform direction of the cisterns, parallel to the cell wall in cells from the pancreas. In both cases there is the same structure, the endoplasmic reticulum, usually shortened to ER. If it can take such different shapes, we are entitled to ask, Is the shape object-specific, i.e., do root-hair cells always have an irregular reticulum, and do the pancreas cells always have it in parallel layers? Can all the possible kinds of cells be characterized by their unchangeable ER structure? Both questions must be answered in the negative. The form of ER that is visible in a section is never unique and unchangeable. Quite the opposite. As the young cell grows up, when it is in full activity, when it grows old and its working powers decline – at each stage its ER is also changed. To begin with, only a few, irregularly distributed cisterns are present; they multiply, occupy a significant part of the interior of the cell, and take up a definite arrangement. In old cells, on the other hand, the ER in some parts of the cell appears moribund, and becoming functionless, loses its coherence.

Furthermore, the ER can change completely *during* a particular stage of development of the cell. We have at least one example of that. Figure 84 shows cells from the pancreas *in a starved condition.* Compare this with Figure 85. These are the same cells, but this time, after starving, and before being fixed, they were supplied with food for one hour. The parallel arrangement is still present, but the elongate cisterns are now no longer straight, but are curved into concentric circles – or more correctly (though to see this we must examine not a single section, but a complete sequence of sections in three dimensions), they are tightly packed together into balls (hollow spheres).

Not all these spheres are tightly closed; on the contrary, many have the cisterns interrupted, so that the spheres have 'holes' in them. It might be expected that the spheres would assemble around the nucleus of the cell, but this only rarely happens. Much more often there is to be found in the middle a mitochondrion (see p. 161), or else a 'secretory granule,' or 'secretory droplet,' which is a granular or drop-like collection of matter which the cell itself produces and passes out to the exterior – into the blood stream, for example, or into the lymphatic system. (cf. fig. 164).

The ER is thus comparatively labile, and is constantly being rebuilt and reorganized, and that very quickly as a rule. What can be seen under the electron microscope is therefore only a 'snapshot,' a momentary picture of a continuously changing process. We must be careful not to attach too much importance to any momentary state of the ER, however regular and 'beautiful' it may seem.

Obviously it would also be unwise if we were now to go to the other extreme and label the ER, in view of its constantly changing structure, as 'dubious' – as an accidental effect, or even as an artifact produced by the preparation of the object for inclusion in the electron microscope. (The question of artifacts in thin sections will be discussed later on.) On this view it would often happen, and happen by chance, that the cisterns would get tangled together into a kind of 'canal system.' Even the arrangement of hollow spheres, all closed up together and changeable though it is, makes us suspect that the cell is not merely a container into which several million enzymes and ribosomes are thrown at random, but that all their individual locations add up to an organized structure – that the cell displays an 'architecture,' which is not any less to be admired because it is constantly changing in incomprehensible fashion.

Within this architecture, the principal function of the ER, without a doubt, is that of bringing things into order. As we have seen, the cisterns themselves are 'optically empty,' that is, they show no particular structure, and no contrast that is effective under the electron microscope. On the outside they have the surface layer, the ER membranes, which, in contrast with the interior, are beset with ribosomes. Outside the cisterns there are also the mitochondria, the plastids, the dictyosomes (we shall become acquainted with these later), and of course, the cell nucleus. In this way the ER membrane divides an *extracisternal phase* (with mitochondria, etc.) from an intracisternal (or 'empty') phase, and the intracisternal phase threads its way through the entire interior of the cell, according to laws which we hardly yet understand, but which are obviously very strict, like all laws of architecture.

Admittedly this tells us little or nothing about how the ER actually works, or what purpose it serves. In this respect we are still entirely dependent on speculation, which, unfortunately, has to reconstruct phenomena from snapshots, and hence is an extremely troublesome and often completely hopeless undertaking. Who, indeed, would try to work out how a motor car operates, from a series of single snapshots of a moving automobile? Pictures of the body work would be of no use, and we should have to make sections of the car, longitudinally and transversely, and then photograph them. Who

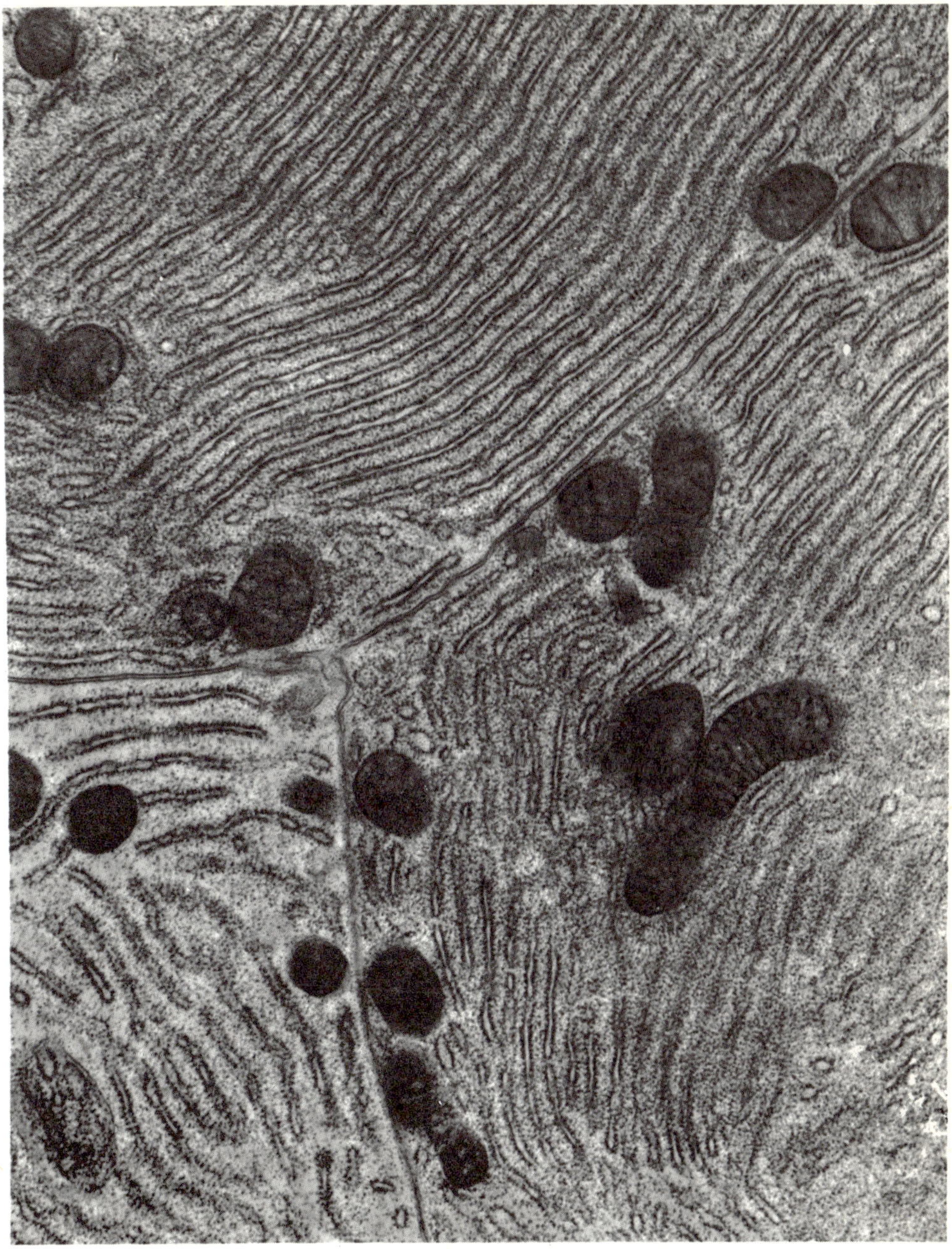

84. Three adjacent cells from the pancreas, in which the ER cisterns are especially numerous, and mostly run parallel to the cell walls. The dark, egg-shaped masses are mitochondria (see fig. 91).

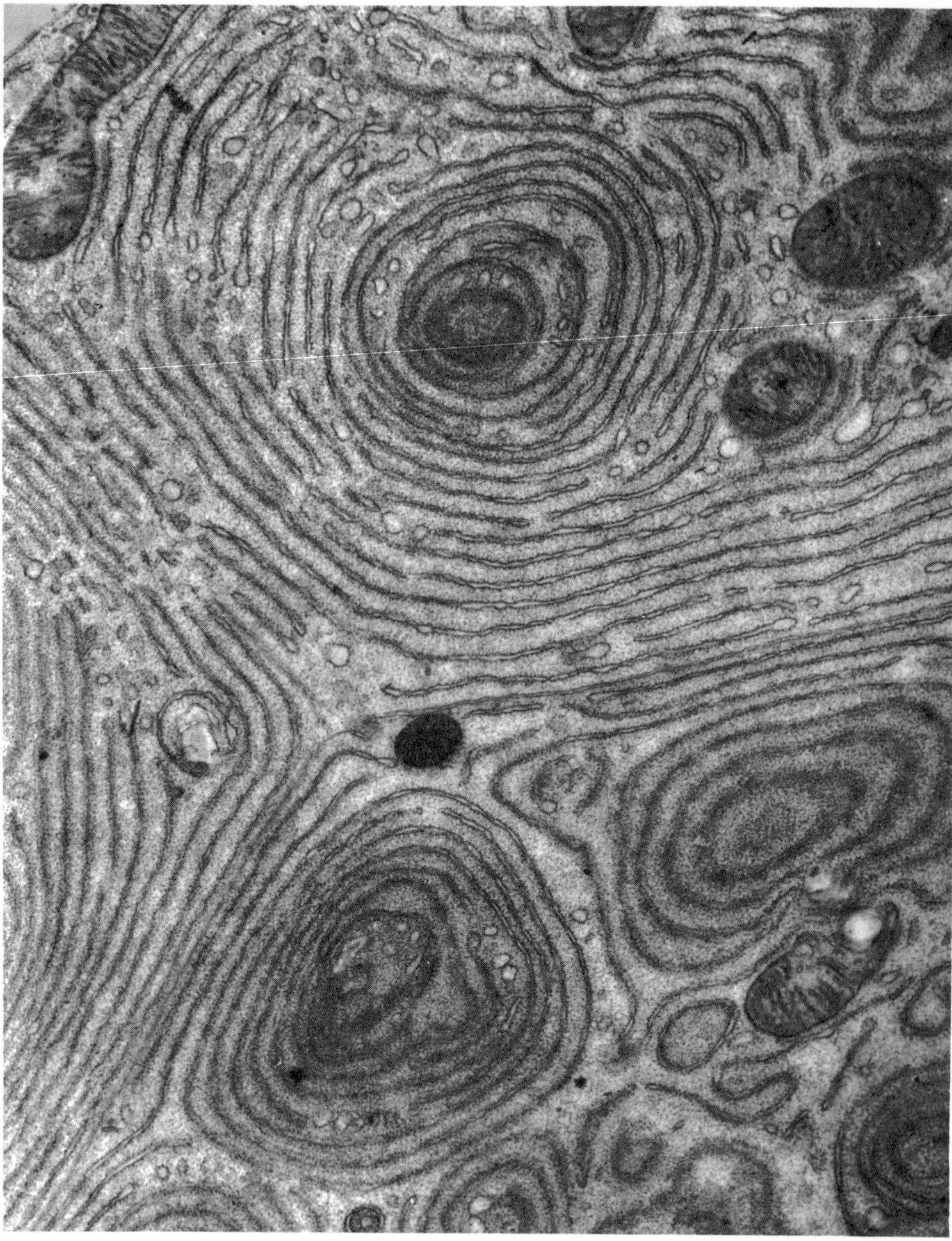

85. Pancreas cells, as in the previous picture, but this time well-fed ones. The cisterns of the ER appear to form approximately concentric rings, though in three dimensions these are actually hollow spheres.

would guarantee that the engine block, after we had sawed through it in order to take a snapshot, would play its part in the operation just as well as before, so that the motor as a whole could go on working? Here, once again, we come up against the problem of artifacts!

There is a difference in the fact that the cisterns of the ER have *a transport function.* Indeed, because they have no trace of any objects that have a structure visible under the electron microscope, but appear 'optically empty,' we must conclude that they contain predominantly small molecules of nutritive substances, protein-building materials, and so on, molecules which are in transit from the places where they are being produced to the places where they are needed; the sites where these substances are being produced and consumed, themselves lie in the extracisternal phase. In that case the cisterns would be canals, transit routes along which substances could be transported, probably more quickly than if they had to be dragged with difficulty through the turmoil of ribosomes, mitochondria, dictyosomes, etc. The substances would then have merely to pass through the walls of the cistern, the ER membrane, either from inside to out or vice versa.

Some such explanation immediately comes to mind, even though, so far, proof is lacking, but this will always be so. To explain *structure* on the basis of *function* is just as dangerous – and unlikely to be correct – as to carry out the reverse process of explaining function, in this case the movement of materials, on the basis of the underlying structure. And this is indeed the business of cell and molecular biology – surely no easy matter: to find out how structure and function fit together, how they mutually depend on each other, even to demonstrate that they are really only two different aspects of one and the same phenomenon – i.e., of life.

It is doubtful, however, if the purpose of the ER is *exclusively* the transport of materials. A new angle appeared when it was discovered that the ER is often linked with the outer surface of the cell body, the so-called *plasmalemma,* and this outer surface touches that of the neighbouring cell (or in plant cells is in contact with the solid cell-wall). That is to say, the outer surface of the cell is, so to speak, a continuation of the interior, and by means of the ER, the cell obtains for itself a considerable area of *internal* surface. This is interesting because almost all the processes – metabolism and so on – take place *on boundary surfaces*, i.e., on outer and inner surfaces. We need only remember the ribosomes, which lie so thickly on the external surface of the ER membrane. In fact, the increase of surface area is a principle that is met with in all organisms, and in all cells, and now we can see that it is not only within the ER that this increase of surface plays an important role.

Nevertheless this interpretation – that the ER serves to make the surface area grow – is unfortunately only intuitive, and is not, or not yet, experimentally proved. No doubt we shall come to definite conclusions about the functions of ER when it is possible to isolate the entire ER, or at least a substantial part of it, by homogenizing and centrifuging the cell; we will then be able to obtain a more or less pure fraction of ER, in the same way that we obtained a pure fraction of ribosomes on p. 45. For the time being, this is obviously impossibly difficult, because the first things to be destroyed when a cell is smashed up are the links between the cisterns, which are left isolated. It seems doubtful whether from the ruins any reliable conclusions can be drawn about the functions of the intact ER system. Nevertheless, experiments are in progress with the object of carefully isolating the ER, and is there any reason why one day they should not succeed?

3.05 Outer and Inner Surfaces
Unit Membranes Sometimes Separate Things and Sometimes Join Them Together

Whatever may subsequently come to light about the functions of the ER, the *membranes* will participate in some way or other, since they separate things and join them together at one and the same time. They separate the intracisternal (transport?) phase from the extracisternal (operative) phase, and by the same token they assist and possibly control the passage of materials. So the dividing membranes have been studied in detail, in especially thin sections, and with especially efficient microscopes. It was not at all difficult to find out that they were by no means homogeneous, but were really built up from units. They were found to be laminated, in layers of astonishing regularity; a clear, inner line was bordered on each side by two dark lines, but in reality, of course, these lines were only the cut ends of different layers.

This may seem rather superfluous and incomprehensible for anyone who is interested only the structural picture, but most electron microscopists are also biologists, and they will remember a time when the so-called permeability investigation was in full bloom, from 1920 to 1935 approximately. *Permeation* comes from the Latin *per*, through, and *meare*, to go; thus to *permeate* is to go through – in this context, through a membrane; the *permeability* of the membrane is the extent to which it allows a dissolved substance to pass through; the membrane is *permeable*. From innumerable experiments with different substances and different kinds of cells, it was proved that in living cells small molecules are taken up more quickly than big ones, and that substances which only dissolve in water (and not in fats or fat-like solvents) infiltrate into the cell more slowly than those that dissolve readily in fats (lipoids). It is not absolutely necessary to go into details – for our purposes the only important conclusion is that the membrane that regulates the transit of materials through the cell surface (i.e., the plasmalemma) must contain:

1. Lipoids – fat-like substances which repel water.
2. Proteins – albuminoid substances which mix with water.

As far as the lipoids are concerned, it is known that these are not pure fats like plant oils or butterfat, but a so-called phosphatide. This means that the lipoid molecule is joined to yet another molecule of phosphoric acid (as well as certain nitrogen-containing compounds such as choline and cholamine – the formulae are of no importance). The well-known substance lezithine is a phosphatide.

If a drop of lipoid is put on a water surface – which is a boundary layer between water and air – it does not mix with the water, but spreads out over the entire water surface. If the drop is very small, however, and the water surface

86. A unit membrane in section has a laminated structure, with an inner, pale strip bounded by two dark lines.

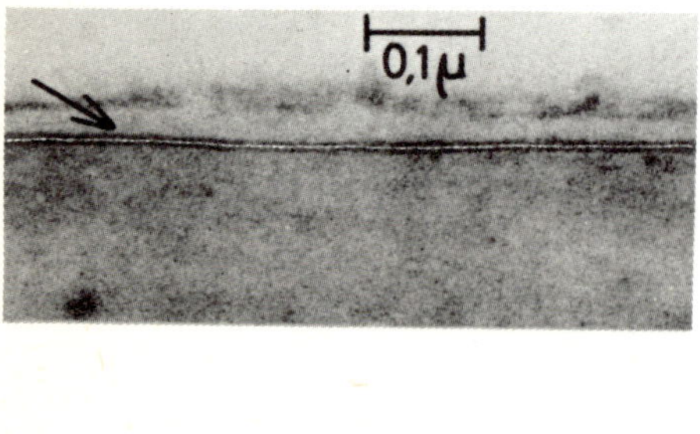

very extensive, the quantity of lipoid is not sufficient to cover the entire surface – and then the lipoid itself has an edge. The area covered by the lipoid can be measured, and the volume of lipoid is known (it has been measured beforehand). It is then a fairly simple matter to work out how 'thick' the lipoid film is – just as thick as the length of a lipoid molecule. Thus the lipoid covers the

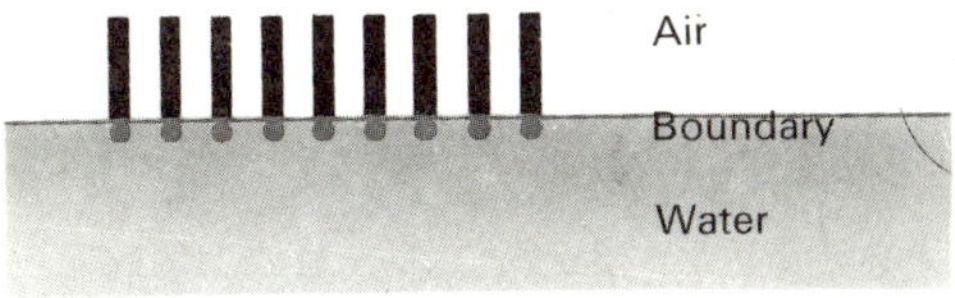

87. A monomolecular lipoid film at an air/water surface. The hydrophobe ends of the lipoid molecules project into the air.

water in a *monomolecular layer* (Greek *monos*, single). In this layer the molecules of lipoid stand neatly side by side, like soldiers, with their feet in the water and their heads in the air.

It may sound rather strange to talk about a molecule having a 'head' and a 'foot,' but this has its justification. The foot is the position in which the phosphoric acid is attached, and this is water bound, water loving, or 'hydrophile' (Greek *hydro*, water, and *philein*, to love). At the head end, in contrast, are attached the stems of the fatty acids, and these are water repellent, or hydrophobe (Greek *phobein*, to shun). So it is no wonder that this end removes itself as far as possible from the water surface.

Such a lipoid film can exist, however, not only at the boundary between water and air, but anywhere where two watery phases come together, for example between the cell contents (or protoplast) and the cell wall, which contains cellulose and at the same time is hydrophile. Since the hydrophobe heads

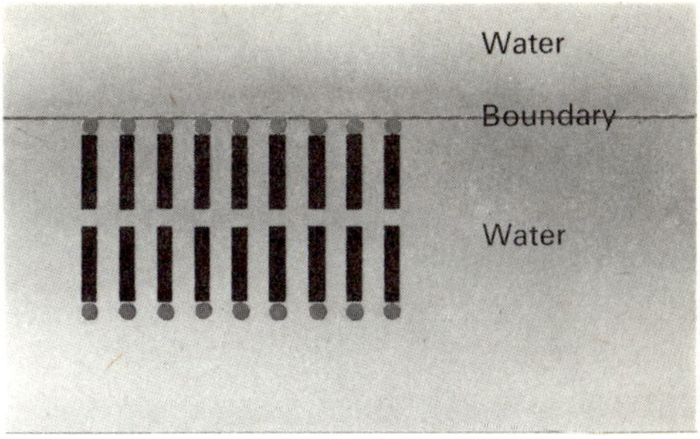

88. A bimolecular lipoid film at an interface between two aqueous media. The hydrophobe ends of the molecules are now opposed to each other.

of the lipoid molecules cannot make friends either with the protoplast or with the cell wall, they turn around on the spot towards each other, and put their heads together, thus forming a double film (a *bimolecular* film, from the Latin *bis*, double). The hydrophobe sides of the two monomolecular films lie against each other, and the hydrophile sides are directed outwards.

A bimolecular lipoid film is obviously not particularly stable. The molecules easily slide over each other, and the film disintegrates. The lipoids alone are not sufficient for

89. A bimolecular lipoid film with a protein layer on each side of it. Natural unit membranes are built on a 'sandwich structure' like this.

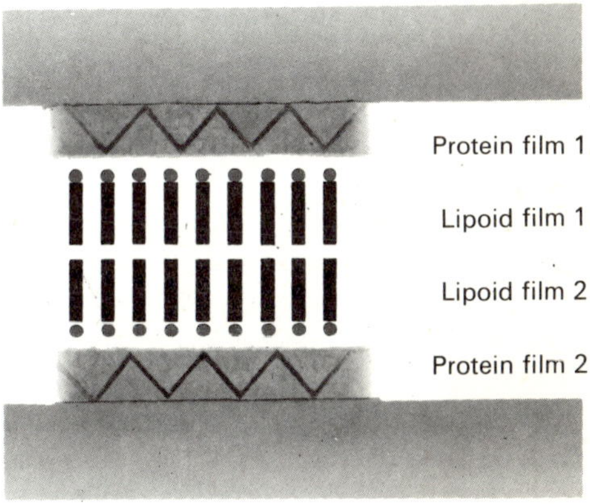

the erection of a boundary layer or membrane that is to last for any length of time (e.g., an ER membrane) – hence proteins, which are hydrophile, must also be present. These also can make monomolecular films, and these can lie for a time alongside the hydrophile outer surfaces of the bimolecular lipoid film, though it would seem that these monomolecular films cannot be measured as accurately as those of lipoids, because the protein molecules are sometimes stretched out, sometimes coiled into a secondary structure, and sometimes folded into a tertiary structure (see p. 30). Now the membrane is complete, just as it ought to be according to the findings of permeability science, and it has the appearance of a bread-and-butter sandwich; inside is the butter, the double lipoid film; outside are the two pieces of bread, the protein films. Americans call this, very appropriately, a *sandwich structure*.

Whatever the permeability scientist has deduced, or has called for, the physical chemist can produce artificially. Under the right conditions actual double films form in water which contains lezithine; these have even been 'fixed,' i.e., made solid, and thus made more durable – so durable that they could be examined under the electron microscope. And what had been expected happened: the films when they were seen in cross-section had exactly the same appearance and exactly the same thickness as the middle part of the ER membrane!

It is even more difficult to coat these bimolecular lipoid films on *both* sides with a film of protein, but that is actually no longer important, since in the meantime it has become evident that the membranes that are visible under the electron microscope have a genuine sandwich structure. One reservation must be made: what we actually see in the electron microscope as membranes – namely, the pale middle layer and the two darker stripes – are not exactly identical with the lipoid film and the two protein films. The darker 'parts' are made more contrasty by artificial means, just as the ribosomes were. This is best done with potassium permanganate or osminic acid (osmium tetroxide, molecular weight 254). This substance, however, does not 'colour' the protein film, but forms a layer along the boundary between lipoid and protein (just where the butter ends and the bread begins!). The complete permeability-controlling membrane is thus in reality somewhat thicker than the 70 to 100Å than it seems, and even measures, under the electron microscope.

This finding reinforces the previous interpretation that ER membranes not only separate two phases from one another, but at the same time, promote and control the exchange of materials between the two phases. Furthermore, this makes it readily understandable that the ER can easily change its form – not so easily as a film of pure lipoid, since the ER membrane is stabilized by its protein film, though the term 'stabilized' does not mean, in a hard-and-fast way, 'concreted.' Quite the opposite. Thus, just as artificial double films of lipoid 'grow' by further addition of lezithine and not only grow in thickness, but also in area, so the cisterns of the ER can also grow and shrink, can merge together, or separate.

That is not yet everything, however, that we know about these membranes. They are not only formed with really astonishing regularity around the cisterns of the ER, but we shall come across them again in many different forms. They are part of the 'iron ration' of many cells – of the mitochondria, for example, and of many others. They constitute such a non-interchangeable 'building unit' that the expressions 'elementary

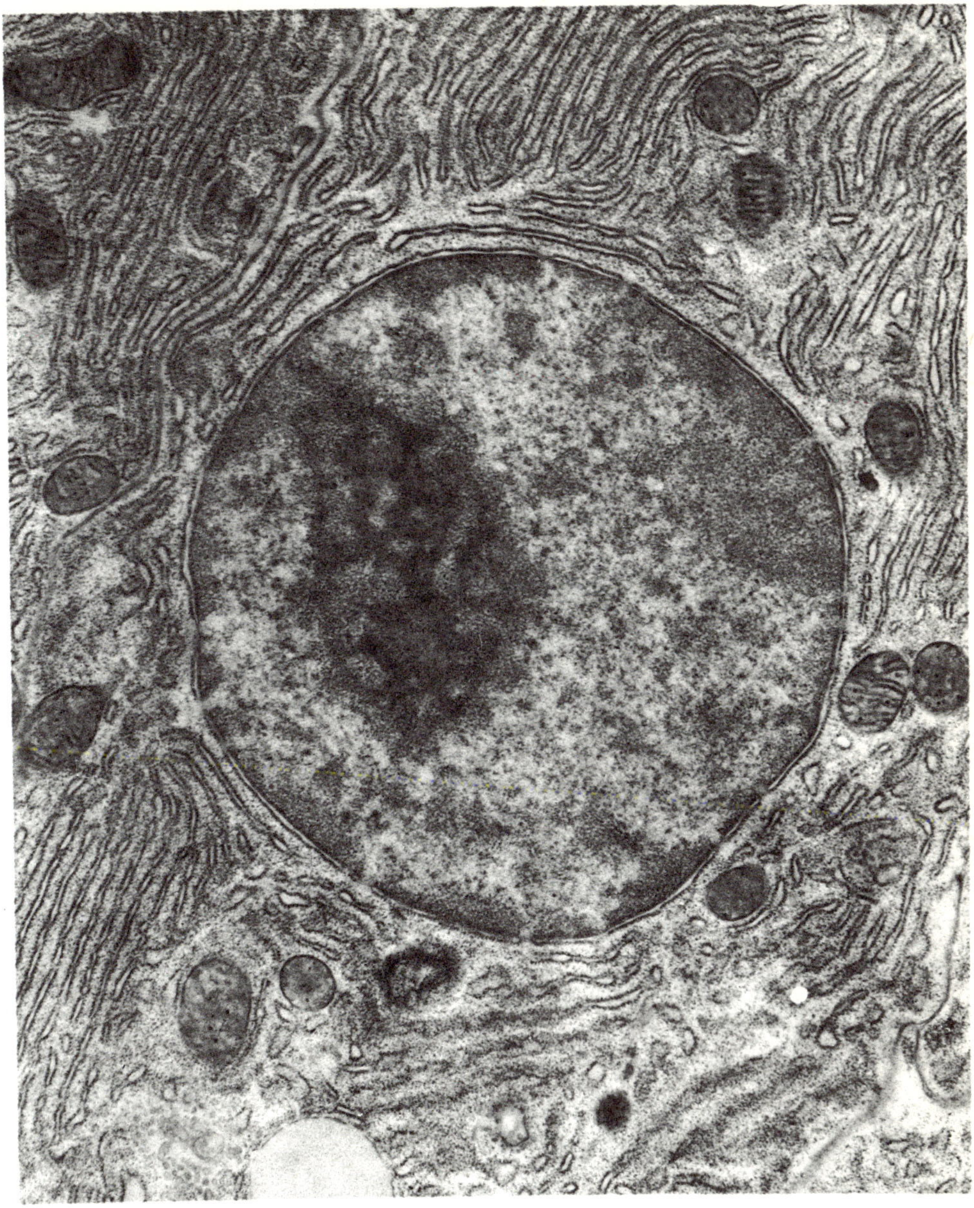

90. Section of a cell, with nucleus and nuclolus (centre), nuclear membrane, and cisterns of the endoplasmic reticulum.

membrane' and 'unit membrane' have come into use for them.

3.06 The Cell Nucleus

An Important Constituent of the Cell That Remains Obscure

To round off our studies of the ER, let us take another look at Figure 90, a section of a cell from the pancreas of a bat, once again made more contrasty. In the middle, and reaching almost to the edge of the picture, gigantic, and therefore conveniently visible under the light microscope, is the nucleus of the cell. Around this are cisterns of the ER, some of them arranged parallel to each other, others changing their direction, sometimes flowing around small egg-shaped enclosures which (on p. 161) we learned to know as mitochondria. The picture does not seem to show anything that is new for the ER system, but after detailed examination something comes out of it: the cell nucleus is enclosed in a 'double envelope,' and

1. This nuclear envelope is not completely closed, but interrupted in many places, so that the envelope has 'holes.'
2. The envelope, and even particular parts of it, looks exactly like parts of the ER system.
3. In a few places it can be seen that the cisterns of the ER open into the nuclear envelope.

May this nuclear envelope, then, be nothing more than a part of the ER system? However unlikely this idea may seem at first, it fits in. What is recognized under the light microscope as the 'nuclear membrane,' and is spoken of as if it were the outer part of the nucleus, really forms no part of the nucleus, but has brought us already into the ER. It is that part of the ER that envelops the nucleus and screens it from the surrounding cytoplasm – and that is why the nuclear envelope can be so easily dissolved at the time of cell division (see p. 77).

Certainly the nucleus is by no means *hermetically* sealed off from its surroundings by the ER system, and both nucleus and cytoplasm belong to the extracisternal phase. For an exchange of material, perhaps for the passing on of genetic information, for messenger-RNA, and the like, there are pores present, holes which admittedly are not always free passages, but which can be temporarily closed by a kind of stopper, made of a dense substance the nature of which is unknown.

We have found out at least some of the services rendered to the cell by the ER; now let us pass on to other components of the cell that are at the very least as important as the ER. Fig. 90, in which the cell nucleus occupies nearly half of the area, is again useful. What does the electron microscope tell us about the nucleus? Nothing that is particularly impressive. The ground-substance is irregularly and unevenly granulated, so that areas that are densely and less densely sprinkled with granules lie in a variegated pattern, and no order can be seen. There is just one, larger, somewhat lens-shaped fleck in the lower half of the nucleus, which stands out by reason of its dense granulation, and of the more or less uniform distribution of the colourless enclosures. This body is obviously the nucleolus (cf. p. 77), although, admittedly, it is no more strongly contrasted against the surrounding medium than are the distinct contours that can be seen inside the nuclear membrane. The nucleus is entirely lacking in unit membranes, which are so numerous everywhere in the cytoplasm.

From the fact that a nucleolus is visible as well as a nuclear membrane (formed from the ER!) we must conclude that the illustration shows an 'interphase nucleus,' i.e., a nucleus that is not in the process of dividing, because during cell division both nuclear membrane and nucleolus are dissolved. In this phase, compact, contracted chromosomes are not to be expected; on the contrary, the chromosomes in the interphase nucleus are so strongly twisted together that under the light microscope no boundaries between individual chromosomes can be seen (cf. p. 73). All the same, the chromosomes must be there in some form, because during the prophase of the next cell division the chromosomes become visible again. Their individuality has persisted, and so they must have remained separated from each other.

The hope that the boundaries between the chromosomes could be seen at least under the electron microscope has apparently not materialized. Why is that? Why does the cytoplasm, which surrounds the nucleus, contain such an overwhelming abundance of distinct structures, while the nucleus – which carries the genetic information and is thus the principal seat of government of the cell – appears to be so extraordinarily disorderly? Is this difference significant, or is it only apparent? There is a suspicion that the methods that have to be employed to prepare a specimen for examination under the electron microscope have played a trick on us, and suspicion naturally falls particularly on the techniques used for killing the cell. It is unavoidable that the cell must be killed, dehydrated, embedded in a synthetic resin, sectioned, and finally enclosed in the vacuum of the electron microscope. 'Killed' in the strict sense means 'fixed' – that is, the structures are brought into such a condition that subsequent stages of preparation will not change them any further. The great question is whether the condition in which things are fixed is, in all instances, the condition in which they were present in the living cell or whether it is only a facsimile of it – and if the latter, how close is the resemblance?

A reagent used in fixing must both penetrate into the cell and get on with its work there as quickly as possible – it must overwhelm the cell in such a way that the cell has no time, and no opportunity, to alter its structure. This means that, in the first place, all *enzymes* – and therefore all proteins – must be rendered ineffective; otherwise, between them they could, during the time that they remained intact, go on working, building up and breaking down substances without being under control, and as a consequence of this unrestrained activity, they could do things that never happen under normal conditions.

The process of rendering protein compounds ineffective is called 'denaturing,' because they are changed from the 'natural,' active condition to an inactive one. As a rule, the denaturing process is sufficient to fix not only the proteins but also other groups of compounds – for example, lipoids and fats – though these others are usually not denatured, only preserved in their original form. The great danger is that during the subsequent stages of preparation – such, for example, as treatment with fat-dissolving chemicals – the fats may be dissolved out and thereby lost. In the case of proteins this is hardly to be feared. Denaturing makes them adhere together in a network, and this makes them insoluble. This very process has another danger; for the proteins are not only denatured, and thus to some extent suddenly congealed, but they also lose a great deal of the water that formerly bound them together (since proteins are all hydrophile) – which means a loss of volume, i.e., shrinkage.

We will not go into unnecessarily confusing details; it is enough for us, first, to establish that a cell that has been fixed is never exactly like a living cell, and so an electron microscope picture never shows exactly what living cells look like. We always have to allow for the artifacts of fixation that is, with artificial things. For a long time, therefore, electron microscopists have debated with each other in argument and counterargument whether certain structures that appeared in their pictures were real and not some kind of artifact. This happened most of all if the same details (or very similar ones) could be seen persistently after using different methods of fixation and different objects. As an example we might take the unit membranes: their stratification, their thickness, and their shape are all convincingly constant. Only – and we must be very clear about this – we never make the 'real' structures visible; what is visible is only something corresponding to what is in the living cell, i.e., something *equivalent to the real thing*. 'Equivalent' means having the same value, but not identical.

It follows from this, moreover, that the ideal method of fixation does not exist, and never will. Consequently we must make use of different methods of fixation, according to whether we want to study the ER, or the ribosomes, the nucleus, or the dictyosomes; at one time we use potassium permanganate, at other times osmium tetroxide or formol – to name just a few chemicals.

Furthermore, there exists a method of fixation, suitable for many special problems, that does works not by chemical means, but physically: the method of *freeze drying*. The material being studied, is frozen in small portions, as quickly as possible, either after immersion in glycerine, which does not congeal even at $-20°C$, or else by treatment with carbon dioxide 'snow' ($-78°C$) or liquid air ($-190°C$). After freezing, the cell or block of tissue is placed in a high vaccum, for days or even for weeks. The effect of this is that bit by bit, extremely gently, the water and any other volatile substances are vapourized; what is left behind is the air-dried, water-free 'skeleton' of the cell, which can now be studied more fully. The method is fairly tedious, but gives good results. Many of the results agree exactly with those obtained by totally different, chemical methods, and this fact is very consoling. Freeze-dried sections, however, show some things that are not seen by any other method.

Two examples may be chosen. In cells that have been fixed chemically, the ER membranes appear dark on a pale ground, but when the cell has been freeze dried this is exactly reversed. Probably, therefore, important constituents of the plasma are lost when the cell is fixed chemically. As a second example, take the ribosomes: in freeze-dried sections they are *not seen at all* as granules, yet it would be wrong to draw from his the conclusion that the ribosomes that are to be seen in 'normally' fixed sections are complete artifacts. Ribosomes can be isolated from freeze-dried material by homogenizing and centrifuging it, and these are fully functional; so they must be there all the time, but just not visible under the electron microscope. On the other hand, our satisfaction over the otherwise good agreement between the results obtained by different methods of fixation should not lead us to overlook the fact that the electron microscope is giving us merely pictures of 'equivalents' and not real structures.

How, then, shall we answer the question on p. 159? Is the unorganized appearance of the nucleus during interphase a result of using unsuitable methods of fixation, or does it indicate that orderly structures are absent?

The first possibility cannot be dismissed offhand, but it is not very likely; for thirty years we have known about methods, good and even very good, by which all the other constituents of cells can be preserved. So the second possibility – the persistent lack of any orderly principle which can make an impression on the electron microscope – carries more weight. The reason for this lack is that the arrangement of genetic information is to be found *at molecular level*, in the sequence of nucleotides, or triplets. The orderliness is so strongly established at this level that it does not need to be reinforced by a higher (which means a coarser) organization, such as that of lamellae, membranes, and so on; these are superfluous and do not exist.

Note that this applies only to the nucleus at interphase. When the nucleus is *dividing*, the genetic information must be converted into the 'transportable form,' and this means chromosomes in their contracted state – when they are so compact that they become visible even under the light microscope. We must postpone a study of the morphology of chromosomes until after we have discovered, in the next chapter, how the transfer of information functions and is regulated. In this field, structure and function are so closely linked together that the one cannot be understood without the other.

3.07 Mitochondria

Power Plants of the cell

Although Figure 90 shows us little of the structure of the nucleus, we will take one last look at it. Apart from the ER there are other extracisternal structures to be seen, spherical or egg-shaped, with good contrast; we have referred to these already as *mitochondria*.

Mitochondria (Greek *mitos*, thread, and *chondros*, grain) are thread-like bodies. The name was given to them around the turn of the century – long before the beginning of the electron microscope era. It applies to the thread-like structures which were to be seen under the light microscope, occasionally in living cells and sometimes in cells that had been fixed. For a long time their existence was controversial, since they could by no means always be found; moreover, they changed their shape, disintegrated, disappeared, then came into being again. Above all, no one knew how they should be related to the physiology of the cell. Had they any function, and if so, what? Should we believe those authorities who said they had to do with the respiration of the cell?

Here the electron microscope (also, as we shall see later, the centrifuge) came as a blessing. With its help, it was conclusively proved that mitochondria formed a regular item in the constitution of *every* cell – every nucleated cell, that is; bacterial cells, for example, which have no genuine nucleus, but only a genome in the form of a double helix of DNA, nevertheless contain objects that resemble mitochondria, but of extremely primitive design.

All the same, mitochondria are usually not thread-like, but spherical or egg shaped. Their maximum length is then around 1μ (1/1,000 mm.) and is thus right on the limit of resolution of the light microscope. Only if several mitochondria coalesce, do they form threads that are visible under the light microscope. It is little wonder that when these separate into individual mitochondria they seem to disappear.

Figure 91 shows about half a dozen individual mitochondria, characterized by a peculiar cross-banding. They are not really very distinct – though this may be because this particular section was made specially to show the ER; the methods of fixing and embedding that happened to be chosen are, in our opinion, not necessarily the most suitable for showing up the mitochondria.

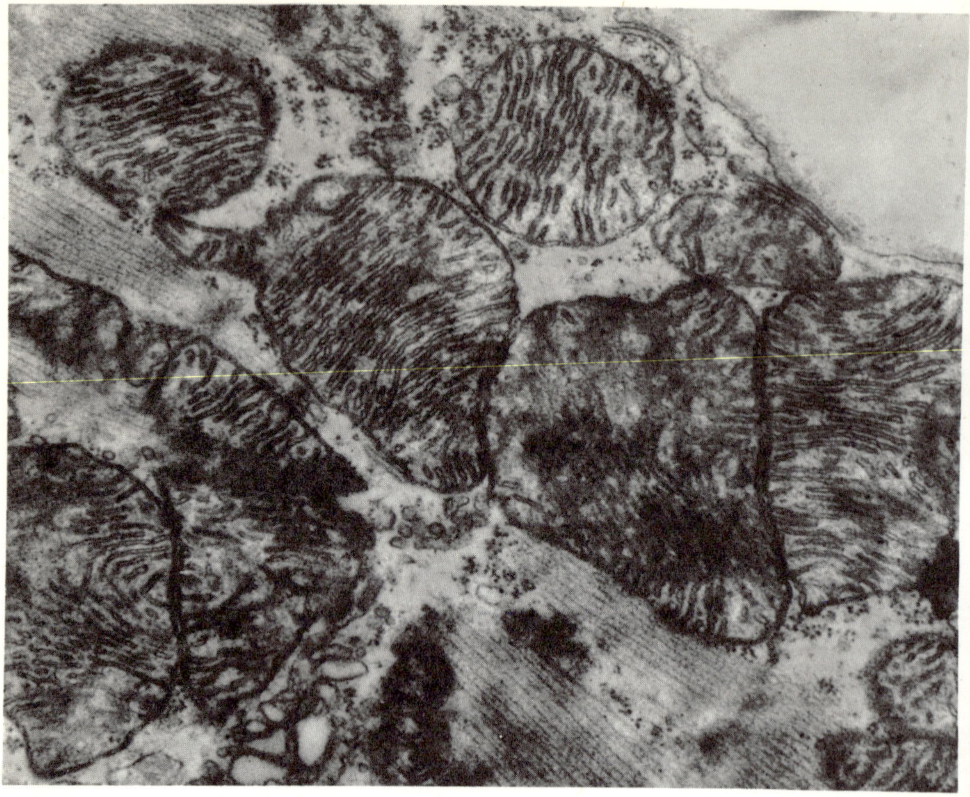

91. Mitochondria of the 'crista' type, from heart muscle.

Figure 93 shows substantially more detail, even though the mitochondria are to some extent pressed against each other, and so their round contours are lost. It is not the outline that is important, however, but the internal structure. Inside them, embedded in a ground-substance, lie narrow, double-contoured stripes; one part of these ends freely in the ground-substance, but the other part continues to the periphery and merges with an envelope which isolates the entire mitochondrion from its surroundings.

Now, a picture of a cross-section still does not allow us to make out details of the *spatial* arrangement of what are obviously real spherical and egg-shaped mitochondria. To do that, we need to examine one and the same mitochondrion in successive sections, and so reconstruct its interior. Figure 92 shows schematically what comes out of this. It now appears that the 'stripes' are really flattened structures, which extend right through the mitochondrion from side to side, running more or less parallel to each other, though it is rare for a section to be aligned so accurately that it shows a complete stripe; this is why we sometimes see 'short' stripes, which fade away in the middle of the section. The envelope of the mitochondrion is a double wall. (Note that this is not the same as the unit membrane of p. 158. Each individual membrane is a unit membrane, and the space between two unit membranes is, broadly speaking, an ER cistern.) The stripes, however, are produced by a folding, or invagination, of the inner membrane. These are called *septa,* or more frequently, *cristae mitochondriales* (Latin *crista,* a comb). This type of mitochondria is called the crista type after these structures, and we shall see more of them shortly.

The interior of a mitochondrion (which is also, confusingly, called a matrix; cf. template or matrix RNA) is furrowed and indeed even fissured by septa that project from the envelope of the mitochondrion. The envelope itself has both an inner and an outer membrane, and the septa, being invaginations of the inner membrane, are in turn separated from the matrix by a membrane on each side.

Thus the electron microscope has not only confirmed the findings of the light microscope but has also made visible a mass of the finest detail, which the light microscope would never be able to show. Finally, a basic membrane is only about 70Å thick; the space enclosed between the membranes of a crista about 60Å. The boundary lines of a septum therefore lie only a maximum of 200Å = 20 nm. = 1/50 μ apart.

The mitochondria that are shown in Figure 91 come from cells of heart muscle. Figure 93, which is a cross-section from a slipper-animalcule shows something rather different. Although generally similar in outline and in size, the interior shows a tangled mass of sausage-shaped processes, with circular figures among them. The 'sausages' seem to be attached to the outer envelope, just as in the crista type of mitochondria. This is in fact so: a three-dimensional reconstruction shows that in this case what grow out from the inner membrane (actually 'grow into' the matrix) are not septa, not cristae, but long tubules, which are coiled around each other many times. If these happen to be sectioned lengthwise, we see them as sausages; when they are cut across, they appear as circles. This type of mitochondrion is called the tubular type, after the tubular shape of the invaginations.

The reader is likely to object at this point that these may not really be two different kinds of *mitochondria,* but only two quite different kinds of cellular *organs.* The two types have, indeed, much in common: size, outline, construction from double membranes, invaginations of the inner membrane, matrix, etc., etc. It will be realized that the

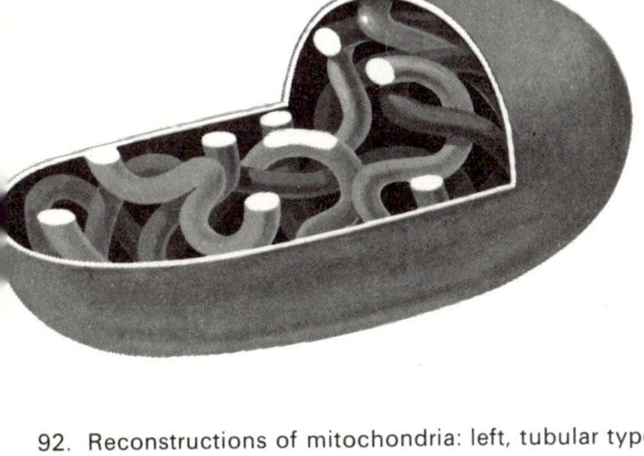

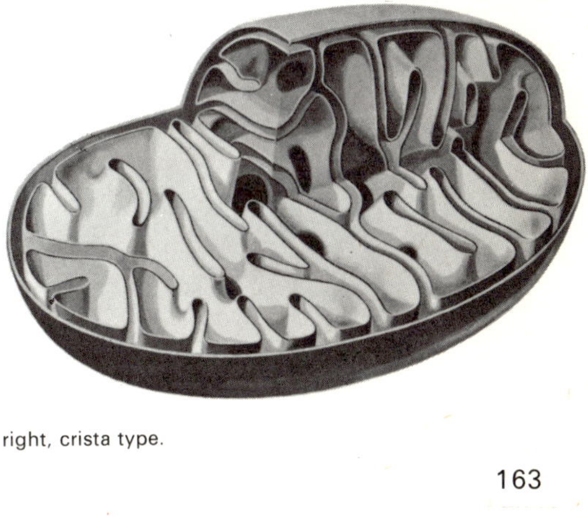

92. Reconstructions of mitochondria: left, tubular type; right, crista type.

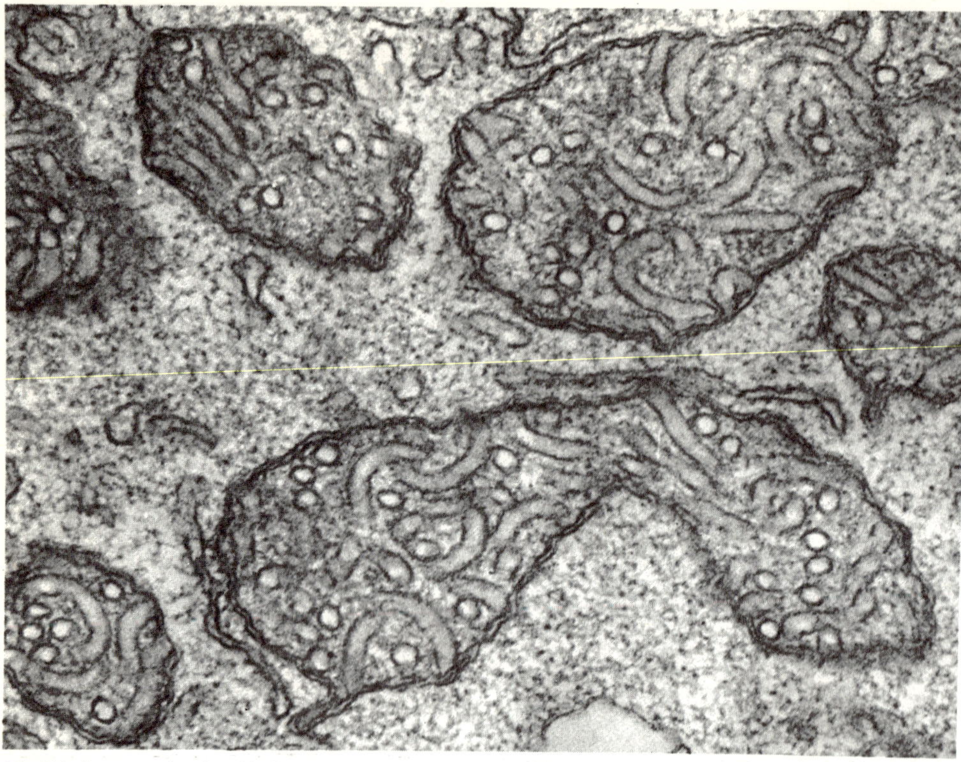

93. Tubular-type mitochondria from a slipper animalcule.

tubular structure produces a quite stupendous increase in the internal surface area, and this is the principle of its construction. We have already seen the same thing in the ER and its connection with the outermost layer of the plasma – the plasmalemma (see p. 153). An increase of surface area also takes place in other things besides mitochondria; so why cannot the crista and tubular mitochondria just be essentially different organs?

This objection is thoroughly justified, and if we had to depend entirely on electron-microscope pictures, we should have difficulty in overcoming it. With this object we must try to demonstrate that the two kinds of mitochondria develop out of the same kind of elements – but we must defer the story of the development of mitochondria for the present. One thing is certainly puzzling. We never find crista and tubular mitochondria mixed together in one and the same cell at the same time. Indeed, tubular mitochondria are mostly found in highly organized unicellular animals, the so-called whip-infusoria, or flagellates. Crista forms, on the other hand, are formed only in the higher plants and animals. Moreover, the plants sometimes, if only rarely, have yet a third type of mitochondria, the sacculus type, so-called because their invaginations are sac-like, widening out, but not drawn out like the tubules.

The fact that the types appear as alternatives to each other is still not sufficient for them to be interpreted as one and the same organ working in different ways. Fortunately, further evidence exists, and now the centrifuge that was only so recently laid aside comes into action again. What was not possible in the ER, namely to obtain it in undamaged form from homogenized cells, can be carried out for mitochondria, successfully and with little trouble. At a particular speed of rotation, between 10,000 and 20,000 r.p.m., a fraction separates which contains almost nothing but mitochondria; by repeated washings and renewed centrifugings, this can be refined into a pure fraction. This process can be well controlled under the electron microscope.

If, now, pure fractions of the different types of mitochondria are compared in the first instance with regard to their chemical composition, it is always found to be 85 per cent water. But this does not tell us much, because the water content of all cells is at about this level. A more important point is that the dry substance consists of half protein and half lipoids, and 90 per cent of the latter is those phosphatides that were described on p. 154. Related to this is the fact that the mitochondria – on account of the invaginations – are extraordinarily rich in unit membranes; these consist of a bimolecular lipoid film and two protein films, which are applied to it on each side, and this gives them a ratio of roughly 1:1. Other organs of the cell, but above all the 'ground plasma,' have many fewer lipoids. Nucleic acids are present in mitochondria in tiny, evanescent amounts, if at all.

Every doubt about whether tubular and crista mitochondria are identical vanishes if we now examine their functions. The best way of doing this is to test whether the mitochondria possess enzymes, and if so, which enzymes; from this it can be deduced whether metabolic processes go on in the mitochondria, and if so, which processes. The mitochondria actually contain many enzymes, in all circumstances, and in great numbers; indeed, they contain all the enzymes that take part in the respiration of the cell.

3.08 Respiration

Burning Fuel and Producing ATP

Expressing it more or less precisely, though not down to the last chemical detail, respiration is the burning, or oxidation, of grape sugar ($C_6H_{12}O_6$) to give carbon dioxide ($6 \times CO_2$) and water ($6 \times H_2O$). If this took place suddenly, in one step, it would release such a large amount of energy that the cell would be quite unable to cope with it, and would itself be burned up. The complete reaction therefore takes place in more than two dozen stages, and each of these is brought about by at least one enzyme.

The first group of about ten stages is designated *glycolysis,* because they involve the breaking down of sugar; they also take place without oxygen, i.e., anaerobically. This group of steps leads from glucose to pyroracemic acid (which we have already met on p. 57).

Many organisms, such, for example, as many yeasts, are content with this. From the pyroracemic acid they only split off carbon dioxide (a process known as decarboxylization), and add hydrogen (which has been separated in an earlier stage of the reaction, and 'stored up'), and thus produce alcohol, in fact ethyl alcohol: alcoholic fermentation. In addition by this means they gain one more molecule of energy-rich phosphate, ATP.

This part of respiration, 'anaerobic glycolysis,' does not take place in mitochondria, but the succeeding stages do. These are known as the 'citric acid cycle' and the 'terminal respiration,' and all their chemical details are known. In this book we need to know only the principles. The pyroracemic acid is gradually split up into CO_2 and H_2, and finally the hydrogen combines with the oxygen of the air to form water; this part of the respiratory process is thus *aerobic*.

Since the end-products of CO_2 and water are valueless 'energy-wise,' the chemical energy that previously resided in the pyroracemic acid must still remain somewhere else. A small amount is converted into heat, but most of the energy is to be found in a new chemical compound: adenosine triphosphate – which we have already met (on p. 57) as ATP, the universal source of energy of the cell.

The production of ATP can be regarded as the aim and purpose of respiration. ATP is produced, however, whenever any ADP present takes up a third molecule of phosphoric acid, when the ADP is 'phosphorylated' to ATP. Since this phosphorylation involves the consumption of oxygen, it is known as oxidative phosphorylation. The mitochondria, whether they are of the tubular or crista type, are the places where oxidative phosphorylation takes place. They provide the ATP for every possible process that requires energy, and so they may rightly be called the power plants of the cell.

If this was as far as we could go in demonstrating the presence of all the enzymes of respiration, it would certainly be convincing, but not entirely conclusive. The results of techniques of fractionating go much further. It can be shown quantitatively that mitochondria will also promote respiration *in the test tube*, take up phosphate and oxygen, and produce CO_2 and ATP. They need to be supplied with just enough substrate and phosphate, as well as with a whole lot of so-called co-factors, e.g., ADP! Fortunately we do not need to go through the whole list, as we did when we were dealing with the synthesis of proteins *in vitro*. Only one prerequisite need be briefly mentioned.

If the mitochondria, after centrifuging, are flushed with fresh water, they receive the same kind of 'osmotic shock' that was given to the bacteria on p. 112; they swell up and burst; and then, naturally, there is no more respiration. A liquid suspension of saccharose is therefore added, in a particular concentration; this must be measured to give as exactly as possible the same osmotic pressure as that prevailing in the interior of the mitochondria. If this agreement is not exact enough, or if one of the co-factors is missing, a gentle swelling of the mitochondria will set in. They will then take up more oxygen, and release CO_2, but ATP production ceases. Oxidation and phosphorylation, which previously went hand in hand, are now 'de-coupled.'

The process of respiration is now 'running empty,' not achieving its object, which was to produce ATP, and so the mitochondria will soon wear out. This 'de-coupling' is important in the understanding of many biochemical processes, but to talk about it here would be to wander from our subject. On the other hand, we can and will take the existence of oxidative phosphorylation as evidence that the mitochondria are intact and functioning normally in the test tube as well as in the cell. The fact that the mitochondria probably promote other biochemical reactions (e.g., the incorporation of amino acids) can be left as just a marginal note.

Although it is important to have undamaged, intact mitochondria in the test tube, for many purposes it is equally good to work with shattered mitochondria, or even with fragments. When this is done, the phosphorylation reactions are understandably omitted, but the enzymes, which break down the pyroracemic acid step by step, have other work to do. If we take only *very small* debris from the mitochondria, then these particles are only capable of carrying out one or two parts of the whole reaction, whereas bigger fragments can carry out several or even many successive steps. From this it has quite justifiably been concluded that in both crista and tubular mitochondria the enzymes which work together are also *placed close together*, and that the substrate upon which they are to operate is passed along from one enzyme to the next just like a conveyor belt. Thus, just as the enzymes that work together were produced by a corresponding gene structure in the operon (cf. next chapter, p. 266), so the enzymes themselves lie in sequence on the mitochondria – an arrangement that has been only speculative for a long time, but which could not be clearly demonstrated.

After all that science has brought to light about mitochondria, with the aid of the electron microscope and the centrifuge, it does not surprise us any more that these structures should be *found in all cells*. They are simply essential for every cell. ATP is plainly a general requirement, sometimes needed very quickly, and in great quantities; hence it is not feasible for a multicellular organism to have only one powerhouse from which to supply all the cells with ATP; transport would take far too long.

In any case, cells such as yeast cells, which supply their ATP demands by alcoholic fermentation, can exist without any mitochondria. In fact, however, they do have mitochondria, even if these are few, and not fully developed. As soon as they change over from fermentation to respiration – and they usually do this if oxygen is available – they have numerous, fully functional mitochondria.

On the other hand we would expect that mitochondria would be particularly numerous in those cells that have an extremely high demand for energy, such as the muscle cells of the human and animal bodies. To contract and relax a muscle, a muscle cell depends on the contraction and relaxation of protein molecules, which on their part lie by the hundred, bundled together into muscular fibrillae, or myofibrillae.

We have already seen (in Chapter 1) that polypeptide chains can change from the shortened, α-helix into the still shorter β-helix, and vice versa. The fact that this contraction requires energy is obvious to sportsmen and others. If we have to lift a basket of potatoes off the floor and onto a table, we have to do a good deal of work in a short time, and hence the individual muscle cells need a lot of ATP, *all at the same time*.

In this connection let us consider Figure 91, which shows the heart muscle of a rat. Now, heart muscle has to work hard and continuously. Many mitochondria with cristae can be seen, closely packed together; many more than in the slipper animalcule shown in Figure 93. The direction in which the heart muscle was sectioned was not specially chosen for effect; if we take the trouble to cut the section exactly parallel to the muscle fibers, then we see a really incredible orderliness (see Fig. 94). The fibrillae, which are pale, with longitudinal and transverse stripes, alternate regularly with dark, tightly packed, frequently block-shaped mitochondria, inset with cristae.

This is shown even more distinctly in a diagram built up from a large number of individual sections (fig. 95). What a large number of individuals this diagram shows,

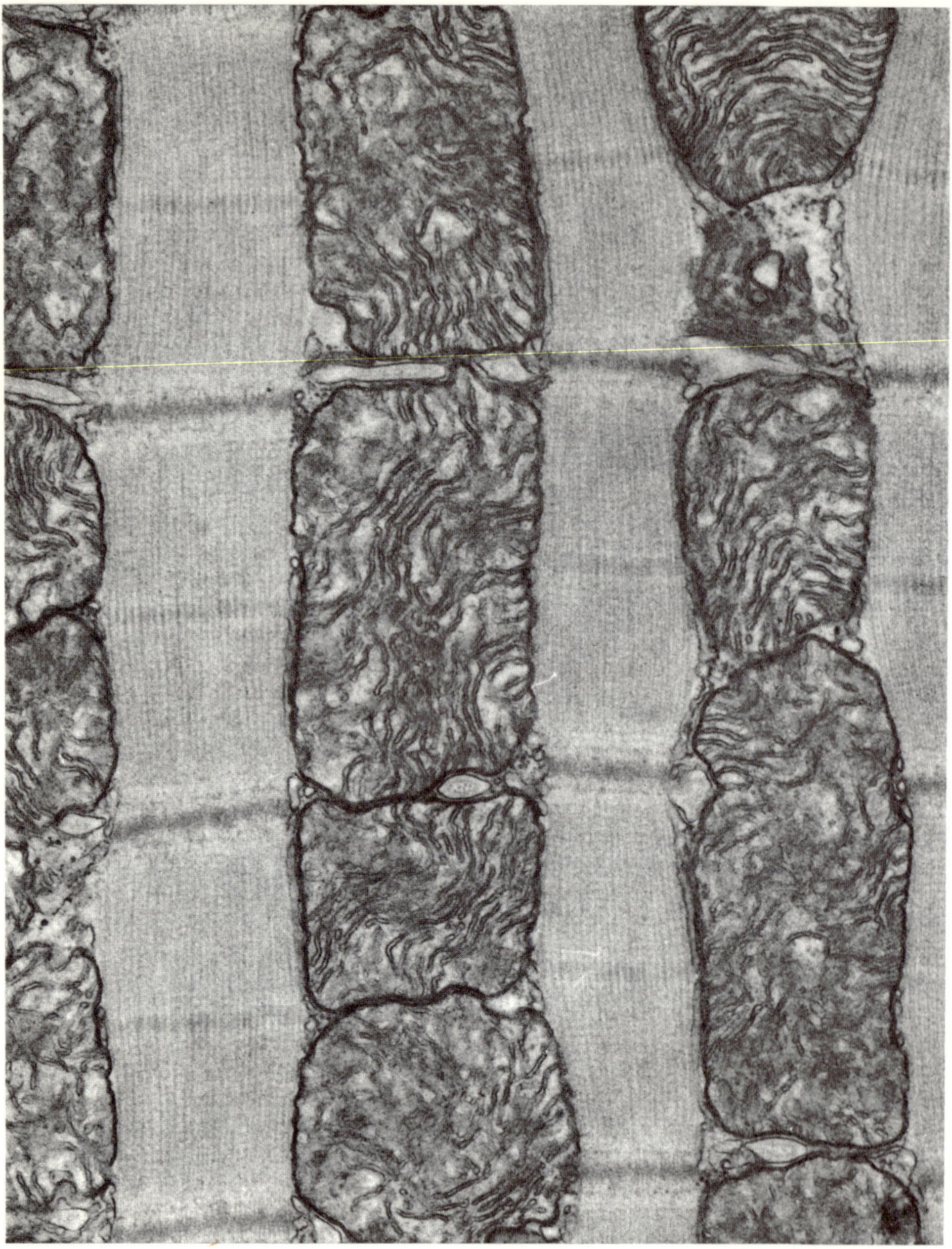

94. Section through a cell from heart muscle, as in Figure 91, this time parallel to the muscle fibers. The fibrillae, pale, with longitudinal and transverse stripes, alternate regularly with dark, closely packed mitochondria.

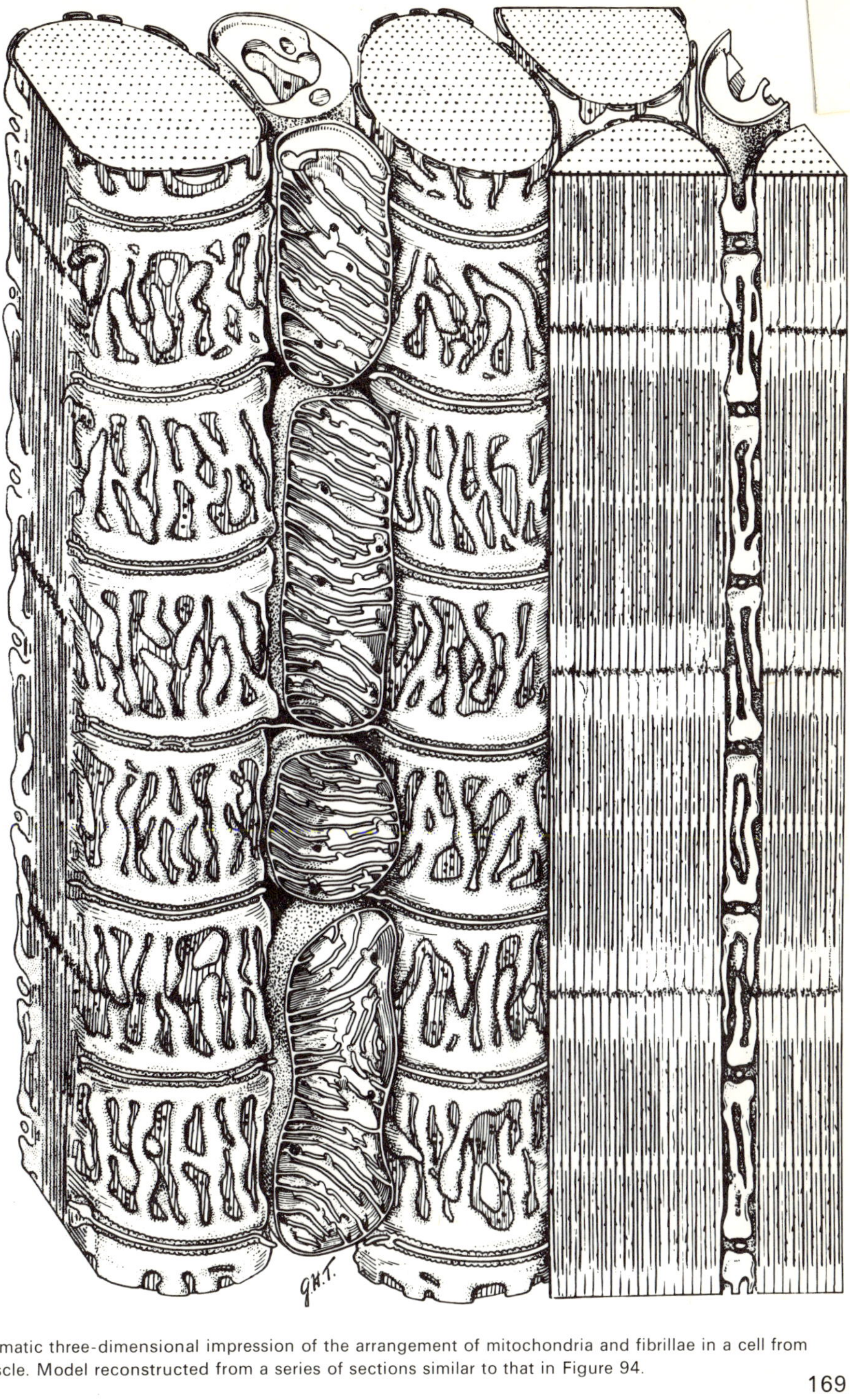

95. Schematic three-dimensional impression of the arrangement of mitochondria and fibrillae in a cell from heart muscle. Model reconstructed from a series of sections similar to that in Figure 94.

and what a high degree of organization of the structural elements! It is believed that these tightly packed mitochondria hold the reserve of energy necessary for the contraction of the fibrillae, and that this type of arrangement is especially suited to the purpose, in particular because it provides a very short path for the ATP between the place where it is made and the place where it is consumed.

3.09 Dictyosomes

Vesicles and Cisterns

So far, when new problems, structures, and so on had to be introduced, we have as far as possible avoided talking about their historical aspects. It seemed in the best interests of the reader to talk less about the errors of the past and more about the inventions and discoveries of modern times. At this point, however, we shall make an exception and introduce a little history, because otherwise the naming of the structures that we are about to discuss cannot be understood. Dictyosomes are called 'net-bodies' (Greek *diktyon,* net), but they have not the smallest resemblance to a net.

In the year 1898 the Italian anatomist Camillo Golgi reported that he had observed in certain nerve fibers an '*apparato reticolare interna,*' which seemed to him to be very significant. His contemporaries, who were already, even then, addicted to abbreviations,

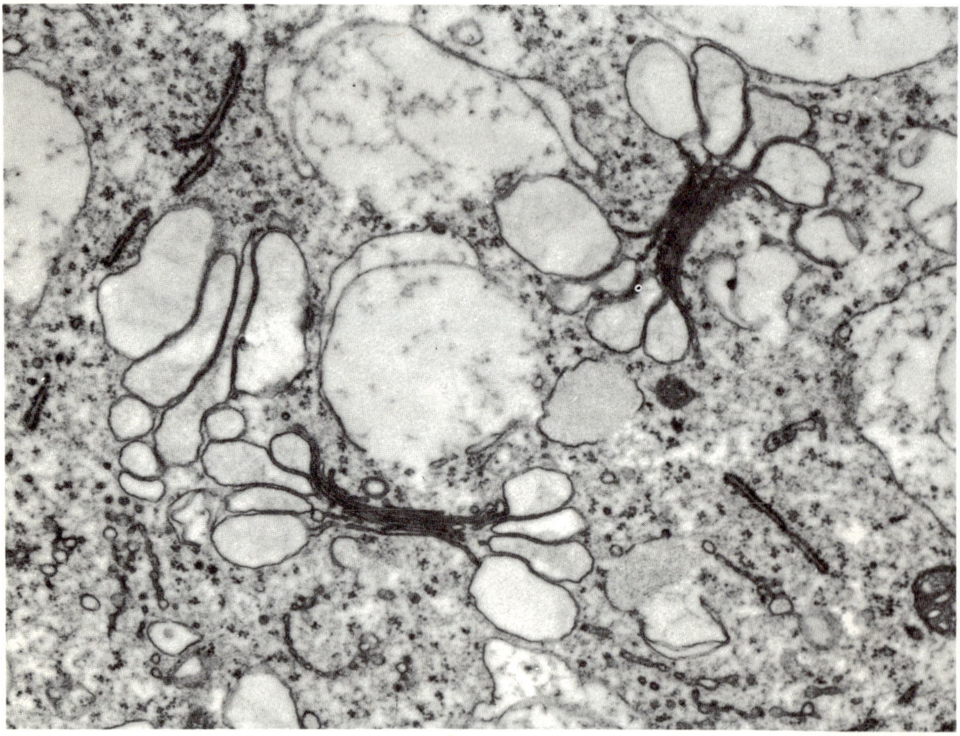

96. Cells from the root surface of a maize plant, showing dictyosomes (right and below).

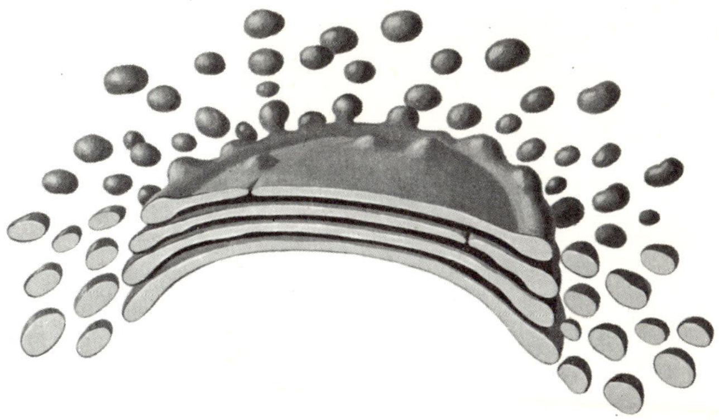

97. A reconstructed picture of a dictyosome, with numerous Golgi vesicles nipped off from the outside.

although not so badly as are the scientists of today, called this apparatus for short the Golgi apparatus. It was found in many animal cells, and later also in cells from plants, particularly often in glandular cells. It seemed to have something to do with secretion. Occasionally it appeared as if the net-like Golgi apparatus broke down into independent small units, to which the name 'dictyosomes' was given. Everything else remained obscure for a long time.

Here, too, the electron microscope brought clarification. Dictyosomes really exist, but they do not join on to each other, at least not in such a way that we could call them a network. Nowadays the term 'Golgi apparatus' can still be used as a collective name for all the dictyosomes of any one cell. Unfortunately, however, this has not prevented some authors from speaking of a single dictyosome as a Golgi apparatus.

Figure 96 shows a section of a cell from the root tip of a young maize plant. This picture clearly shows two dictyosomes sectioned; the rest of the picture is occupied by large and small *vesicles*. According to this, dictyosomes consist of a thick bundle of three to five flat sacs, or bursae, surrounded by a membrane. On the edges of the bundle these are inflated into vesicles. Close by there are usually vesicles of similar form which are not (any longer?) linked with the sacs, but which may be supposed to have separated off from them quite recently.

A solid diagram, reconstructed from serial sections, allows these 'interrupted attachments' to be better understood. First, we see that the sacs, known as 'Golgi cisterns,' are almost exactly circular, very much flattened, and arranged very closely parallel to each other; near the edges they broaden into spherical cavities. The curving of the sacs that is often observed has been ignored in the diagram so that the spatial arrangement can be understood better. Many of these however, are already branched off, and distributed around the periphery as 'Golgi vesicles.' Only very rarely is it possible to gain an insight into the function of a structure such as this from pictures of sections, i.e., from mere snapshots. It is quite obvious that something is produced in the Golgi cisterns, carried to the margin, and from there passed into the cell by the pinching off of the vesicle. But what substance is involved?

A botanical subject that has been much studied is a plant, *Drosophyllum lusitanicum*, which is closely related to the familiar sundew *Drosera rotundifolia*, and which similarly has many 'tentacles' on its leaves. These are

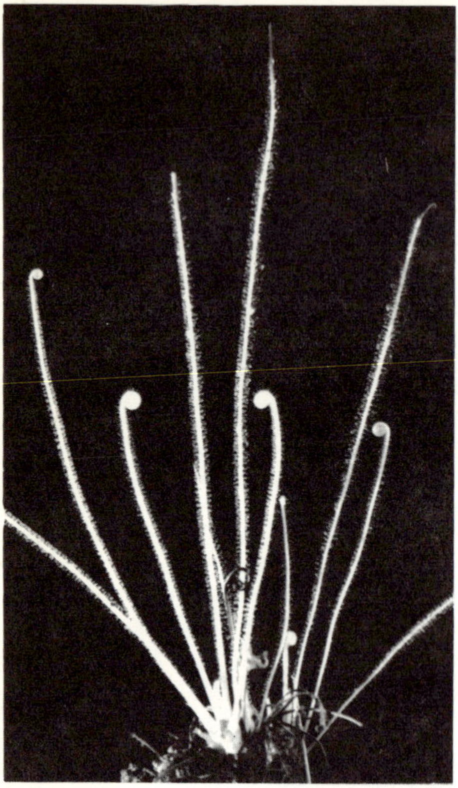

98. *Drosophyllum lusitanicum*, a relative of the sundew, with numerous fine 'tentacles.'

stalked bodies, with many glandular cells. The glandular cells produce a sticky secretion which oozes out and spreads over the surface of the knobs. Each knob glistens like a dew-drop in the sun, hence the name 'sundew.' The secretion is quite extraordinarily sticky, so sticky that any insects that touch it become inescapably trapped. Enzymes that are subsequently secreted dissolve the contents of the bodies of the insects and make them available as an additional source of nourishment for the plant. Lusitania was the name of the Roman province which occupied more or less the same area as modern Portugal, and even today a potted *Drosophyllum lusitanicum* plant is often used in Portuguese homes as a living flytrap – so powerful is the sticky secretion.

As might be expected, the glandular cells in which the sticky secretion is produced are especially rich in dictyosomes and in Golgi vesicles. Moreover, their abundance varies in the same rhythm as that which can be observed in the production of secretion. There is no doubt that the secretion is produced in the cisterns of the dictyosomes, or at least accumulated there. Now, we already know that nearly all chemical syntheses need energy, in the form of ATP; therefore in the glandular cells, close to the dictyosomes, are to be found a particularly large number of mitochondria.

On the other hand, if respiration is impeded – and an effective respiratory poison is one that uncouples the oxidative phosphorylation (see p. 166) – then hardly any more Golgi vesicles are nipped off, and new secretion can no longer be produced. Removal of the respiratory poison allows production to start up again; it soon reaches its original level again – and, once again, every dictyosome can nip off up to three vesicles per minute.

It is easily possible to collect enough of the sticky secretion to allow it to be studied chemically. It consists of carbohydrates, sugars of high molecular weight, and sugar-like substances which are polymerized to macromolecules and excreted in this form.

It can be shown that dictyosomes build up polysaccharides in other circumstances too. In the fine root-hairs of the maize plant the ground-substance of the cell walls produces so-called *hemicelluloses*, which are carbohydrates very similar to those in the sticky secretion of sundews. Golgi vesicles containing hemicellulose show up in particularly strong contrast (see Fig. 98), and thus their

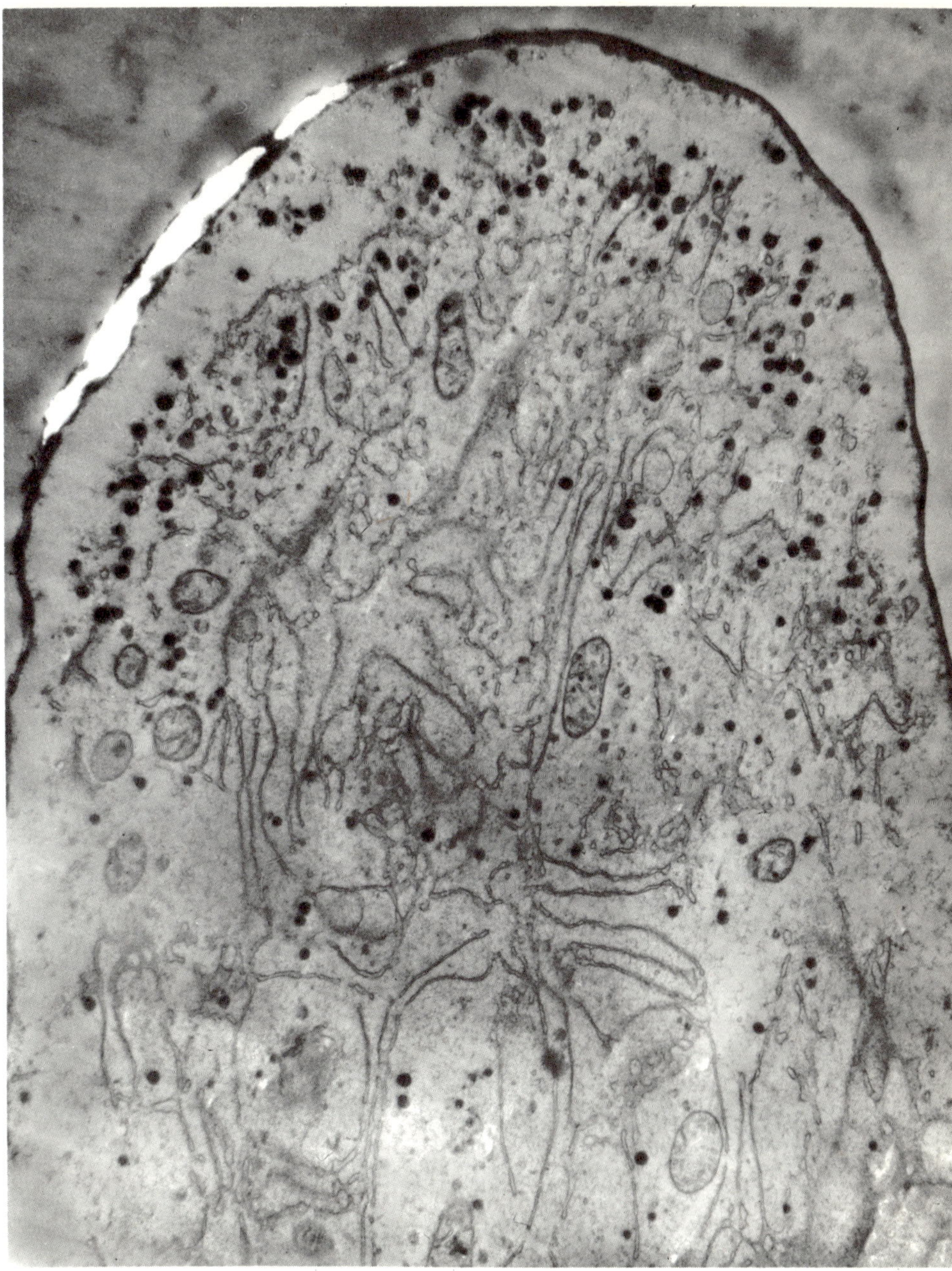

99. Section through a root hair from a maize plant. Many dark Golgi vesicles, which were formed in the interior of the cell, move towards the growing tip.

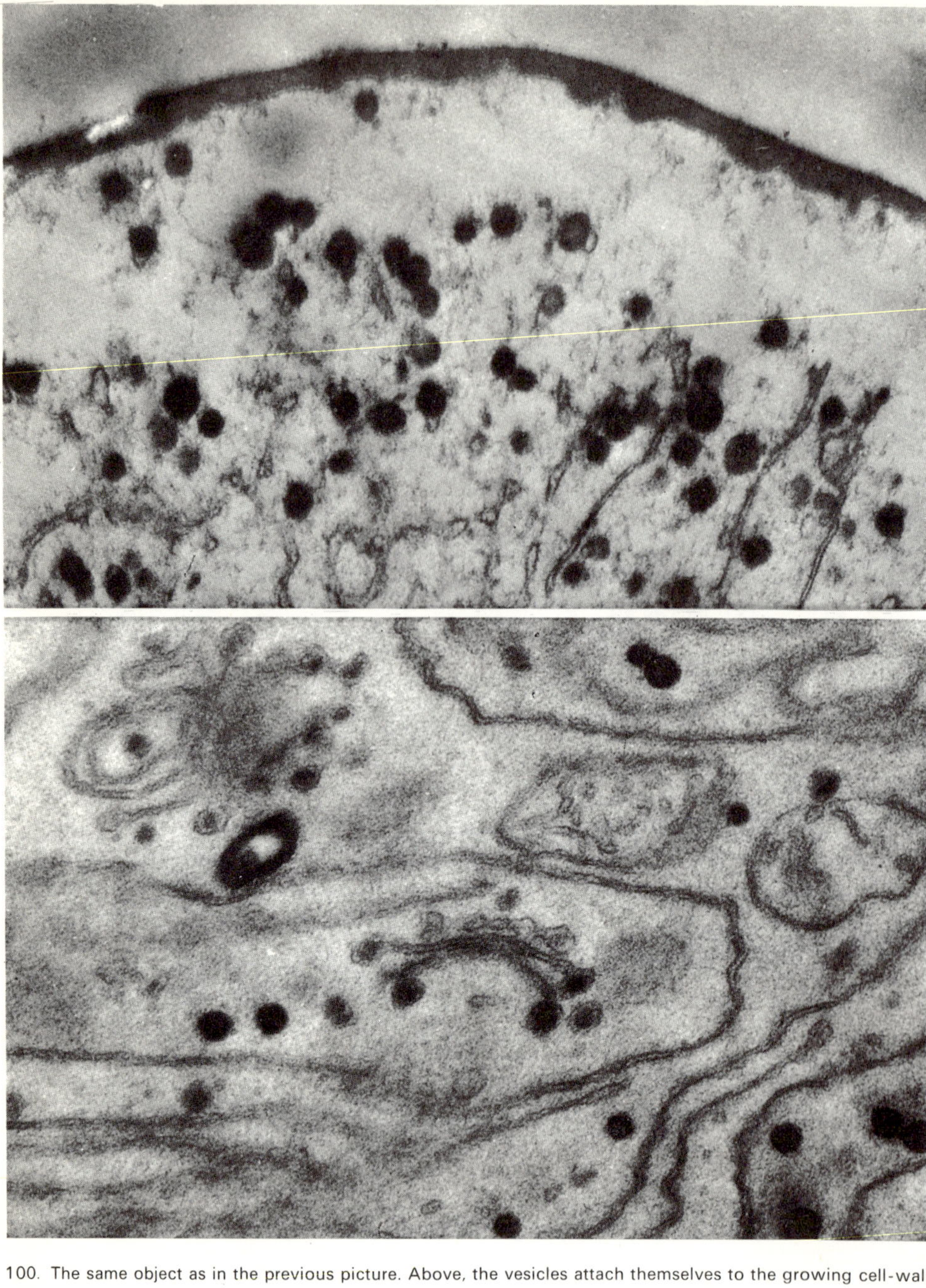

100. The same object as in the previous picture. Above, the vesicles attach themselves to the growing cell-wall and 'go along with it.' Below, dictysomes with divided dark Golgi-vesicles.

movements can easily be followed. They migrate to the growing tip of the cell – that is to say, the point at which the cell wall is enlarging; they attach themselves to this, and eventually merge with it.

In the same way the material which forms the cell-walls of silicaceous algae (diatoms) is polymerized in the dictyosomes; only, this time, it is not a hemicellulose that is involved, but a polymer of silicic acid. This endows each individual alga with a hard, durable armour. In other cells, glandular cells again, the dictyosomes manufacture an essential oil, as they do, for example, in peppermint. Thus the dictyosomes are comparatively versatile in respect of the kind of substance that they can produce. There is only one kind of secretion that the dictyosomes seem to be unable to make, and that is a secretion containing proteins or enzymes such as that produced by the salivary glands or the pancreas.

Up to now we have spoken almost exclusively about glandular cells, but this must not give the impression that dictyosomes do not occur anywhere else. The opposite is true. There exists hardly any kind of cell in which dictyosomes are not to be found, and such as there are, are unable to excrete anything. What their function may be is still not known, but perhaps we shall be helped by the studies of cell-wall construction after nuclear division, which we start on p. 200.

To conclude, we still have to complete the anatomical details of the dictyosomes. The membranes which surround the Golgi cisterns and vesicles are, once again, unit or elementary, membranes, and their thickness and distance apart are known. The cross-section of the disk-shaped cisterns is usually somewhere between 0·2 and 1·2 μ, and thus lies close to the limit of resolution of the light microscope – hence the uncertainty that surrounded the earliest findings about the 'classical Golgi apparatus.'

3.10 Profit and Loss with the Electron Microscope

Phagocytosis and Extrusion

One thing more needs to be brought out about dictyosomes. We need to go a little further into how the cell discharges its contents, i.e., its secretory product. The Golgi vesicles are little bladders with a unit membrane. The cell body is likewise enclosed externally in a unit membrane running right round it, the plasmalemma. If a Golgi vesicle, in its wanderings, arrives at the surface of the cell, with which its outer unit membrane makes contact, then the two membranes fuse together in such a manner that the point of cohesion becomes a 'hole' (see Fig. 101). The bladder hangs like a spherical jug on the plasmalemma, but the mouth of the jug opens through the plasmalemma to the exterior. Continuous contraction of the jug causes the contents to flow out, if indeed they are not expelled suddenly, and soon there is nothing to be seen except a small indentation, which eventually disappears. Nothing more is left of the Golgi vesicle, and its contents have been expelled by the cell. In the case of the hemicellulose in the root hairs of Figure 99, the secretion passes into the growing cell wall.

It is only rarely that the corresponding electron-microscope pictures are as clear as diagrams (see Fig. 102), and even more rarely can one be found that shows the secretion actually being pressed out, i.e., shows *extrusion*. There is a good reason for this, and that is why the caption for this particular section has been worded like a balance sheet. The entire process can be 'read' *in the reverse direction* – i.e., the plasmalemma at the cell surface forms an indentation, which enlarges

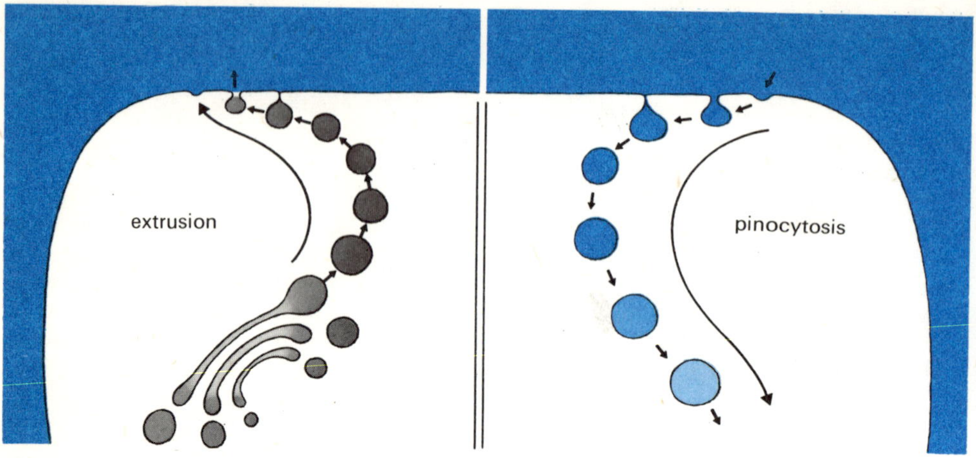

101. Left, *extrusion,* with the Golgi vesicles releasing their contents into the medium surrounding the cell. Right, *pinocytosis,* transference of water and dissolved substance into the interior of the cell.

into a spherical invagination, the sphere cuts itself off and migrates about the interior of the cell as a vacuole or vesicle. The cell has 'taken a drink' – hence the name *pinocytosis* (from the Greek *pinein,* to drink) for this process.

It is fairly certain that the cell can take up water in this fashion, at least so long as it has no firm cell wall. It is very difficult to tell from a photograph whether an invagination is the end of an extrusion or the beginning of a process of pinocytosis.

What is true for water is true for many dissolved substances, though not for all. Potassium and sodium salts, for example, may certainly provoke a pinocytosis, and the same is true of such macromolecules as those of gelatine, ribonuclease (we meet with this in connection with molecular memory, on p. 233), and gamma globulin (which plays an important part in the biology of immunity; see p. 241). Other macromolecules, in contrast, cannot be taken up by pinocytosis; strange to say, the carbohydrates and nucleic acids belong to this group.

Pinocytosis can therefore help us to solve – or at least to understand better – two problems to which physiologists were still looking for answers thirty years ago:

1. How to introduce into a cell a molecule to which a bimolecular lipoid film (i.e., a unit membrane) is impermeable.

2. That many substances are taken up 'actively,' that is, that the cell expends energy in the process. It has been known for a long time that many salts, for example, only enter into a cell if the cell's respiration is intact. If the respiration is interrupted ('de-coupled'), then the intake of salts is also interrupted. At the present time hypotheses (unproved ones) about how substances are taken up are numerous. Pinocytosis, which certainly cannot proceed without expenditure of energy, is a model that may be accepted as valid, at least in some circumstances.

It may now be asked whether pinocytosis is a rare occurrence, or whether it is a necessary part of the process bringing appreciable quantities of material into the cell. This certainly cannot be given an answer that is valid in all circumstances. Nevertheless, it has been possible to show that starved

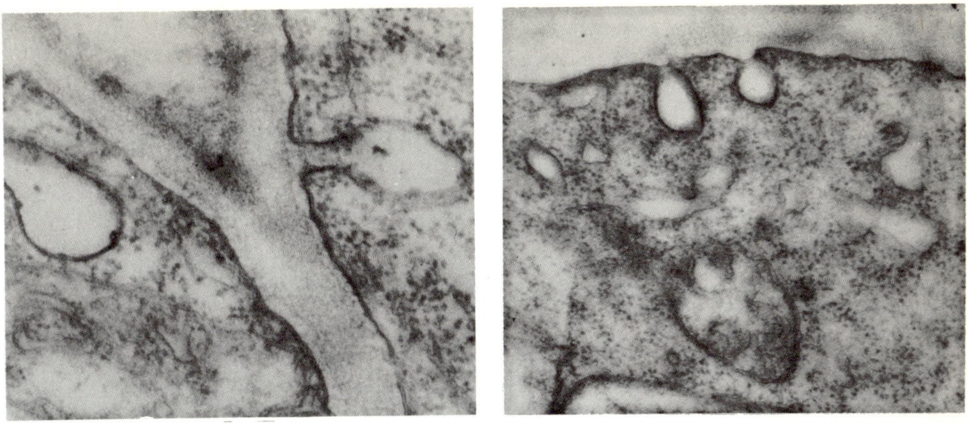

102. Electron-microscope pictures showing jug-shaped invaginations of the outer surface of the cell. These could be either the beginning of pinocytosis or the end of an extrusion.

amoebae, cells that have not been mutilated, can take up from 1 per cent solution of globulin inside two hours, 30 to 40 per cent of their own volume. This is quite a respectable performance.

Let us stay for a moment longer with the problem of dissolved substances. When it concerns ribonuclease (molecular weight 13,000), or even more, gamma globulin (molecular weight 16,000, sometimes even 1,000,000), it is doubtful whether these molecules are really genuinely in solution in the same sense as common salt (NaCl, molecular weight 58) or alcohol (CH_3CH_2OH, molecular weight 46). In spite of this, these molecules are taken up. And if we offer the cell still bigger molecules, or even solid particles which are certainly not dissolved, but only in suspension, to many cells it still makes no overall difference: they devour these, too.

The Greek for 'to devour' is *phagein*; the process of taking up solid particles is called *phagocytosis*. This has been known for a long time, since it can be observed clearly under the light microscope. Basically there is no difference in operation between pinocytosis and phagocytosis; invagination, dissolution of the plasmalemma, and migration of the particle into the interior of the cell occur in both processes. Naturally the size of the bladder-like cavities must be considerably bigger than that of the Golgi vesicles. Small,

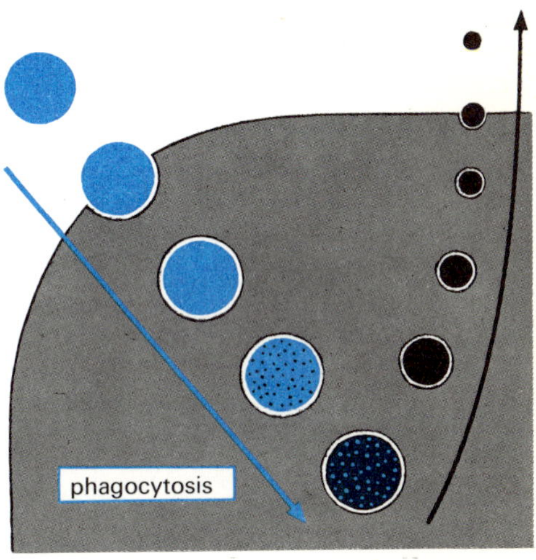

103. *Phagocytosis*, the process of taking solid particles into the cell. The digestible parts are digested, and the unusable residue can be returned to the outside by the process of extrusion.

indigestible balls of synthetic materials (polystyrene, 2,200A across) have proved suitable, and are readily taken up by the cell.

Indigestible – now we have come to the last catchword of this section! What actually happens to the bubbles inside the cell during pinocytosis? Obviously it is not very likely that pinocytosis and phagocytosis are completely the reverse of extrusion – that is to say, that the bubbles end up by merging into the Golgi cisterns of the dictyosomes. Instead of doing that, they remain for a shorter or longer time lying in the interior of the cell. Any water that they may hold is given up gradually to the surrounding plasma, and finally the unit membrane is dissolved away. The same thing happens to materials of a lower molecular weight, which can be used either as building materials or as food. Bigger molecules, so long as the appropriate enzymes are present, are broken down and digested, and sometimes the surrounding basic membrane is digested too. From coarse, solid particles the usable substances are extracted, and the indigestible parts are left behind in the vesicle – which now starts the process in reverse. During the process of extrusion the indigestible residue is passed out to the exterior.

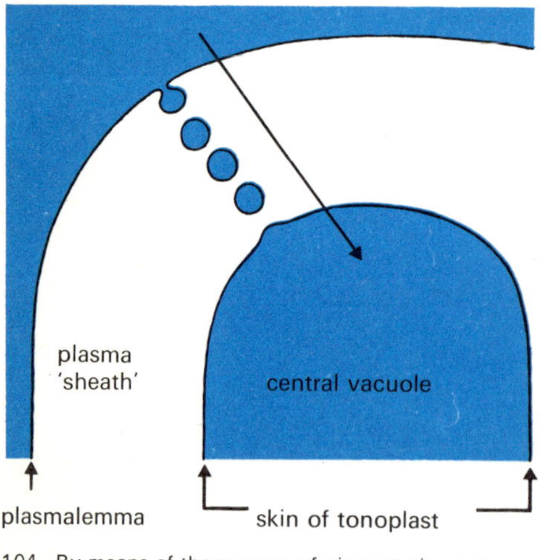

104. By means of the process of pinocytosis, water also can be transported into the interior of the cell.

Another possibility remains to be considered, namely that the vesicle, as long as it contains water and dissolved substances, does not remain in the plasma, but migrates to the central vacuole, the excretory space of the cell, and is incorporated into it – but this belongs to the next section.

3.11 Vacuoles

Keep the Cell Clean and Store Waste Materials

The various bodies, vesicles and little organs, that have been mentioned so far all belong to the non-expendable equipment of the cell, but this is not true of the vacuoles. The occurrence of these is almost exclusively restricted to plant cells that have germinated and become differentiated. Here they are so conspicuous, and clearly so significant, that they cannot easily be ignored.

Anyone who has seen under the microscope a section through living plant tissue – a piece of a green leaf, for example – knows that inside the cell there is an immensely large excretory chamber. Sometimes it occupies as much as 90 per cent of the total volume of the cell, and this means that the cytoplasm, the nucleus, and all the rest of the cell structures are pressed against the cell wall, leaving only a thin 'skin' of plasma surrounding the vacuole. This central vacuole has no apparent structure, but is optically empty; if it is cut open, almost pure water comes out.

It is fairly difficult to prepare such vacuolated cells for the electron microscope. It is almost a regular occurrence for the boundary between the viscous cytoplasm and the very fluid contents of the vacuole to become stretched and torn, and for this reason, good

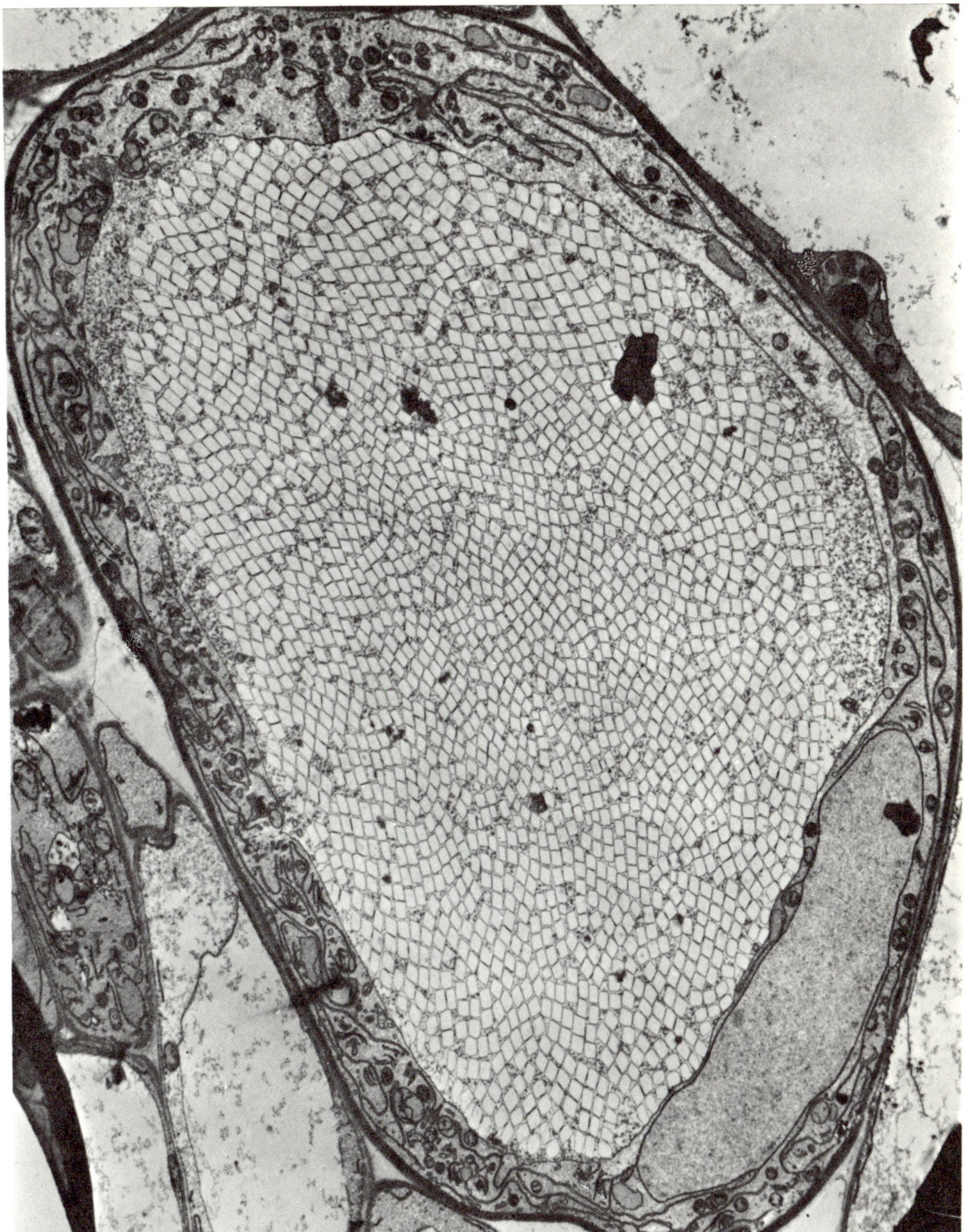

105. A growing cell with a big central vacuole – this large space is entirely filled with cubic crystals of calcium oxalate. All the structures of the cell – nucleus, mitochondria, ER, and so on – are compressed into a narrow space between the vacuole and the cell wall. On the boundary between vacuole and plasma the skin of tonoplasts constitutes a unit membrane.

pictures of vacuoles are rare. Even in good pictures, it is always found that the excretory vacuole, or at any rate the space in the cell where this existed before fixation, appears empty under the electron microscope.

An exception appears in the latest electron-microscope picture, shown in Figure 105. Here the excretory vacuole contains not almost pure water, but large amounts of calcium in the form of calcium oxalate – so much so that it has crystallized out. It now exists as a large number of needle crystals, lying parallel to each other. In the figure these have been faultlessly sectioned, without splitting them. Now, crystals of oxalate are special formations, by no means universally found in plant cells. The picture is not reproduced here for that reason, but because the boundary between the vacuole and the surrounding plasma remains beautifully defined. This makes it a unit membrane, exactly like the plasmalemma.

Once again modern terminology is historically inaccurate, even contradictory. Originally the complete vacuole – that is, the excretory space plus its surrounding membrane – was given (by Hugo de Vries in 1885) the name *tonoplast*. A certain carelessness in use has resulted in most scientists of today applying the name only to the envelope of the vacuole. This makes tonoplast and plasmalemma homologous: the one, the plasmalemma, divides the interior of the plasma from the outside world; the other, the tonoplast, divides the plasma from the central vacuole.

Now, the name 'tonoplast' itself is not entirely well chosen. It might refer:

1. To a moulded shape (Greek *plastos,* moulded) surrounding a cell organ (organelle).

2. To a structure that controls the tonus of the cell. *Tonus* is the Latin term for tension. In the physiological sense this might apply to the tension or pressure obtaining in muscles, vessels (e.g., blood vessels), tissues, or cells.

Many plants do not have the benefit of a supporting structure, as animals do; such plants have neither an external skeleton, like the armoured integumen of insects, nor a bony skeleton such as that of mammals and man. Plants can do something to help themselves by strengthening the outer envelope of each cell, the cell membranes. This is done in the cells of wood, and in the shells of hazelnuts. The hazelnut is particularly hard, and consists of *dead* cells. The thickening of the cell walls and the deposition of woody or corky substance, however, has its consequences. Thicker cell walls make it more difficult to exchange materials between the cells; also, they absorb light – which is particularly unfavourable for leaves, which need light to be able to manufacture sugar and starch from carbon dioxide and water.

Leaf cells – as well as many others – are held taut (turgescent, from the Latin *turgescere,* to swell) by the tonoplasts. According to the laws of osmosis, which we shall not go into here, they take up water, which is conducted along water channels. As a result, their volume increases; the skin of plasma that surrounds the tonoplasts is pressed against the cell walls, and exerts pressure, so that the cell wall is stretched. The cell wall is elastic for a time, but the more it is stretched, the more strongly does it press back – putting the cell *under tension.* Continued increase in volume of the tonoplasts ultimately makes the pressure so great that it succeeds in inhibiting any further intake of water by the tonoplasts. In a given period of time the amount of water expelled by pressure from the cell wall is equal to the amount that is taken up by osmosis; the inward pressure of the cell wall holds the turgor in balance. The pressure inside the cell keeps it turgid, and no more water is taken up.

Everyone has at some time seen the converse of this. If a plant is not watered, then the tonoplasts do not get enough, the turgo-pressure falls off, the cell walls are relaxed, and the leaf, or even the whole plant, wilts. Eventually it dries up, unless it is supplied with water in time. It is evident from this that it is a necessity of life for plants, or at least for many of them, to have vacuoles or tonoplasts in their cells.

The ultrastructures, which can be seen in fine detail under the electron microscope, are the tonoplasts and the vesicles of pinocytosis, with only a narrow dividing wall between them. They contain water and occasionally substances dissolved in it, such as salts and organic compounds, and they are surrounded by a unit membrane. This relationship is further confirmed by the phenomenon that was described at the end of the previous section: the vesicles of pinocytosis, starting out from the plasmalemma, transverse the cytoplasm and then give up their contents to the central vacuole (see Fig. 104).

In all probability the tonoplasts have other functions besides producing and controlling turgescence. For example, the red and blue colour of the blood in many organisms, as well as the red colour of the copper beech, is produced because certain coloured substances are released in great quantity into the vacuole. These are the so-called *anthocyans* (from the Greek *anthos,* flower and *kyanos,* blue), carbon compounds of complicated structure which can change their colour according to the salt content of the water in which they are dissolved. The unusual colour of the copper beech, which is particularly intense when the tree is young, is not altered in autumn by reddening or yellowing of the leaves; these pigments are produced quite differently, and are not to be found in the vacuole; more about this on p. 187.

Generally a large number of substances are dissolved in the fluid of the vacuole, depending on the metabolism of the cell; but these are substances that cannot be utilized any further: tannins, mineral substances, and many others. Many research workers therefore consider the vacuoles to be refuse dumps, which are necessary because plants, unlike animals and man, do not have the benefit of an excretory system – but these are speculations, which can hardly be avoided occasionally, at least not where electron-microscope studies are concerned.

3.12 Chloroplasts

Manufacture Fuel for the Power Plant of the Cell

If we now consider the plant cell in a little more detail, we shall immediately be involved with yet another kind of cellular organ, which is not found anywhere else. To be more precise, this organ is not merely confined to plants, but occurs only in *green* plants, though its significance extends beyond these, to all plants, and in the long run, to all living creatures. We are talking about the grains of chlorophyll, or *chloroplasts.* The Greek *chloros* means green, and *phyllon,* leaf; chlorophyll, or leaf green, is the colouring matter of leaves which reveals the presence of chloroplasts. With the help of this pigment, which absorbs the energy of sunlight, and through the operation of a very characteristic fine-structure the chloroplasts utilize carbon dioxide and water to make glucose, an organic substance upon which feed not only plant cells but nearly all living creatures found on the earth.

This process, without which life is hardly possible, is known nowadays as *photosynthesis.* At one time it was called *assimilation,* but this term is now considered not precise enough. 'Assimilation' implies no more than making two things similar, taking a substance that is strange, or 'foreign' to the organism, and converting it into other substances, which are at home there – and this may take place between quite different materials. Thus, for example, there is a process of nitrogen assimilation, and so in this case we ought properly to speak of carbon-dioxide assimilation. Yet this expression is still much too general. To assimilate carbon dioxide (which is valueless energy-wise; see p. 166) requires energy to be supplied. What goes on in the chloroplasts, however, is *assimilation of carbon dioxide with the help of light energy* – and this is really very complicated to say; the term *photosynthesis* is shorter and more to the point. A whole range of lower organisms carry on the process of CO_2 assimilation just the same, using not light energy but energy from chemical reactions; logically this kind of assimilation of carbon dioxide should be called chemosynthesis.

Now, the 'glance into the cell' that we were going to take in this chapter with the help of the electron microscope will teach us nothing about the process of photosynthesis, but it may tell us something about the fine structure of the chloroplasts. We won't start right away with the big guns, but begin with the light microscope, since chloroplasts are so big that they can conveniently be examined under the light microscope. In the cells of higher plants the chloroplasts are all fairly similar, more or less disk shaped or lens shaped, with a cross-section of about 5 μ. Rarely is there just one in a cell (as, for example, in a few mosses); as a rule, there are ten to twenty of them in the equipment of an assimilation cell; sometimes there are even more than twenty.

106. A light-microscope photograph showing chloroplasts with a grana structure.

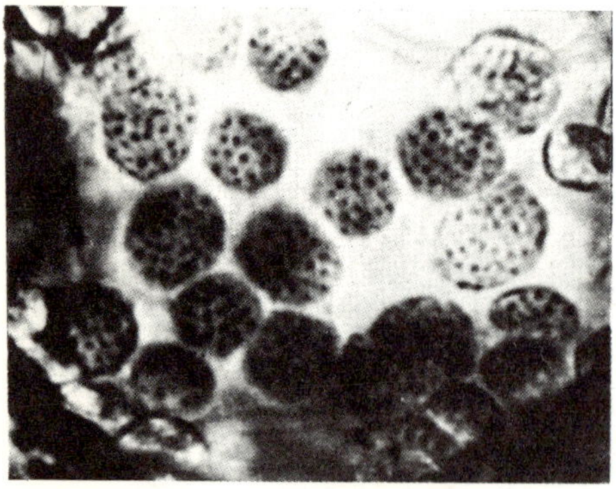

Algae differ considerably from this, at least in part, in the form as well as in the number of their chloroplasts. Many have only a single, large, plate-like chromatophore (Greek *chroma,* colour, and *phorein,* to carry), while others have them band-like, or crescentic, in a network or tattered. This is only mentioned in passing, in order that we can refer later to the great variety that exists in the living world; unfortunately we must dispense with details. In what follows we must concern ourselves mainly with the chloroplasts of flowering plants.

Indeed, not very much more can be found out by means of the light microscope; we must bring into play the so-called *grana.* These were described as long ago as 1883. Seen in the flat, chloroplasts are sometimes not uniformly green, but granular – i.e., green grains, or *grana,* lie embedded in an almost colourless substance called the *stroma.* To begin with, nothing was really known about these grana; many research workers thought that they were not an original structure, but an artifact, a product of disintegration or something like that. Over fifty years elapsed before they were rediscovered (see Fig. 106); soon after that the first electron microscope pictures were published.

107. Section of a chloroplast seen 'in the flat,' under the electron microscope. The dark spots are the grana, which are just as poorly defined as the mesh or network in between them.

Certainly the side views were not very impressive under the electron microscope (see Fig. 107). The objects actually to be seen are more or less circular in outline, and these agree well in diameter with the grana that can be seen under the light microscope. These are certainly the same grana, but they are noticeably blurred in outline, and we miss the distinct layers, basic membranes, ribosomes, and so on that were so clear in the electron-microscope pictures that we saw previously. What lies between the grana is even less

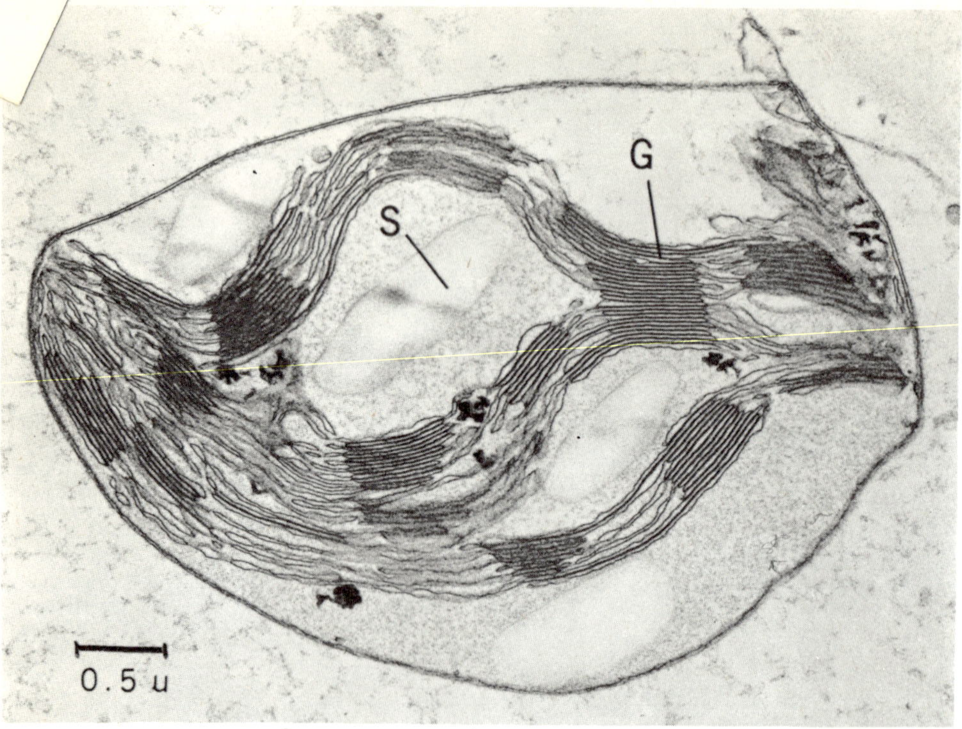

108. A chloroplast sectioned in a direction at right angles to that pictured in Figure 107. The surrounding envelope of the chloroplast (a double membrane!), stroma-thylacoids, and grana stores (indicated by G) can be clearly seen. S is a starch grain.

distinct. All that one sees is a peculiar diffuse grating, or meshwork, with perforations of varying sizes, which appears very much as if it were a consequence of damage to the cell. Is it possible that the skeptics are right in their opinion that the grana are artifacts?

A completely different picture is obtained if a section is taken *at right angles* to the plane of the chloroplast, so that we are looking at the edge of a lens-shaped body (see Fig. 108). Here once more are all the layers and the basic membranes, and this time as densely present as the organs we mentioned above, and even denser than the cisterns of the ER membrane in Figure 83.

After detailed examination the following points may be made:

1. The entire chloroplast is enclosed in an envelope which completely surrounds it. This takes the form of a double membrane, composed of two unit membranes, just as in mitochondria.

2. The envelope encloses a ground-substance, which is called the *stroma,* which shows no characteristic structure beyond a fine granulation.

3. In the stroma lie several big grains (S), which are completely structureless, and so appear white. These are starch grains; they can be given this name with confidence

because before fixation the chloroplast was fully active in photosynthesis, producing glucose, which was immediately converted into insoluble starch.

4. Many *lamellae* run through the stroma, sometimes reaching from one end of the chloroplast to the other, though occasionally they end blindly. Above all, they are always paired, and joined together at their ends, so that at times they constitute a cistern. In this case the cistern is notably flattened, but otherwise it is completely analogous to the cisterns of the ER or of the Golgi systems, the more so since each individual membrane can clearly be regarded as a unit membrane. The term thylacoid (from the Greek *thylakoides*, sac-like) is generally applied to such an exceptionally flattened cistern, complete with its pair of surrounding unit membranes. Thylacoids usually occur in groups, and then they run parallel in places.

5. At certain points the (stroma) thylacoids have between them still smaller thylacoids (G), and then these are usually numerous, and piled all on top of one another. In this pile the number of unit membranes is naturally particularly high, and the preparation unusually dense. Now, if this picture is compared with Figure 107; it will become clear that the pile of short thylacoids, together with the stroma thylacoids running right through them, *are none other than the grana*, seen end-on. The short thylacoids, which we can now call grana thylacoids, must consequently be circular in plan-view. Since they lie on top of each other, like a heap of coins at the bank a granum must seem more compact than it really is. Actually grana thylacoids are obviously not quite as uniform as coins, and so the outline of such a heap is never as sharply defined as the outline of a single thylacoid; and since, in addition, stroma lamellae run right through the pile, it is not surprising that the outline should be so indistinct.

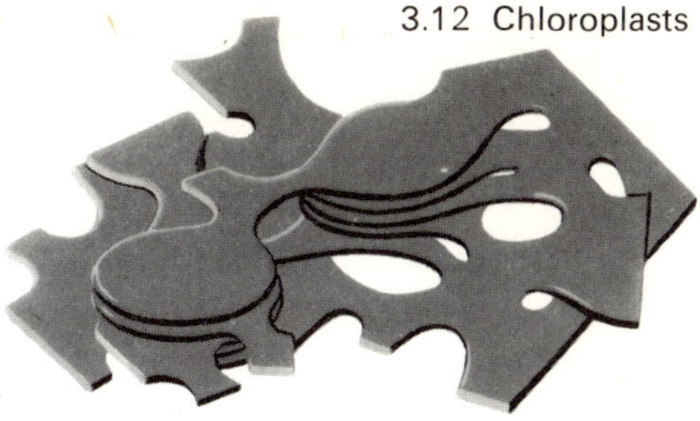

109. A grana heap is formed when irregularly piled stroma thylacoids flow laterally into excrescences, which are individually almost circular in outline, and are piled on top of each other like coins. (After Wehrmeyer.)

Sometimes the stroma lamellae remain attached to the envelope of the chloroplast,

110. Reconstruction of a piece of chloroplast.

and it will be realized that for this reason, at least some of them become obliterated, just like the tubuli and cristae of the mitochondria; later on, however, the connection is broken. The purpose of these structures is obviously the same as that of the mitochondria – namely, an increase in the internal surface area.

But where do the grana thylacoids come from? Originally it was thought that they arose in special places, where they were released by an unknown factor, and that they were built up simply by doubling of the unit membranes and splitting of the stroma thylacoids. The view that is held today differs in many ways. The stroma thylacoids do not run through the entire area of the chloroplast, but are patchy and beset with large and small pores. They also form lateral 'excrescences,' more or less round in outline, as shown schematically in Figure 109. Such outgrowths arise repeatedly from adjoining sites, and press against each other, and this accounts to some extent for the irregular outline. This representation gives a good explanation of the lateral view shown in Figure 107.

Whatever method of origin really produces the appearance of grana, the enormous increase in internal surface area is obvious. What purpose does this serve? Certainly not to absorb as much light as possible, otherwise the grana thylacoids would be arranged to lie one behind the other and not side by side. For all that, they must have something to do with chlorophyll. If plants are left to grow in the dark, they do not become green, but remain white, or somewhat yellowish, like potato tubers. Their chloroplasts contain the ingredients of chlorophyll, but the final step in synthesis cannot take place in the absence of light. Leaves and stems like this, which show 'chlorosis,' have no true chloroplasts in their cells, but only the *precursors of chloroplasts,* the so-called *proplastids* (see p. 193), and these have no thylacoids. So the action of light not only forms chlorophyll but also produces thylacoids.

This strongly suggests that the colouring matter lies in the thylacoids themselves, either in the unit membranes or in the cisterns: and this idea receives further support from the fact that the heaps of grana which appear dark under the electron microscope are identical with the green grana that are to be seen under the light microscope. Naturally these conclusions are not absolutely certain, and this goes, too, for certain calculations that have been carried out several times. The number of molecules of chlorophyll in one chloroplast has been estimated. Since the chemical formula has been known for a long time, it can also be calculated how many plates would be needed for all these molecules if the plates were arranged side by side. At the same time, it is possible to discover from electron-microscope photographs what is the total surface area of all the thylacoid membranes. Both figures are surprisingly good, on the whole. Such calculations, however, are always subject to a big uncertainty factor if we take into account that:

1. Chlorophyll really exists in two forms: chlorophyll a and chlorophyll b. These differ from each other in a short side-chain. The proportion of form a to form b varies from one sample to another. Furthermore:

2. In addition to chlorophyll, there are other yellow or reddish yellow pigments, of different chemical composition, such as the so-called carotin (which is named after *Daucus carota,* the carrot, the thick root of which contains very large amounts of this pigment).

Thus the final evidence for the view that all the colouring substances of chloroplasts are actually present in the thylacoids is still lacking, and the electron microscope cannot help any further at the present time. Under the electron microscope we cannot even see any colour, nor can we, with rare exceptions, even identify any single molecule.

Nevertheless no scientist today any longer doubts that this view is correct. Admittedly there are more than half a dozen hypotheses as to where and how the molecules of pigment are actually held. Probably they lie on the boundary between the bimolecular lipoid film and the protein films that lie on each side of it. This can be deduced from the fact that the chlorophyll molecule has a hydrophobe stalk and a hydrophile head, whereas in carotin, which is completely hydrophobe, the arrangement must be different. It is certain that the molecules of pigment are arranged very regularly, and parallel to each other, like the phosphate molecules in a lipoid film. This conclusion, however, is not a result of observations made under the electron microscope, but follows from physical measurements.

Moreover, it is still completely uncertain how necessary carotinoids are in photosynthesis. They are obviously not immediately involved. Possibly they, too, take up light energy, but instead of utilizing this directly in photosynthesis, pass it on to the chlorophylls. In any event, it is noteworthy that the yellow and red carotinoids absorb just those colours of light that the green chlorophyll cannot take up.

Normally the amount of carotinoid is inferior to that of the chlorophylls, so that the former is completely eclipsed, but of course this is different in plants of temperate countries in autumn. Before the leaves fall off, they become golden, but the chlorophyll has not turned yellow; it has been decomposed, and as a result, the carotin, which was previously hidden, has become visible. The decomposition of the chlorophyll is understandable when we remember that every molecule of chlorophyll contains one atom of magnesium and several of nitrogen, both elements that are always in demand. Of course, this is true only for plants; animals and humans generally have an excess of nitrogen, so much of an excess that they get rid of some every day in their urine.

Carotinoids, on the other hand, are carbohydrates, consisting only of carbon, hydrogen, and a small amount of oxygen, and the plant can obtain a sufficiency of these at any time.

We have now accumulated a few facts about the chemical composition of chloroplasts, and we can go a little further towards completing this. Chloroplasts contain a little RNA, and (less) DNA, as well as even ribosomes, so it is not to be wondered at that a fairly active protein synthesis takes place in the chloroplasts. We must now add a little to what we have already learned; our views about what goes on in the plastids are still by no means clear. At most, we know something about their fine structure, and for that we have to thank the electron microscope. We also know for certain that the fine structure is decisively important for the process of photosynthesis, but which structure and which function go together, and co-operate, will remain obscure for some time yet. To throw light on this, we need newer and still more precise methods. So for the time being, the dream of the researchers on photosynthesis will have to remain unfulfilled: the dream of making synthetic thylacoids from chlorophyll, carotinoids, lipoids, and proteins, and then carrying out artificial photosynthesis, first in the test tube, and then as a full-scale technical process. Of course, this is no reason for giving up, but rather a spur to work harder than ever. The reward: of one day being able at one blow to relieve mankind of all nutritive problems.

urther Steps

ctural Elements Which Still Need estigation

In order not to forget what we have learned, this is perhaps a suitable moment to make a short recapitulation. We already know all the important constituents of a cell: the nucleus, the mitochondria and chloroplasts, the dictyosomes, the ribosomes and the ER, some of them with a highly characteristic structure, even if this is not always completely clear. Furthermore, we know that these structures have a common factor insofar as they are equipped with – or at least, surrounded by – unit membranes. On the other hand, we associate them at times with different functions.

Ribosomes carry on the synthesis of proteins.

ER transports materials (among other things).

Mitochondria manage the respiration of the cell.

Dictyosomes synthesize material for the cell walls, and similar things.

Chloroplasts carry on photosynthesis.

The nucleus is the bearer of genetic information.

Usually these functions are mutually exclusive, but occasionally there is overlapping: protein synthesis occurs not only in those ribosomes that lie on the membranes of the ER, but also in chloroplasts. It has also been established with equal certainty that the division of labour inside the cell involves a sharing of the available *space*; the cell is divided not only into vague 'regions' but into distinct and different compartments and into structures in which from time to time different functions are organized. The term 'compartmentization' is often used for this.

Now, one thing is indeed certain: although it is true that with photosynthesis, respiration, inheritance, protein synthesis, sticky secretion, and transport we have covered important functions of the cell, these are by no means all the phenomena of life. Even if we take into account the fact that sometimes a single one of the structural elements known to us may tackle several problems, there still remains a substantial residue of functions for which the related ('correlated') structures have not been seen under the electron microscope.

If on the strength of this we investigate cells of various kinds, of course we shall still find a multitude of other sorts of structure – small vesicles, granular structures, lamellae and bundles of lamellae, fibrous structures, drop-like enclosures, and so on. They exist in different kinds, and in different combinations, are often completely absent, and conjointly and severally they are difficult to characterize. In some cells it has been thought that a regular repetition of particles could be detected, and since they may perhaps be responsible for some special function, they have been given names like lysosomes, phragmosomes, quantasomes and many others. However, in this field, investigations by means of the electron microscope are still in a complete state of flux, and for the present the link with physiology has gone no further than the planning stage. It is still too early for us to commit ourselves to anything about this, and we will not meddle in the discussion that is in progress between people concerned in this field. We shall know the answer better in a few years.

Probably the lysosomes that have just been mentioned will soon take a firm place in the picture of the cell, and therefore we might mention them, at least briefly. If we consider once again the works which have been carried out on the various constituents of the cell, as mentioned above, we shall establish that the essential constructive processes are:

1. Production of carbohydrates (photosynthesis).

2. Synthesis of proteins.

3. Synthesis of constituents of the cell wall, and of secretions, sticky and otherwise.

The mitochondria are an exception, since they break down glucose. Now, true respiration cannot take place just by the disintegration of carbohydrates, but the disintegration of the carbohydrate can produce ATP. ATP, however, is synthesized from ADP and phosphate. So without more ado it can be formally stated that the activity of mitochondria is *the synthesis of ATP*.

It is clear, however, from first principles, so to speak, that the life process cannot depend solely on one discontinuous synthesis, but that it results from a very finely regulated (and continuously adjusted) balance between synthesis and disintegration. Enzymes of decomposition keep construction in check – or stand by ready to come into action when they are needed. In practice there is a whole range of such enzymes of decomposition, the most important of them being the so-called phosphatases, i.e., enzymes that split off phosphoric acid from compounds which contain phosphate (phosphorylated compounds). When circumstances are right, and not until then, they bring about a pause in the production of the universally important ATP: as ATP-ases they split up the ATP into ADP plus a free phosphate. It is evident that such a lytic ('releasing') enzyme must remain under control, so that it does not interfere indiscriminately in the process of synthesis. To all appearances, for most of the time this enzyme is kept isolated in special vesicles, which are called lysosomes for this reason. Only under strictly defined conditions do the lysosomes liberate their enzyme, and these conditions are dictated by the rate at which synthesis is going on. Here is yet another indication of the close correlation between structure and function. On the other hand, these relationships are complicated by the fact that structural elements can obviously influence each other in contrary ways. Whereas, up to now, the regulation of complete cell-systems has been studied from the point of view of enzyme chains (more about this in the next chapter), it becomes obvious at this point that even structural elements can control one another – a field that is just beginning to be investigated.

Before starting to put the cell together from its individual components, we must take another quick look at the 'ground material,' or ground-plasma. We have discovered that the ER system runs through all the cell like a system of canals, and that its membranes separate an intracisternal phase from an extracisternal phase. Structures that belong to the extracisternal phase include ribosomes, mitochondria, plastids, the cell nucleus, and so on, and these are not juxtaposed without separation, but have a dividing space between them (these spaces can even be seen in the heart muscle shown in Figure 81), and are embedded in the ground-plasma.

The definition can also be formulated the other way round. The ground-plasma (or matrix, not to be confused with the matrix of mitochondria) is then that which remains when all the definite structures, from ribosomes to ER system, have been removed. Naturally during the light-microscope period it was not possible to draw any picture of structures within the ground-plasma. All that was known about these – and even that was not very much – was deduced indirectly. The ground-plasma appears to be fluid – not so fluid as a watery solution, but viscous, slimy, rather like a gelatine solution. It was

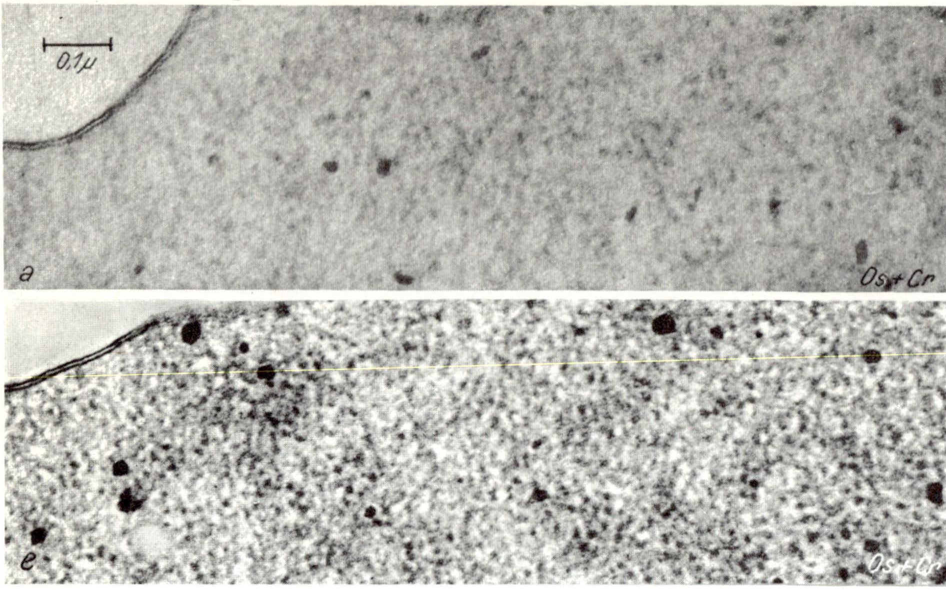

111. The cytoplasm of an amoeba: above, fibrillar structure; below, granular structure.

thought to consist of either fine grains (granular structure) or of the very finest fibers, or fibrillae (fibrillar structure). The physiologists had demonstrated that it contained proteins, and among them, many enzymes of decomposition, of a kind that were not attached to the basic membranes, as they are in mitochondria. In particular, there was found the same enzyme that converts glucose into pyroracemic acid (glycolysis; see pp. 165–6).

When the electron microscope is used on the ground-plasma it is unfortunately the same story as it was for the cell nucleus: hardly any features are made visible which can be definitely said not to be artifacts produced by fixation. Among the best pictures are those taken by Professor Wohlfarth-Bottermann of the ground-plasma of an amoeba (see Fig. 111). They confirm what the light microscopists had thought, but of course up to the higher resolving power of the electron microscope. The ground-plasma can be *both* granular *and* fibrillar in structure. The two states pass into one another, which suggests that the fibrillae may be merely a number of granulae strung together like pearls. Hardly anything more can be said at the present time; so if we want to get on with reconstructing our complete cell, we can disregard the fine structure of the matrix, or ground-plasma.

3.14 Reconstructing a Cell

Analysis Is Followed by Synthesis

The 'reconstruction' is of course not to be taken literally. We have indeed made the acquaintance of all the various bits of the cell, but – except for the cell nucleus, of which there is always only one – we do not yet know *how many* of each we must have. By taking a

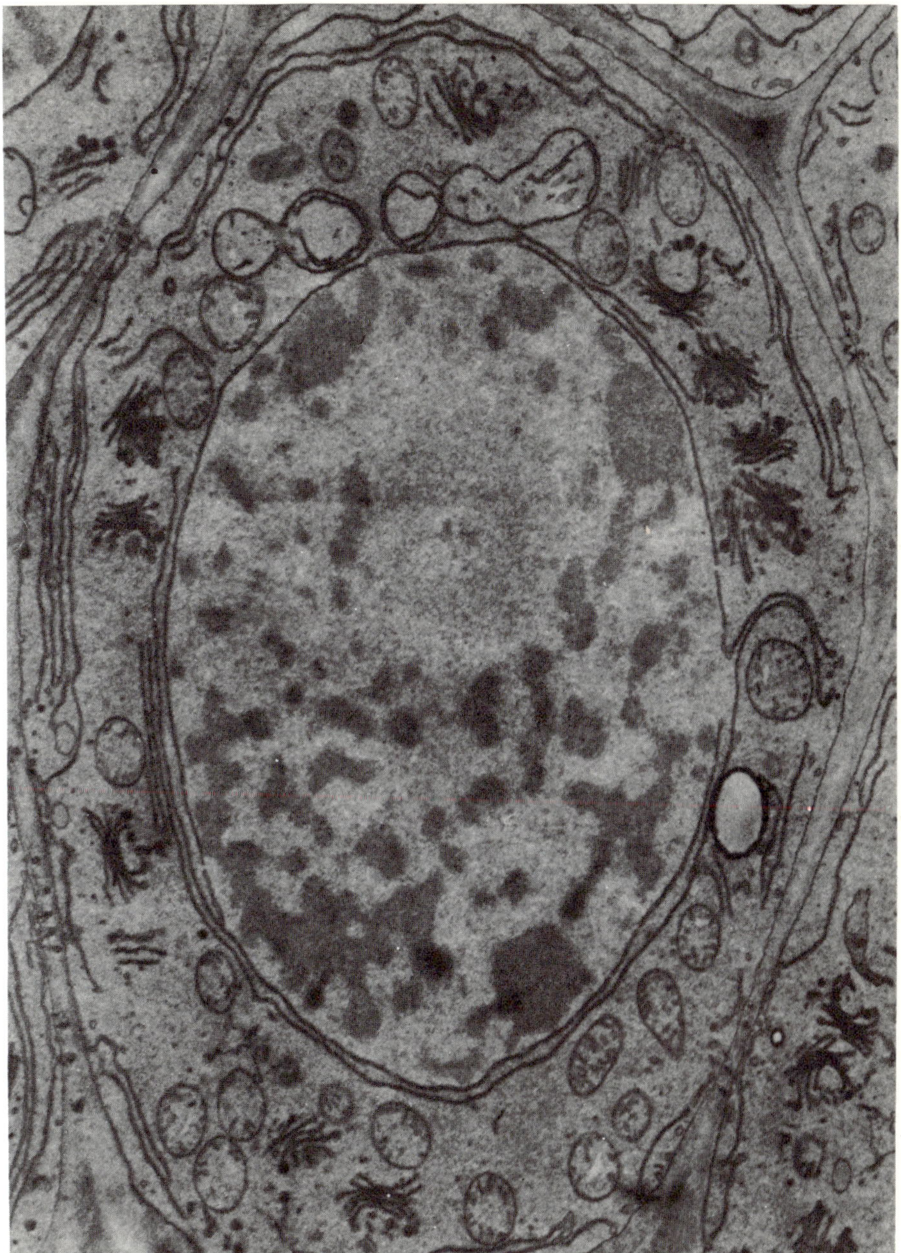

112. One of the younger cells from a maize root. In the middle is the large nucleus, and between this and the cell wall lies the cytoplasm, with strands of the ER system, dictyosomes, proplastids, and promitochondria.

series of sections through one and the same cell, we must find out how many dictyosomes (for example) are present. If we do this, it will make things simpler for us, because we can then study one good, complete section through a cell with a reasonable hope of finding what we are looking for.

Of course this is more easily said than done. A normal cell, washed out from a homogenate – and that is what we always end up with, so far – is usually so big that it cannot be fitted into the field of the electron microscope. Even if we select a smaller cell, the meshes of the supporting net will annoy us by cutting across the picture just at the point where it is particularly interesting. So it is not surprising that a glance through the relevant literature will show that there are only a few photographs in existence which show an isolated cell in its entirety.

Thus we are not quite free in our choice of the cell that we want to study. The best thing would be to decide in the first instance upon a young plant cell which is not yet completely developed (differentiated), even if there is a risk that one or another cell element will be lacking or not yet fully formed. Figure 112 shows such a cell, from a root of maize.

The first thing to notice is of course the large oval structure in the middle of the cell, which can be identified without difficulty as the nucleus (n) by its indistinct, granular structure. As we know, it is surrounded by an envelope, the clear outline of which connects at one point with the lamellar system of the ER. The nuclear envelope is single in some places, double or treble in others, and in several places it can be seen how the inner envelope is broken by nuclear pores (on the upper margin).

The nucleus is surrounded by a broad zone of ground-plasma, which in its turn is bounded externally by the cell wall. This will be dealt with later, but first let us take a look at the vesicles that appear in the ground-plasma, since the plasma itself, as expected, shows no particular structure.

The dictyosomes are the easiest to find. The sinuous Golgi cisterns and the Golgi vesicles, which are divided off from them and show here in strong contrast – cannot be mistaken. There are about a dozen of these dictyosomes in the entire cross-section, and they are distributed more or less uniformly throughout the plasma.

The ER is also easily recognized, even if its cisterns are not nearly as numerous as they appear in Figure 83. The ER runs all around the nucleus, with much branching. It is not very easy to reconstruct its spatial arrangement, but this is superfluous for our purposes.

One observation must be made about the ER. Its strands often lie together in pairs, so that the inference might be made that these strands are unit membranes. This is an illusion, because they do not enclose an intracisternal space, but end blindly in the plasma. In a few places it can clearly be seen that these strands really function as double membranes, themselves surrounding ER cisterns, which in this case are particularly narrow. This becomes even more clear if we examine the extreme edge of the ground-plasma, that is, where it touches the cell wall. Here there is a dark line, which runs all around the cell, produced by the plasmalemma. The plasmalemma, however, consists of *only one unit membrane*; consequently, it is, at most, half as thick as the narrow, contiguous double membranes of the ER.

The remaining particles are somewhat more difficult. A few of them, about twenty in number, are certainly mitochondria; they have a well-formed envelope, whereas the cristae are obviously still at an initial stage. Structures that seem to be entirely lacking are chloroplasts, but this is not surprising. The finished cell, after all, is a root cell, which never sees the light and therefore will never be able to synthesize chlorophyll or form any grana thylacoids.

In place of complete chloroplasts, with their piles of grana, we observe here the precursors, the so-called proplastids. There are several of these in the upper part of the cell. Although the envelope of the plastid is somewhat irregular in outline, it is nevertheless clearly visible from the individual thylacoids which sprout from it. The junction with the envelope is still recognizable here, but later on, at the 'ripe' stage, it will be lost. In the maize root, of course, the proplastids develop not into chloroplasts, but a dense accumulation of leucoplasts.

The ribosomes are not easy to see, but the main reason for this is that the object was fixed by a method designed to show up the basic membranes, and such a method would usually tend to make the ribosomes less visible.

No vacuole can be seen, the cell being still too young for this.

So on the whole our expectations have not been disappointed, and we have been able to find in the complete cell virtually all the cellular organs (organelles) that we described separately. Furthermore, we have been able to assess their number and the comparative amount of each, to a first approximation, of course. On the other hand, we have discovered that in a young cell many organelles are not yet fully formed, or not yet fully functional.

Now we come to the question, What is to be seen in a fully developed cell? Obviously no general answer can be given. Young cells are usually very much alike, but in later life cells go their own way and develop (differentiate) in different directions. At the end of their journey they must necessarily be completely different in structure, while their functions also are no longer the same. A nerve cell can hardly still be comparable to a cell in a connective tissue, or to a cell that forms part of the parenchyma of a plant and is active in photosynthesis, either in shape or performance.

To describe all the different kinds of cells – there are hundreds of them – would fill several volumes, but we are neither able nor willing to write a textbook of the fine structure of the cell; our aim is simply to state a few of the principles of modern biology.

We mentioned on p. 178 the vacuoles and tonoplasts that are characteristic of well-differentiated plant cells, Figure 105 should be examined again. The first point to be established is whether the great increase in size that the vacuole has undergone is mainly caused by the withdrawal of water from the cell. The cell nucleus has grown little or not at all; the plasma has grown a little; the proplastids have become chloroplasts, and their number has increased in the process; and so have the mitochondria (the precursors of which are called promitochondria).

The number of dictyosomes may be reduced if the cell wall is eventually dissolved – as may happen if the cell is one that later becomes part of a sticky secretion; also, the ribosomes become smaller if there is no longer a need to synthesize proteins. In such cases a big vacuole comes into existence in the middle of the cell, where young cells have their nucleus. Everything else has been squeezed together into a thin layer of plasma, which is hardly thicker, thinner even, than the space that is left in the young cell between the nuclear envelope and the cell wall (as we already know, however, that this space is not present in animal cells).

One thing will still puzzle the reader. All the organelles were already present in the young cell, even if some of them were in an incomplete state. Where do they come from? It is sufficiently well established that a nucleus can arise only from another nucleus, but what about dictyosomes, mitochondria, and chloroplasts? Do they arise only from their own kind, or can they come into existence *de novo*?

This is a very good question. It has been known for certain that plastids (which include not only chloroplasts but also reddish yellow chromoplasts and colourless leucoplasts) and mitochondria were structures *sui generis,* which always arose only from their own kind. This meant that they arose from proplastids and promitochondria, or perhaps from even more simple precursors. Out of the many results that support this view, one may be selected.

If after cell division each of the daughter cells possesses the same complement of organelles, then the plastids and the promitochondria must have divided at the same time as the nucleus, and they must have been shared equally by the two daughter cells. This is what happens as a rule, though even today we do not know exactly how. There are cases, however, in which, for example, the division of the proplastids takes longer than the division of the nucleus. Then, after successive cell divisions, fewer and fewer proplastids remain in the cells; they become 'thinned out' until, finally, only a single proplastid is left in each cell. If yet another cell division then takes place, only one of the daughter cells can receive a proplastid, and the other daughter cell must remain without. These cells remain without chloroplasts and

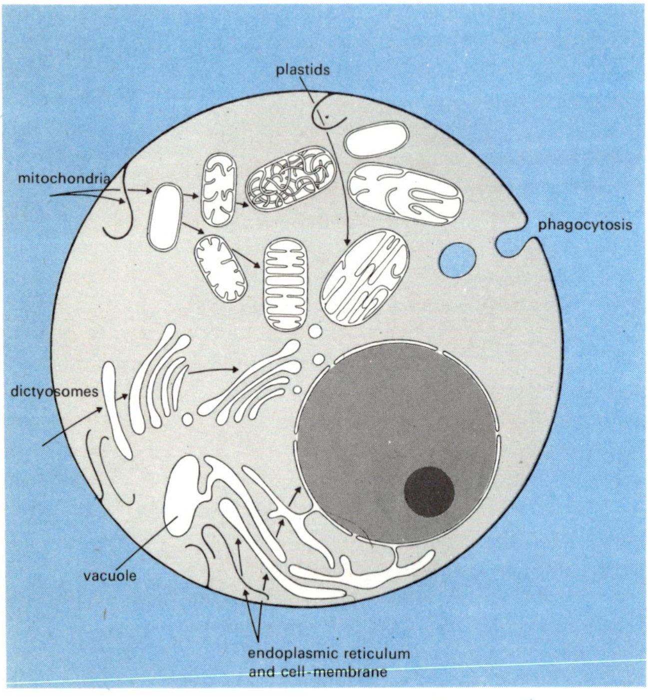

113. A simplified diagram of a cell according to a current, much-discussed hypothesis that derives all the constituent organelles of the cell from the plasmalemma.

so without colour to the end of their lives, showing that they are unable to convert any other organelle into a chloroplast. The same story applies to mitochondria.

Very recently a new hypothesis was introduced into this argument. This was based upon an observation made under the electron microscope, according to which the precursors of both plastids and mitochondria originated as little bubbles cut off from the plasmalemma, similar to the bubbles of pinocytosis, though naturally with a different content. From these developed first the proplastids and promitochondria, and eventually the chloroplasts and mitochondria themselves. Although it is always dangerous, as well as difficult, to reconstruct the course of development from an instantaneous picture taken under the electron microscope, this hypothesis cannot be dismissed out of hand. Arguments from the opposite point of view are still in full spate; once again we must not trespass on the specialist's field, but must await the results of the latest work.

In conclusion let us take yet another look at Figure 105, giving our attention to the cell wall, which has been neglected so far. This figure includes the whole of the cell wall but shows no kind of fine structure. On the other hand, it sometimes appears, for example on the lower right, as if strands of ER run to this wall from neighbouring cells, and bind them together. We shall now examine these cell walls and filaments in a little more detail.

3.15 The Cell Wall

The Skeleton of the Plant Cell

Figure 114 shows a cross-section through the cell wall of a cell from the root surface of a plant of *Rhizinus,* at a higher magnification than the preceding picture. In the middle lies the cell wall, and to right and left of it are the two cells which meet at this point.

First, the cell itself. At times the plasmalemma is to be seen pressed against its surface, i.e., immediately adjacent to the cell wall, and it then appears as a fine line, i.e., as a unit membrane; the plasmalemma then acts as a barrier between the cell wall and the granular plasma. Enclosed in the plasma are some of the cisterns of the ER that are already well known to us, stretched out and thin, lying more or less parallel to the cell wall. In several places, however, they break off and run towards the cell wall. In one place, above the middle, it can be seen how an ER strand from the right-hand cell runs through the cell wall and joins with a bubble-shaped ER cistern of the left-hand cell. In most places, of course, the ER strands end immediately before reaching the cell wall, but in one place at least it is very obvious that strands in the two cells are in line with each other; the cell wall – which otherwise shows no internal structure – in this place appears 'blackened' in the line of the two strands. This can hardly be explained in any other way than by saying that the two ER cisterns, which in this case are plate-like, are also extended in depth. The link between two cisterns in neighbouring cells, on the contrary, is obviously *not* plate-like in form, but is either tubular or thread-like, from which it gets the name of a plasmodesma, from the Greek *desmos,* thread. Of course, it is completely by chance if one of these is cut exactly along its length in the section; usually they are sliced across and appear transparent in section, thus accounting for the blurred contours that occur in cells.

Now, there is nothing new in the fact that neighbouring cells can be pressed into contact with each other. Even in the light microscope it can be seen distinctly how the cell wall is pierced with large pores, called 'spots.' It has been believed for a long time, however,

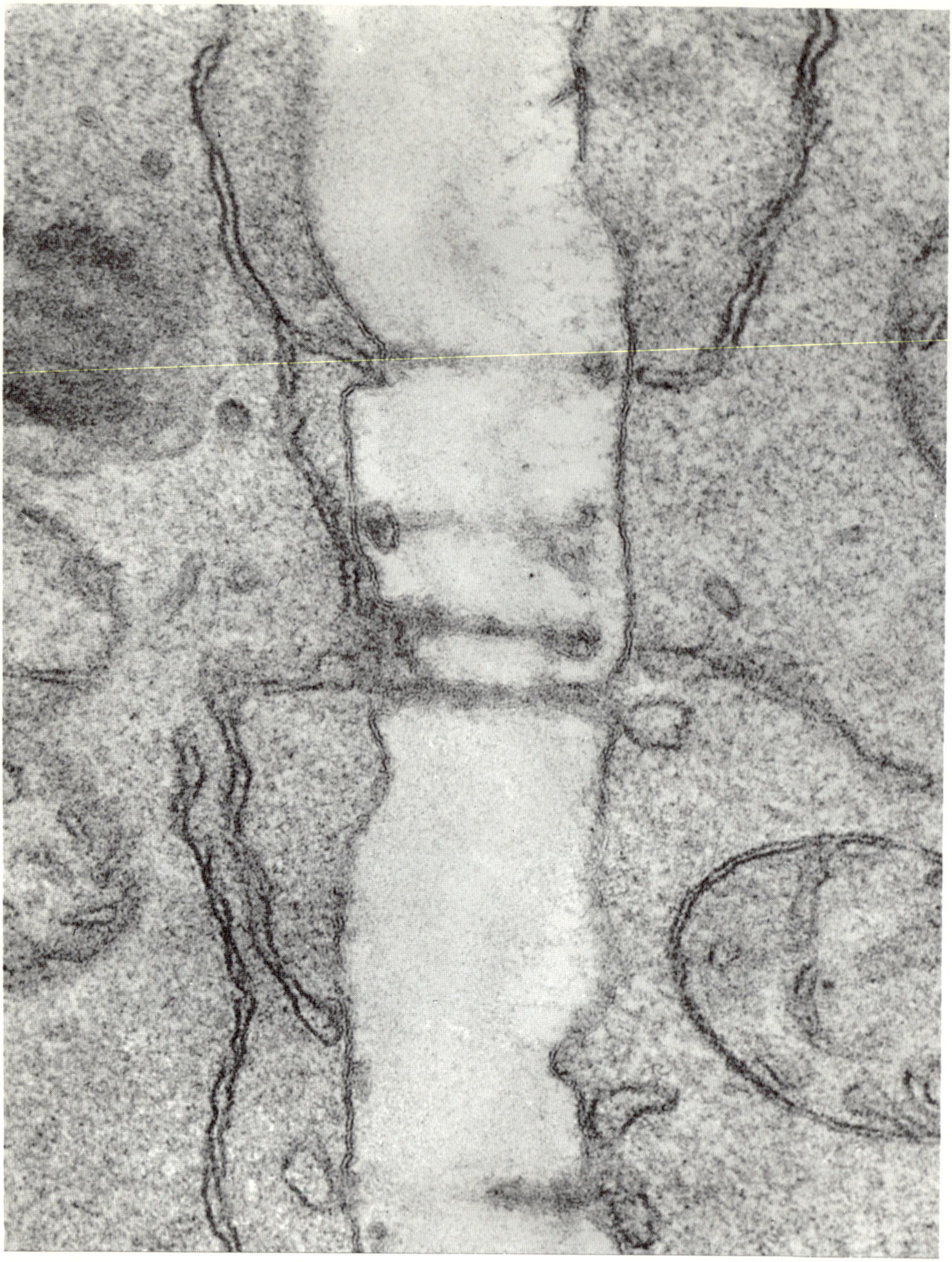

114. The cell wall (pale strip down middle) is traversed by canals from the ER system, the so-called plasmodesmata. Close up against the cell wall on both sides lies the plasmalemma, which is a unit membrane.

that adjoining cells merely touched, with plasmalemma lying against plasmalemma, and that the individuality of the cells was therefore maintained. Obviously this belief must now be corrected. The ER of neighbouring cells goes directly from one to the other through the cell wall, and it might almost be said that in a tissue hundreds of cells possess, as it were, a single, common ER system. This is probably an exaggeration, since we do not know for certain whether the *entire* ER system of one cell is linked to that of another; very likely it can be 'shut off,' either temporarily or over a period. However, it can no longer be doubted that an exchange of materials from cell to cell is possible, without having to pass twice through the plasmalemma.

If we leave the ER tubules out of consideration, the cell wall is certainly 'empty' to the electron microscope, and that is a little strange. During the period of the light microscope, opinion was eventually united in thinking that a cell wall was not a homogeneous body, a complete unit in itself, like a sheet of glass. Certain physical results served only to show that the cell wall is heterogeneous, and must be composed of elongate units, laid parallel to each other and embedded in a ground-substance. These units were not represented as being single, elongate macromolecules, like molecules of cellulose, but as packets of individual molecules called *fibrillae*. If such molecules cannot be seen under the electron microscope, this can only be because they do not produce enough contrast. If they could be made more contrasty in some way, then they would become visible.

Further, the expected increase in contrast was produced, as explained on p. 149, by using a heavy metal – in this case not by using a solution of a salt of the heavy metal to stain the slide subsequently, but by vapourizing the metal itself onto the preparation from a particular angle. This method is called oblique vapourization.

Vapourized metal produces pictures in relief

Let us anticipate one thing: metal can be vapourized extremely well onto a cut surface that is quite flat, and forms a uniformly thick layer of metal which scatters electrons all over with equal strength. As a result, the preparation becomes less transparent, but in a uniform way, without introducing any new differences of contrast. It is different if an uneven surface is coated. Figure 115 shows what happens then, taking as example a sphere standing on a flat undersurface. If a

115. The technique of coating a preparation with metal by 'oblique vapourization.'

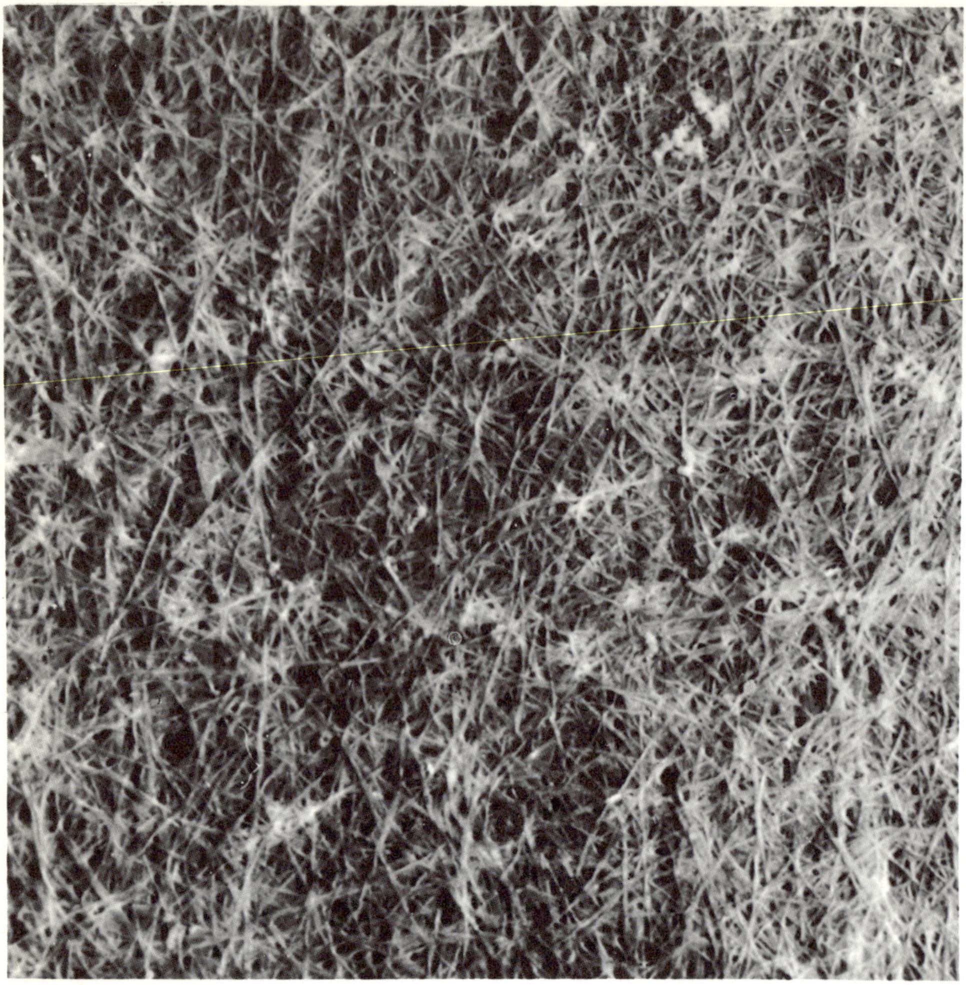

116. The inner surface of a recently formed cell wall of an alga, showing 'straw texture' – i.e., the fibres completely disordered.

parallel-sided beam of vapourized metal is directed at the ball, it deposits equally on the ball and on the surrounding flat; but the reverse of the ball, and any parts of the flat surface that are sheltered from the stream, are not coated. They lie in the 'wind shadow.' The metal that is scattered by the ball strikes the flat plate again at a distance from the ball, and there coats it more heavily.

If a preparation coated in this way is now put under the electron microscope, then those parts that have been coated with metal have become opaque to electrons, and so they record as a clear area on a photographic

117. An older cell wall than that shown in Figure 116. The fibres now lie in crossed parallel texture. At different levels the direction of the fibres is different, sometimes by as much as 90°.

plate, whereas the non-coated areas remain transparent to electrons and so produce pictures, which appear in exceptional relief. An example has already been given in Figure 20. The ribosomes that are shown there had been made visible by oblique vapourization; it would seem that they had been sprayed from the left, and throw a shadow. It is obvious, of course, that the height of each object can be calculated from the length of the shadow that it casts, so long as the angle at which the rays were directed is accurately known. The photographs in Figure 116, 117 show the walls of plant cells that have been treated in this way, and they are of an astonishing clarity, and even beauty. We can see all the thread-like fibrillae in relief, and as distinctly as we could wish. This object comes not from one of the higher plants, but from a green alga named *Valonia,* but that is of no great significance, since similar pictures, if not quite so beautiful, can also be obtained from flowering plants. Nevertheless, it is never *sections* that show up in such good contrast, but always just the *natural surface* of the cell wall, from which all the attached fragments of the cell contents have been cleaned off.

Something else appears in the two pictures. The first shows a cell that is quite young, in which the fibrillae lie tangled together in a disorderly way (called 'straw texture' by the specialist). It is different when the cell wall is older and has become thicker through the deposition of new fibrillae. The new fibrillae are laid down in distinct parallel lines, and sometimes even the old fibrillae can still undergo a delayed organization. In different sections the direction of the fibrillae can vary through as much as 90° – this type of arrangement is called 'cross-parallel texture.'

A wall consisting of several layers may thus be built up exactly like plywood, and have the corresponding mechanical properties.

In this way – incidentally, as it were – we have taken in at one glance a cell wall both being formed and growing. It is known that the growth process is not finished even when secondary and tertiary layers have been laid down: *inlaying* of lignin, a woody substance, into the walls of wood cells; *overlaying* of suberine, a corky substance under the bark of cork oaks, and many other processes may take place, each under special conditions appropriate to the particular cell – but this again must be just an aside.

In conclusion let us look at the first beginnings of a cell wall – and this will really be the last 'glance into the cell.' We saw this momentarily back in Chapter 2, during the telophase of a mitosis. There we saw that between the two daughter cells, in the equatorial plane, blister-like objects appear, which run together to form a kind of cell plate. At first this is more or less fluid, but it quickly solidifies and stretches out to the edges of the 'old' cell. We have assisted at the birth of a new cell wall. We recall equally well the growing cell walls of root hairs (p. 74), where there were Golgi vesicles which migrated to the walls and merged into them. Is it possible that the blisters that merged into the cell – late in the first example – might also be Golgi vesicles?

Let us consider Figure 118, which reproduces the three successive stages of the formation of the 'primary' cell wall. The diagram on the left shows that there is no continuous cell wall, but as will be enlarged upon later, numerous vesicles can be seen, some of which are still isolated, and some of which are linked together in short chains, and leading into each other. In the center diagram the young cell wall is already 'closed,' though new vesicles still migrate about inside it and merge with it, until finally (right-hand diagram) the primary wall is finished, and already equipped with plasmodesmata. However, the vesicles that allow this to take place are actually Golgi vesicles, cut off from the Golgi vesicles of the dictyosomes. It is quite consistent with this that the primary wall should not consist exclusively of cellulose, but contain predominantly the polysaccharide pectin, the same substance that marmalade brings to the palate. First, the secondary cell wall is formed by application of cellulose fibrillae, but these consist not of Golgi vesicles, but of material that arises from the plasma near to the cell wall, the

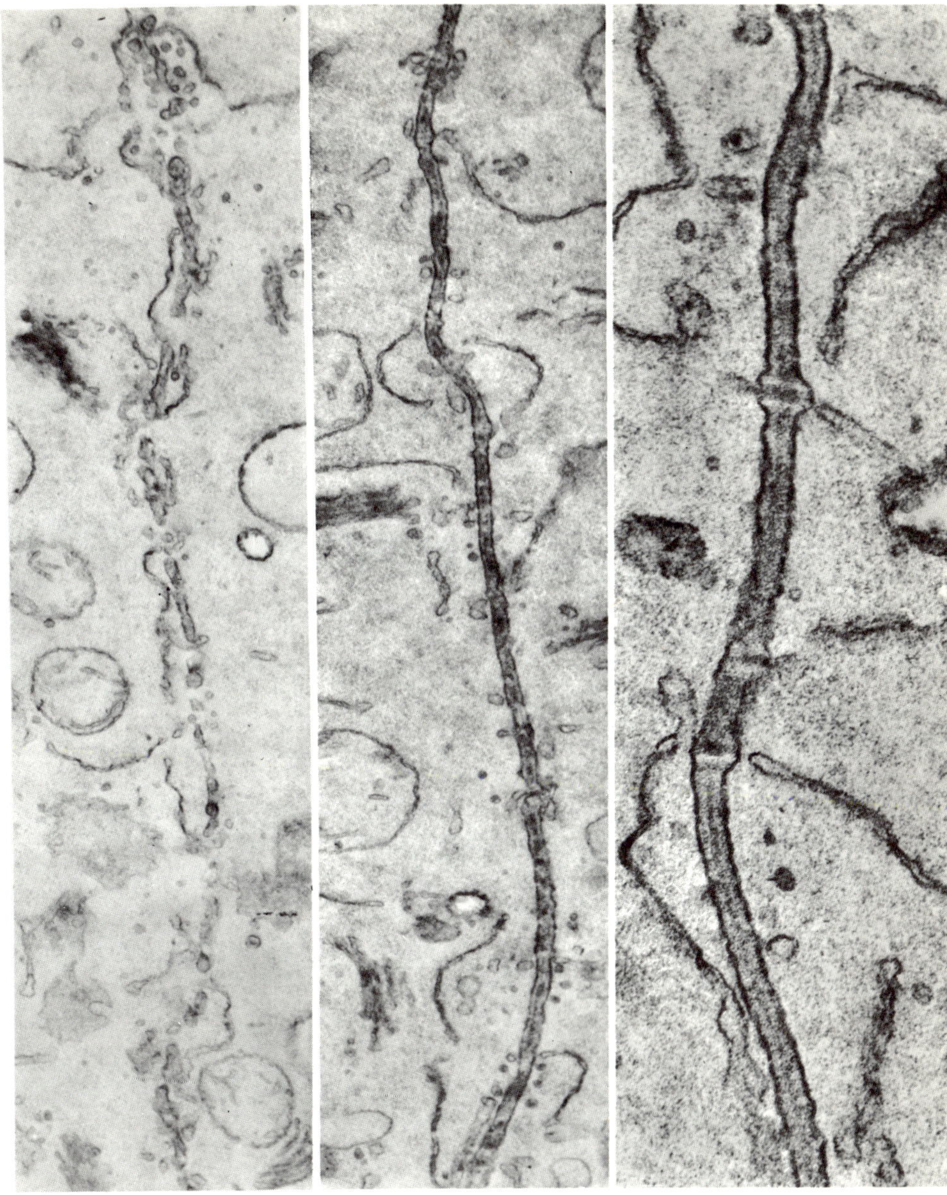

118. Left to right, three successive stages of the formation of a cell wall by Golgi vesicles.

nature of which is still not known with certainty at the present time.

This is the last glance that we take into the cell, with the help of the electron microscope. It would be more correct to say that this *must* be the last glance that we take, because it is unfortunately not possible in this book to review all the units, special structures, and cell types, which are more numerous than would be thought. It is troublesome to master even those that have so far been drawn and photographed, and which were completely new, and generally unexpected; how much more confusing would it be if we were now to try to study nerve cells, renal canals, and cross-sections of flagella, or developmental stages in which geometrically crystalline shapes appear, which then unexpectedly grow into lamellae or into thylacoids. And this in only a part of the new discoveries that have been made recently.

In these circumstances it would be superfluous to say any more – it can only serve to confirm us in two conclusions, which we have to face up to anyway:

One of them is not unexpected. A cell is not just a bag of enzymes, but as we have just reconstructed it, it has an architecture which, wonderful though it is merely as a static thing, is definitely subordinated to its functional relationship.

The other is rather disappointing. The constructural principles are so subtle, even when they do not extend into molecular dimensions, that the 'director molecules' – with some exceptions – still remain invisible.

With this we conclude this chapter of *Modern Biology* and begin another one, which will again be concerned with how things function, but this time with the so-called regulator-mechanisms. The first step will be to demonstrate by new – and sometimes strange – methods that the cell is in fact something more than a 'bag of enzymes.' This demonstration has been delayed in the past by certain other factors; notably by researchers who kept too narrowly within their specialised compartments, thus allowing information to become stale before it was made use of. Admittedly three scientists working in this field, Professors Jacob, Monod and Lwoff, have won Nobel Prizes for their work, but the following chapter was already finished when these honours were announced. In fact, the economy of the cell is not carried on without a plan, but extremely economically, under continual control and regulation. *Planning* does not mean a distribution of food-cards; the author does not mean to imply any resemblance to politico-economic problems, but only those of the oldest living creature on earth – the cell.

4. REGULATION TECHNIQUE WITHIN THE CELL

4.01 Production by Weight

Economy Is the First Priority

We have not merely taken a glance into the cell, but we have studied diagrams and become familiar with the sometimes confusing profusion of detail within a cell. The reader may feel that this is enough for him. Impressive though the architecture of the cell may be, however, it still has to leave space for the real life-processes to go on. So now we must give our attention once again to how things work, especially to those that were included in the first two chapters. They showed us how the genes and their genetic information were passed to offspring and how the genes make enzymes, or in more precise terms, 'press them out.' This may be simplified as

$$1 \text{ gene} \rightarrow 1 \text{ enzyme}.$$

It might be thought that this is all there is to say on this matter. If a cell has all the enzymes, then from its food material it can synthesize every single substance that it needs and build its elements into a structure.

Now, it is true that in practice the simplest of several choices is often the correct one to make, but only rarely is a *simplification of the facts* correct. To say 1 gene→1 enzyme, therefore 5,000 genes→5,000 enzymes, and the cell is finished – this is undoubtedly to oversimplify.

A plant, an animal, a man, moreover, has very many different kinds of cells. The red blood corpuscles contain the red colouring matter hemoglobin, and all the others do not, so they are not red. Only the cells of supporting tissues secrete solid substances and form a skeleton; the others do not. Many enzymes which are essential for digestion are formed in very special glands, such as the pancreas, and never anywhere else.

Is it to be understood from this that the different cells of an organism never have a complete set of genes, the entire genome, but only fragments, different in each type of cell? Then we must look for a mechanism that will break up the genome and get rid of some parts of it; and this must be a mechanism that operates very precisely, according to a carefully worked-out plan. The mechanism must be directed in its turn, and directed *in accordance with some kind of information.* Otherwise it might well happen some time that cells with chlorophyll would turn up in the middle of a tree root, or digestive glands in the eyes.

Many readers may have been successful in growing cuttings from begonia leaves, whereby they have made leaves turn into complete plants, with roots and blossoms. Obviously every cell of the plant body must have contained the complete genetic information, not just a piece of it. This complete information lies, so to speak, 'on call,' but only part of it is actually called, and the rest is left in reserve for any purpose (e.g., for regeneration). The question must now occur, What holds back the genes, and according to what plan is the release of information to the cell regulated?

'Regulated' – this means not merely regulation that is past and done with, but much more: that the cell has at its disposal a complete book of rules. Modern regulation technologists want to take a look at this rule book. How are the rules constructed? And how do they work?

It looks as if the control of the issuing of information were homologous with *differentiation*, the phenomenon that starts from one cell, or a few identical cells, and gradually produces different cells, specialized for certain purposes, while other activities are neglected. In these cases can we study the rule book only in higher, multicellular organisms; unicellular organisms, especially bacteria, must be left out of consideration because they do not have anything that can be called differentiation in this sense? Once again, appearances are deceptive. Certainly differentiation originates as a phenomenon of regulation, but not every regulation leads inevitably to a permanent differentiation. Moreover, it is by no means absurd to look for *regulation mechanisms* in or involving bacteria, and anyway we should be grateful for them because they are so handy for experiments and have given us information about some of the most important aspects of heredity mechanisms.

It is not even very long since the bacterial cell was actually thought of as a 'sackful of enzymes' which had on call the complete information; consequently, all the enzymes associated with the gene were thought to be present, fully developed, and able to carry out all the metabolic processes. Today we know better than that! If the mass of a single enzyme in a bacterial cell is determined and compared with the total mass of protein, in some cases the figure arrived at is 5 to 8 per cent – i.e., from the complete protein content of a cell, up to 8 per cent is earmarked for any particular enzyme. This suggests that such a cell can have no more than about twelve different enzymes – whereas certainly hundreds, if not thousands, of them are needed. Now it is very likely that some other enzymes are not produced in such great amounts, so that up to about fifty different enzymes could form part of the complete protein. All the same, how can all the other life processes proceed in an orderly fashion if such a lot of material (and energy!) must be set aside for one single enzyme? It is not surprising, therefore, that in such cells growth and reproduction are heavily handicapped.

Indeed, the bacteria, with their enormously high rate of reproduction compared with that of higher organisms, are strongly competitive. On account of this, one can speak of a 'selection pressure': every cell that develops some kind of storage mechanism more economical than of its neighbour is better equipped to satisfy new demands and so can grow by taking advantage of changes in environmental conditions. Storage mechanisms have a 'positive selection value,' and so it may be said that in evolution, economical restraint is selected.

The simplest – and most effective – storage mechanism, however, is obviously always that which builds up only those enzymes that are essential for the time being, while the other enzymes are produced either not at all or only in small amounts. The relevant information will thus be called upon to a limited extent, or not at all. In this way the organism does not expend itself unnecessarily, but creates for itself a kind of static reserve, upon which it can call in an emergency.

Naturally such mechanisms can have an evolutionary value only if they can be inherited – that is, if some relevant piece of information is located in the genome, as a so-called regulator gene. Genetic analyses of this mechanism are in progress, and it will not surprise us too much to hear that the groundwork of our knowledge of regulator techniques in the cell was carried out on bacteria. To transfer this knowledge to plants, animals, and man ought not to be very much more difficult.

4.02 The Regulation of Enzyme Synthesis 1

Repression – Restraint Exercised through the End-product

Take protein synthesis. It is indeed not quite as simple as other processes of synthesis, but it has the advantage that we already know a lot about it from Chapter 1. If a cell, for whatever reason, stops producing protein, then the components, the amino acids, remain unused, and if the synthesis of amino acids continues, they even accumulate unnecessarily. A cell in which amino-acid synthesis thereupon comes to a stop undoubtedly works more efficiently and more economically than a cell that is without such a brake.

As an example, let us study how an individual amino acid, arginine, can be slowed down. The synthesis of the arginine from precursor substances (the nature of which is irrelevant here) takes place in at least four steps; consequently four enzymes are needed to accomplish it; the synthesis of these, again, is directed by four genes. The four enzymes always work together, in succession: the first enzyme converts starting substance A to B; enzyme 2 converts B to C; enzyme 3 converts C to D; and finally enzyme 4 changes D into E, the end-product, which in this case is arginine. The first part of this chain is symbolized in Figure 2 (p. 19), which shows a circular saw and a chopping machine. If, now, arginine is no longer needed, then enzymes E_1 to E_4 become superfluous, they are broken down, and their constituents changed in other ways, and at the same time the production of new enzymes is halted.

But how? Singly, or en bloc? The latter would be simpler to bring about. A first hint of this, that the cell uses the comprehensive method, came from the search to find out how the relevant genes lay along the genome (we have seen, in Chapter 2, how such gene maps were worked out). The genes were found to

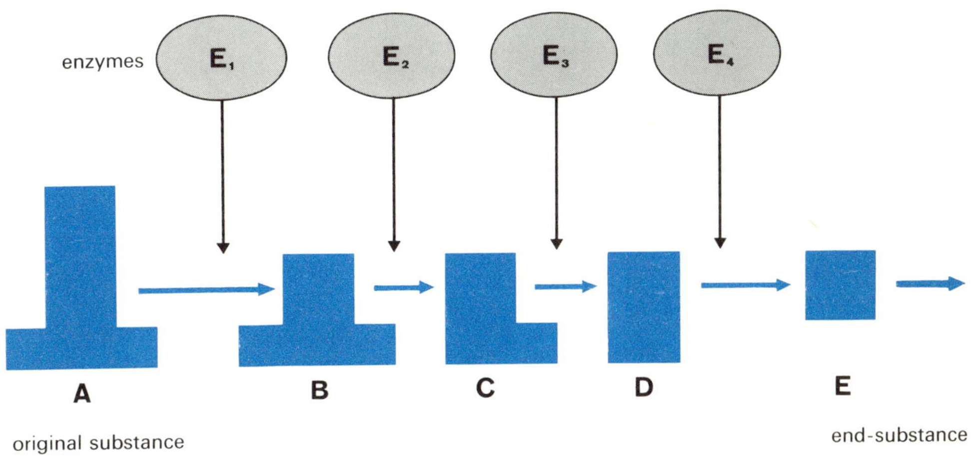

119. Working of an enzyme chain. The original substance, A, is converted into the final substance, E, by the successive action of the four enzymes, E_1–E_4.

lie immediately next to each other, and even in the same order as that in which the en-

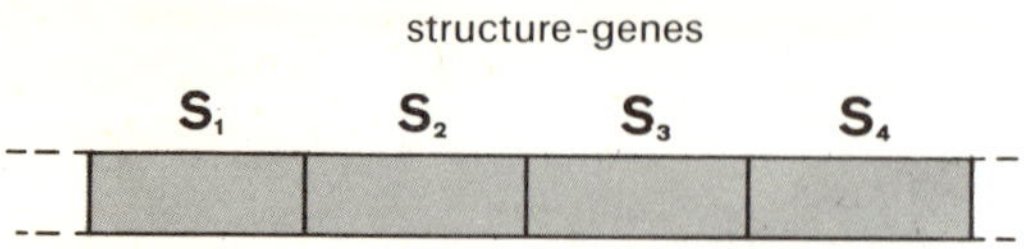

120. The structure genes $S_1 - S_4$ lie along the genome in the same order as the sequence of operation of the enzymes whose manufacture they direct.

zymes went into operation. To avoid confusion let us label the four genes S_1 to S_4, i.e., as *structure-genes* 1 to 4, because they determine the structure of the enzymes E_1 to E_4. The corresponding piece of genome will then appear thus:

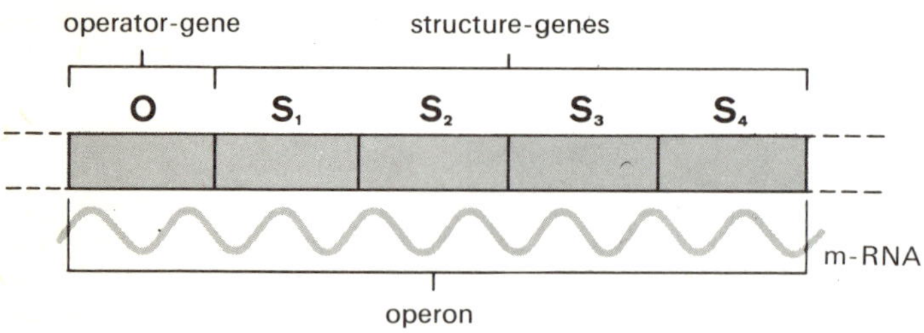

121. An *operon* consists of a sequence of structure genes (such as that shown in Fig. 120) *plus* a special operator gene, which comes immediately before them.

An additional gene, called an *operator gene* (O), ensures that the section from gene position S_1 to S_4 is broken off in one piece. This operator gene probably always lies at one end of the gene-chain, and so to speak, presides over all the structure genes that belong to it; it determines whether they shall be effective, and so produce the appropriate enzyme, or whether they shall remain inactive.

The combination of operator gene and structure gene, is called an *operon*, a term given to it by two French geneticists, François Jacob and Jacques Monod, who received the Nobel Prize for medicine in 1965.

Many uncertainties remain about the nature of an operator gene. *Operator mutants* are known: it is believed that the operator gene either allows or prevents the releasing of genetic information from the DNA (the 'copying' onto RNA; see p. 43), but it is not yet completely certain whether the operator gene is really an independent gene lying near to (in front of) S_1 or whether it is no more than the first part of S_1 itself. One thing is certain, however: the state of the operator gene, whether it is in the open position, meaning release allowed, or in the closed position, meaning release prohibited, is determined *from outside* and not from within the operator gene itself.

Once again, the simplest thing would be if the end-product, arginine, were to act directly on the operator gene and set it into its 'release prohibited' position, but this time

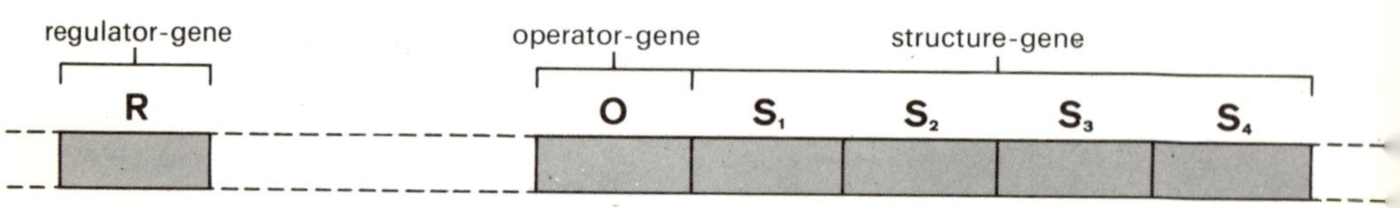

122. The state of readiness or otherwise of the operon is controlled by a *regulator gene* some distance away.

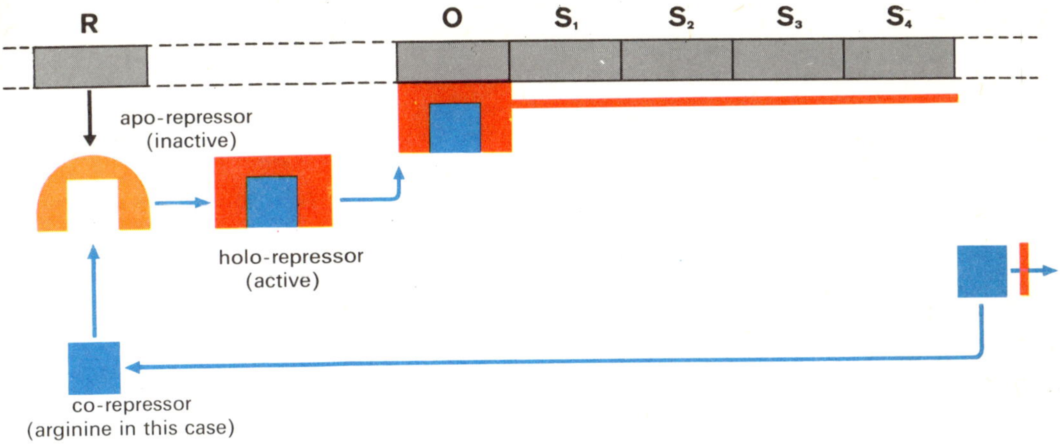

123. Operation of the repressor mechanism. The regulator gene, R, produces an apo-repressor, which is inactive until it is combined with a co-repressor: this may be some substance, such as arginine, which has been stored up. The complete holo-repressor now 'closes' the operator gene, O, and so stops the production of the enzymes E_1–E_4.

things are not quite so simple, for reasons (economical reasons once more) which are again important. So we must introduce yet another gene, the regulator gene, R. The name tells what its function is: to regulate the state of the operator gene, whether open or closed. R, as a gene, naturally lies in the genome; it is, however, not immediately adjacent to O, but some distance away.

And here is another difference between R and O. The operator acts directly on the neighbouring structure genes, whereas the regulator gene, in contrast, releases a substance into the cytoplasm. This substance is ineffective by itself, but has to combine, in the cytoplasm, with a substance of high molecular weight, in this case arginine, before becoming active and causing the operator gene to move into the closed position.

The closing of the operator gene, with the result that no more genetic information can be released and that, consequently, enzyme synthesis ceases, is called *inhibition*, and the eventual conclusion is *repression*. From this it is only logical to call the substance which brings about closing of the operator gene a *repressor* (unfortunately, the nomenclature is not always quite logical and consistent, and

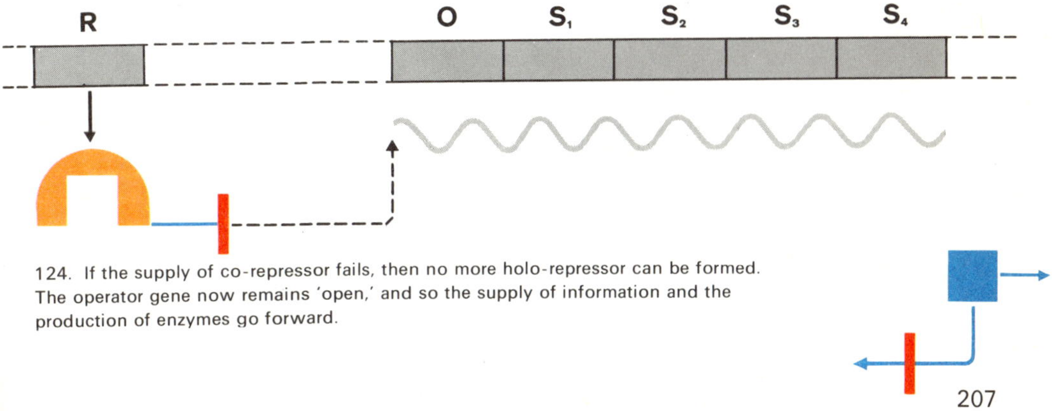

124. If the supply of co-repressor fails, then no more holo-repressor can be formed. The operator gene now remains 'open,' and so the supply of information and the production of enzymes go forward.

so we shall try to get along with as few new names as possible). The repressor, however, is a composite of the substance produced by the regulator gene, R, and another substance, in this case arginine. We can draw a parallel between this and the combination of apo-enzyme and co-enzyme into holo-enzyme, and we can then speak of an *apo-repressor* and a *co-repressor*. R produces the apo-repressor and releases it into the cytoplasm, where it is ineffective by itself; the accumulated end-product of the chain of enzymes E_1 to E_4, namely arginine, combines with the apo-repressor to form a holo-repressor, and this inhibits the operator gene, and thereby the whole arginine operon.

On the other hand, if the arginine, as fast as it is formed by the enzymes E_1 to E_4, is built up into something else – for instance, into particular proteins – it cannot accumulate and so cannot combine with the apo-repressor to form active repressor. In this case the operator is not closed, and the synthesis of E_1 to E_4, as well as the synthesis of arginine, can continue unhindered. Figs. 123, 124 show diagrammatically how the different aspects of this process are related to each other. Apo- and co-repressor are extremely specific. Arginine does not restrain the synthesis of other amino-acids, but works only with 'its own' apo-repressor, which in turn participates only in an arginine operon.

The economical as well as the biological significance of such a set of regulator mechanisms can now be seen. Production is exactly related to need, quantitatively, so to speak. If the need is great, then production continues without interference. As soon as demand slackens, production is throttled back, at source, using the means of production, the enzymes changing over to demolition. This kind of 'restraint through the end-product' (this is again a very formal expression, but it is more accurate than a phrase which is commonly used, 'end-product restraint') is certainly very effective, but it has two drawbacks. One is the complexity, the seemingly roundabout business, of the repressor. In reality this has advantages, because the same mechanism, with only small changes, can be used for other, equally useful purposes (see p. 211).

On the other hand, a real disadvantage lies in the fact that the repressor works a little slowly, and before the enzyme synthesis can be completely shut off, a great quantity of the end-product has been manufactured – which is not needed any more. The repressor by itself is thus only a mechanism for coarsely regulating the process. Is there, in addition, a fine regulator?

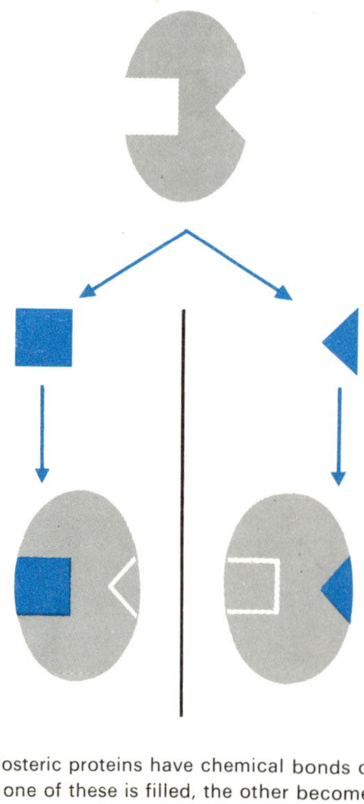

125. Allosteric proteins have chemical bonds of two kinds; if one of these is filled, the other becomes blocked.

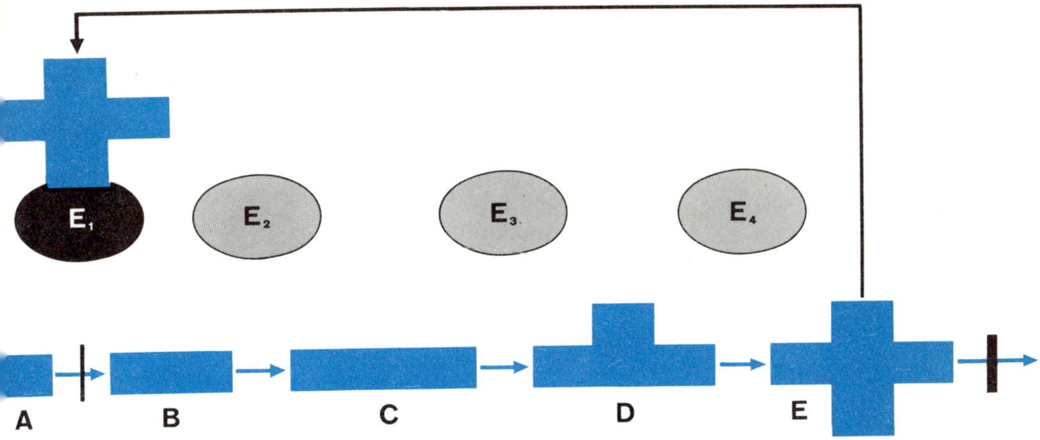

126. Fine regulation of enzyme manufacture. If the co-repressor (e.g., arginine; cf. Fig. 123) accumulates, it can itself act directly on the first enzyme of the chain, E_1, and inhibit it. If this happens, then the whole process stops.

4.03 Fine Regulation

The Influencing of Enzyme Activity

During fine regulation, just as during repression, the complete set of enzymes is split off, and here, too, the end-product is effective, though this time not in combination with anything else. It fixes directly onto the first enzyme of the chain, E_1, and does not combine with a co-repressor or form an operator gene. It combines with the E_1 and inactivates it.

An enzyme protein – at least so long as it stands at the beginning of a chain concerned in synthesis – has two specific bonds, in addition to the place where the co-enzyme fits. One of these is for the substrate, A, on which the enzyme is to work, and the other is for the end-product of the synthesis. The two are *not* identical: it is not simply that the end-product can attach itself in position *in place of* A, and thus exclude A from the surface of the enzyme. The two bonds are directly differently in space. If the end-product could become attached, then the properties of the enzyme would be changed in such a way that the original substrate, A, could no longer be transformed into substance B, the function of the enzyme. Proteins that have bonds differently directed in this way are called 'allosteric' (from the Greek *allos*, different) proteins.

The blocking of the first enzyme of a chain through the action of the end-product is effective immediately, and then at once – as soon as the allosteric effect comes into operation – the first intermediate product (and naturally all the succeeding ones too) is not made any more. This regulator mechanism is nothing more than a back-coupling, or *feedback*. More precisely it is a *negative feedback*, since the effect of the end-product is to inhibit production, not to encourage it (unlike the feedback that can sometimes occur in radio apparatus!). Whence it may be particularly noted that the *principle* of feedback, this comparatively simple technical trick, has existed in living cells for millions, if not thousands of millions of years.

To recapitulate: the regulation of the process of synthesis takes place in two ways – first, through repression, which influences the synthesis of enzymes through control of the release of genetic information; and second, through feedback, which influences enzyme activity. Coarse and fine regulators work together in this way most impressively.

4.04 The Regulation of Enzyme Synthesis II
Induction and the Influence of Nutrition

What we have learned so far about the synthesis of amino acids also applies without any serious limitations to the synthesis of other substances, such as those involved, for example, in growth, cell division, duplication of DNA, and so on. Synthesis, however, is only one aspect of metabolic processes, though it is the aspect for which building materials and energy are needed. The opposite aspect involves energy-releasing processes, which break down the nutritive materials that have been taken up, splitting them apart and thereby obtaining both energy and basic construction materials. Are these processes of decomposition equally regulated, and do they involve regulatory mechanisms similar to those that we have just seen?

Let us put the question a little more precisely. Bacteria obtain nutriment from sugars such as glucose (grape sugar), lactose (milk sugar), and the like, but they can also make use of a whole range of other 'substrates.'

All of these are broken down by enzyme action, usually in several stages. From sixty to eighty different enzymes are needed to utilize a mere dozen or so different foodstuffs. Now, admittedly, it is only rarely that all twelve foodstuffs will be present *at the same time*. We need only think of that 'work horse' of the modern geneticist, *Escherichia coli*, which is present in the human intestine, and also in sewage and so on. How often may the full complement of nutritive substances be present in the vicinity of one bacterium? It would be most uneconomical, and even wasteful, if the cell were to have sixty to

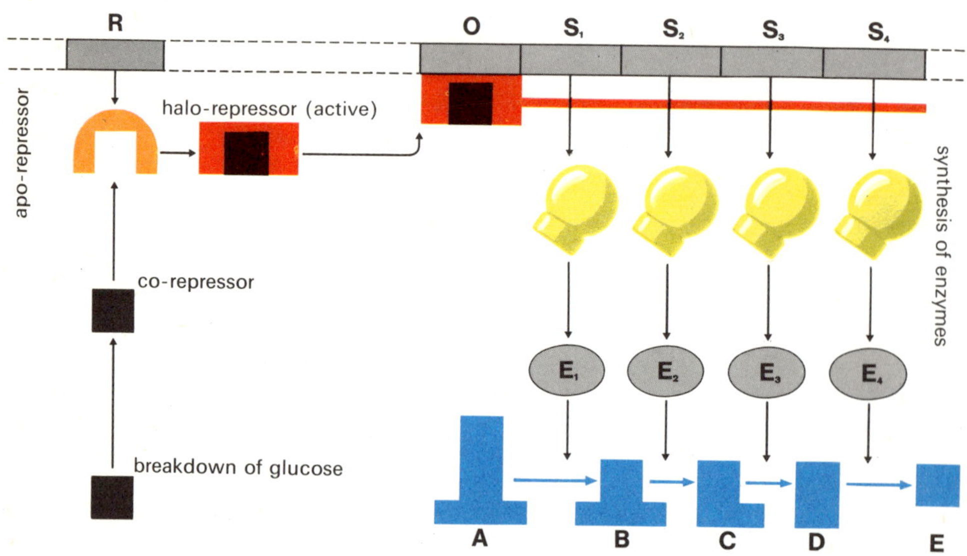

127. Enzyme regulation in nutrition I. A co-repressor derived from the breakdown of glucose combines with the apo-repressor and stops the breakdown of lactose.

eighty enzymes standing by in readiness for 'eventualities' when only some four or eight of them would be actually needed. As we have already seen, these are selected frugally, and mechanisms exist in nature to ensure this; but they must work in exactly the opposite way to those that synthesize amino acids. The product of heavy molecular weight, such as lactose, must *set in motion* the synthesis of the enzyme to be made, not restrict it, as does arginine.

In spite of working in the opposite direction, this is in practice the same regulatory mechanism, though with a slight modification. The roundabout operation of the repressor, which we criticized when we saw it in the synthesis of arginine, makes it possible for the cell to use one system of regulation for two different purposes. Thus it does not need to build up two regulator systems.

Before we turn our attention to this 'slight modification,' we must mention an exception that is worth noting. This particularly concerns the breaking down of glucose. Glucose is not only the preferred foodstuff for bacteria, but also the commonest. For these reasons the mechanism for breaking down glucose is always kept open; the chain of about ten enzymes stands in continuous readiness; and the synthesis of more of these is *not inhibited*. These enzymes, the complete chain of them, are labelled as *constitutive*, for they belong to the basic equipment of the cell. There is another reason why it is important that the mechanism for breaking down glucose should be kept open: most other foodstuffs are changed so much during digestion that at some stage or other they are likely to produce glucose.

Now, however, to a different method of nutrition. Let us take lactose, because the breaking down of lactose, and its regulation, has been particularly closely studied. Again, several enzymes are necessary, and once

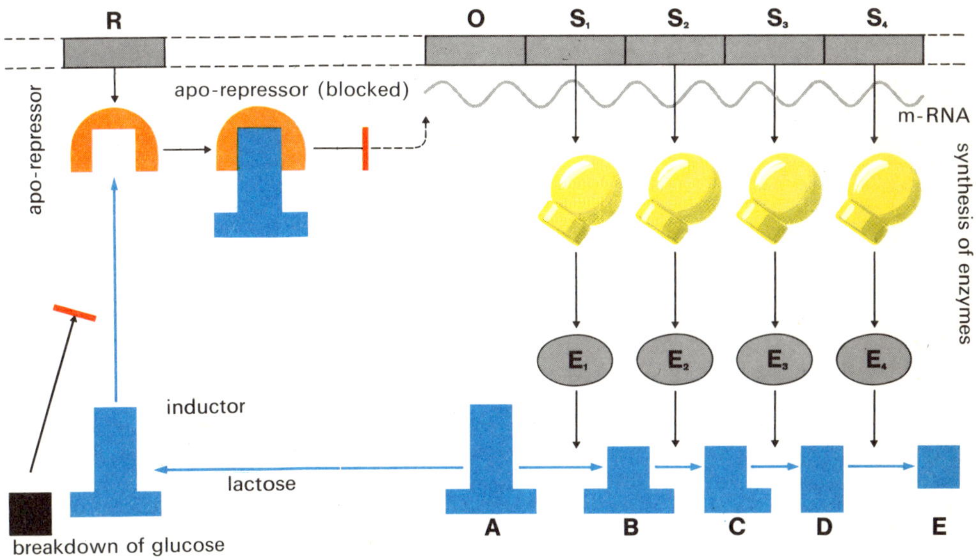

128. Enzyme regulation in nutrition II. If the breakdown of glucose does not supply enough, or any, co-repressor, then some of the lactose present combines with the apo-repressor. The result, however, is not an active holo-repressor, but an inactive compound, and so the breakdown of lactose is allowed to continue (cf. Fig. 131).

again they are combined into one operon, as in Figure 121. Again, a regulator gene is part of the mechanism, and produces an apo-repressor. The parallel with the synthesis of arginine goes a stage further yet. If a co-repressor is present, it combines with the apo-repressor to form an active (holo-) repressor, and this closes the operator gene and prevents the release of the genetic information that lies within the structure gene of the operon.

But now comes the 'alteration': the co-repressor arises out of the breakdown of the glucose. If enough glucose is present, then a great deal of co-repressor and active repressor will be formed, and thus the synthesis of the lactose-forming enzymes will be suppressed. Now, this is really economical, because it will make it difficult for the cell to synthesize the enzyme for lactose at times when it has at its disposal sufficient quantities of glucose, which can be assimilated by means of the constitutive enzymes that are already present.

Now, it must be admitted that there is yet another substance in addition to the apo-repressor; this we shall call the *inductor*. It competes with the co-repressor (produced by breakdown of glucose) for the bond on the apo-repressor; if much co-repressor is present, then naturally the inductor has no chance. It is different if, for example, the glucose has already been broken down completely. Then the concentration of the co-repressor falls so rapidly that the inductor comes into action. If it succeeds in attaching itself to the apo-repressor, then this becomes *inactive*. The combination of apo-repressor plus inductor results in an *ineffective repressor*, which can no longer hold the operator gene in the closed state; the latter is thus de-restrained, or un-inhibited. Hence the structure gene can give up its information, the enzymes that decompose lactose can be synthesized, and the transformation of the lactose is set in train. These phenomena make sense when we learn that *the inductor is none other than the lactose itself*. The lactose itself *induces* (calls forth) the synthesis of the enzymes that are needed for its own conversion (so long as no co-repressor is present from the constitutive glucose-system!).

Let us summarize what we have learned about the induction of enzyme synthesis. It takes place only if: first, the substance that is to be broken down is actually present, and second, no more glucose is present than can break down as part of a constitutive decomposition. If the second condition is fulfilled, then the concurrent repressor is absent (since this must come from the breakdown of glucose); the inductor makes the apo-repressor inactive and releases the operon. If the second condition is not fulfilled, however, then so much co-repressor is present that the inductor lactose is forced away and prevented from attaching itself to the apo-repressor; then co-repressor and apo-repressor produce active repressor, the operon is closed, enzyme synthesis is interrupted, and the lactose is not decomposed, because this is now superfluous.

The correct functioning of this somewhat complicated regulatory mechanism can be traced out step by step by making use of bacteria that have never previously been in contact with lactose – bacteria that, in the words of the microbiologists, are 'non-lactose-adapted.' The expression 'adapted' should not be applied so generally as to suggest that a cell can adapt itself to any substrate it likes – this is true only for nutritive substrates, for the decomposition of which the appropriate operons are already present (laid down in anticipation) in the genome. Suppose we put such non-lactose-adapted

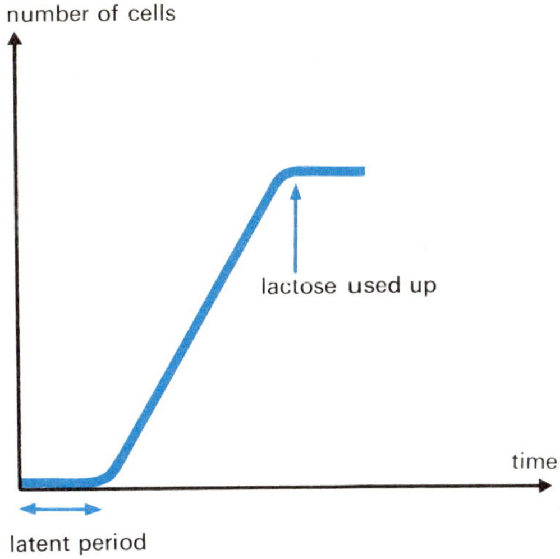

129. Rate of increase of a suspension of bacterial cells. In the presence of lactose, multiplication begins only after a latent period, during which the formation of enzymes is initiated.

bacteria into a nutritive solution in which the only source of energy is lactose and then measure the rate of growth – or better, take counts of the number of cells. As Figure 129 shows, growth does not begin immediately, but only after a certain lapse of time; first, the operon must be released, and before that, some enzyme must be synthesized. After that, the cell count increases steadily up to the point at which lactose is needed, when growth and multiplication are arrested and the curve becomes horizontal.

In a parallel experiment, let us put the same bacterium into a nutritive solution that contains *both* lactose *and* glucose. In this experiment growth begins immediately. It starts with the breaking down of glucose, for which the constitutive enzymes are already present, and because of the consequent lack of glucose, the growth curve becomes horizontal at this point; but now the co-repressor

130. In the presence of both lactose and glucose the bacteria shown in Figure 129 now begin to increase at once, because the enzymes necessary for the breakdown of glucose are already present *(constitutive enzymes)*. When the glucose is used up no lactose can be broken down until the operon has been opened and the necessary enzymes have been manufactured; hence there is a latent period before further growth of the bacteria can take place.

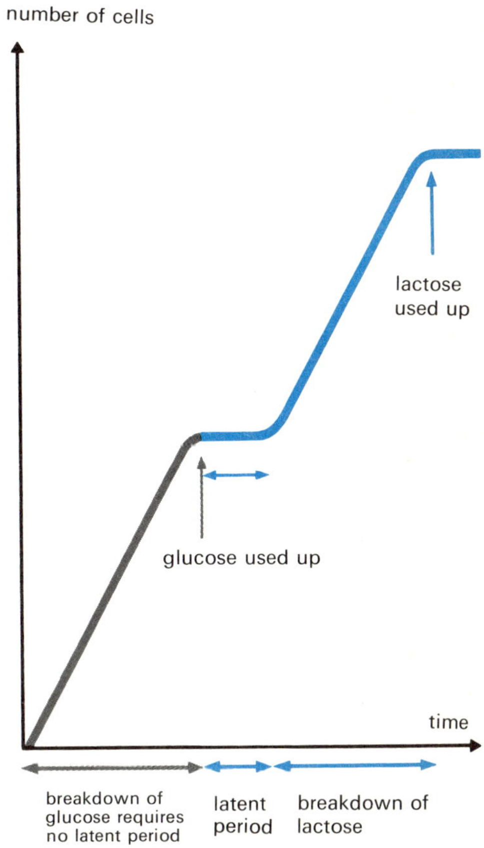

disappears, allowing the inductor lactose to take action – i.e., the inactive repressor mechanism releases the operon. This causes breakdown of lactose to begin and growth to be renewed, though, again, only after an initial delay, necessary for the enzyme synthesis.

We must move on through yet another step, not because it is strictly necessary in an elementary account of the process, but because only then can we appreciate the subtlety of the regulatory processes of metabolism.

During the breakdown of lactose two systems are, so to speak, connected together: first, the actual breakdown of lactose, and second, the constitutive breakdown of glucose. The breakdown products of the glucose are even included in the second system. It is not to be expected that the two systems will be exactly synchronized from the start. If the breakdown of lactose proceeds slowly, that obviously does not matter; the small amounts of decomposition products can be reinforced by glucose decomposition, without more ado. On the other hand, if more lactose is broken down than the glucose system can deal with, then the following 'fine adjustment' comes into operation. The galactose – which is the first decomposition product of lactose – is blocked by the glucose system, and so accumulates. Now it begins to act on its own behalf as a *Co-repressor*, displaces the inductor (lactose) from the apo-repressor, and combines with this to form an active repressor. Then the operator gene is closed, the enzyme synthesis is blocked, the decomposition of lactose is throttled back, until the concentration of galactose has been reduced. Now the tuning is just perfect.

Evidently, the neat division of enzymes into *adjustable* – through induction or repression – and *non-adjustable*, or constitutive, is not quite satisfactory. If the operons, which provide sets of constitutive enzymes, have a completely free hand, then is there not some

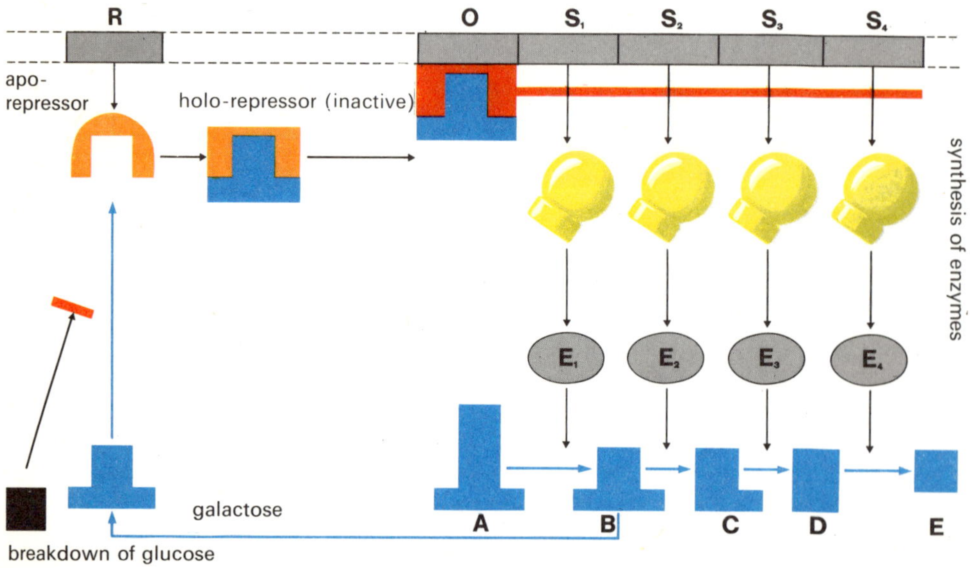

131. Fine regulation of enzyme synthesis. If one of the intermediate products of the breakdown of lactose (in this case galactose, B) accumulates, it may be able to act as an effective co-repressor and combine with the apo-repressor to form an active holo-repressor. If this happens, then the operon is closed and no more lactose is broken down (cf. Figs. 125 and 126).

risk of a serious overproduction of constitutive enzymes? Several biochemists and geneticists have had ideas about this very point. They have supposed – though experimental evidence is still lacking – that in fact the synthesis of the constitutive enzymes is also controlled. Indeed, in this connection, a distinction must be drawn between those enzyme chains that have only an insignificant turnover and those which produce their products in quantity. To the first class belong, for example, the enzymes that build up certain co-enzymes. It is true that every apo-enzyme needs its co-enzyme, but an apo-enzyme consists of about 150 amino acids – i.e., for every 150 amino-acid molecules, only one molecule of co-enzyme is needed. Such 'idle' synthetic chains naturally need only a few specimens of the corresponding enzyme molecule.

Apparently the operator gene in the corresponding operon is indolent by nature, as it were, and produces very little enzyme, except when it receives a very special direction. If at any time there is actually an overproduction of the co-enzyme, this is not usually suddenly reduced in quantity. Probably the construction of a special regulatory mechanism for such eventualities would require more material and more energy than can be spared for this purpose.

It is different, naturally, for the constitutive enzymes that break down glucose, with their higher turnover. Here the enzyme concentration must be high, and must remain high, yet it must not be too high. A regulator is possible if, *first*, inductor and co-repressor are 'related' to each other, either in the sense that the inductor has arisen from the co-repressor, or vice-versa, or in the sense that the inductor and co-repressor arose simultaneously from a common source; and if, *second*, at the same time inductor and co-repressor stand in a constant quantitative relationship to each other (the same thing is true in organic chemistry). Since, however, the inductor always produces inactive repressor, and the co-repressor produces active repressor, these cancel each other out, and so the output of information from the operon remains constant (and constantly high).

However, this is yet more speculation, for which there is no experimental evidence; perhaps the truth is quite different. This much seems to be true, though: the division into controllable and non-controllable (constitutive) enzyme synthesis is not quite correct. It would be better to say that there are chains of enzymes which remain constant at their lowest level (e.g., for co-enzymes), some which remain constant at a high level (e.g., for the breakdown of glucose), and finally some in which the enzyme concentration oscillates between a high level and zero (the synthesis of amino acids by repression, and the decomposition of lactose through induction) – each according to demand.

Probably it is not absolutely necessary that the reader of this book understand the process of regulation in every detail, but it is important that he be at home with the principle involved. Briefly summarized, the principle is this: the process works through repressors; the repressors have a double (allosteric) specificity – the first is for a particular operator gene, and the second for a particular small molecule (of a co-repressor or of an inductor) in the cytoplasm. Now, to go back to the original question – how is release of information regulated? – this double specificity constitutes a double system of transfer of information. C. Bresch, in his book *Classical and Molecular Genetics*, has found such a meaningful formulation of these matters that we cannot do better than quote him.

'Specific small molecules may activate a particular repressor. On the other hand, the repressor reacts with specific operators, and thus puts groups of genes into or out of action. The repressor, is, so to speak, the intermediary between plasma and genome. In the same way that messenger-RNA brings information from the genome into the cytoplasm, so the repressor carries information from the cytoplasm to the genome.

'The reason why the two reciprocal methods of carrying information are so different lies in the nature of the information to be transmitted. In the one case it is a long instruction, whereas in the other it is simply the command "Enzyme stop" or "Enzyme proceed." A second difference is the fact that the information from the genome to the plasma is not directed to a particular target, whereas that from the plasma to the genome is aimed at a particular operon. It is as if the genome sent a message in a bottle, without address, but containing a long letter; whereas the plasma only sent the message "Yes" or "No," but addressed this to a particular individual. Since the plasma cannot write, however, nature makes use of the trick of letting the genome address the envelopes (by action of the repressor), thus ensuring that only news of the availability of a particular molecule is included in that envelope.'

It can also be expressed in these words: the information comes from the genome not as commands, but as suggestions. Only after the plasma has countersigned them with a 'Yes' do they attain the force of law, and come into effect; a negative endorsement rejects them.

Another question arises here. Where do these remarkable repressors come from – can they possibly be amino acids once again? Up to a few years ago it was actually thought that they were amino acids because the first gene-products surely cannot be anything else than molecules of m-RNA; they arise immediately from the DNA during the process of copying (see p. 43). Consequently, one is even more inclined to see them as protein molecules, and even protein molecules of a high molecular weight, of the order of 200 to 10,000. There are two reasons for thinking this:

1. A characteristic of repressors is the allosteric effect – and allostery is predominantly a property of albuminoid bodies.

2. The chromosomes of the higher nucleated organisms do not consist just of DNA with such things as nucleolus-RNA, but they also contain many albuminoid substances, the so-called *histones*. Their molecular weights lie between 2,000 and 10,000, and therefore below those of the enzyme proteins that are actually their own sub-units. It is supposed, therefore, that these histones serve to prevent the premature release of information that is not needed until later stages in development. To this extent they, too, are a kind of repressor – and we shall start to refer to them as such.

If the repressors should actually prove to be proteins, then they could hardly arise directly from the regulator gene, but must come indirectly by way of RNA. Thus our ideas about the *nature* of the repressors are still very uncertain, and hypothetical.

In contrast to this, the *existence* of the repressors is quite definitely unmistakable, as we know from genetic research. Unlikely as it may sound, all the component parts of the regulator mechanisms that we have discussed so far have been discovered in the course of recombination experiments in bacteria. Tens of thousands of investigations have been made, in series, into the phenomena of conjugation, transduction, and sexduction, introducing precisely known fragments of genome into bacterial cells, in addition to the bacterial genome that was already present there (heterogenotes; see p. 120). One cannot sufficiently praise the thoughtful work that has gone into both the planning of the experiments and the interpretation of the results.

The genetic method fails to throw any

light on the chemical constitution of repressors. And since probably a single molecule of active repressor is sufficient to release a specific operon, chemical methods, too, let us down, because chemical methods are based upon a large number of molecules of the same substance. A measurable result can be obtained only as a summation of many tiny separate effects. For the present, neither physical nor chemical measurements can be based upon single molecules, so the genetic method holds out most hope.

Repressors do not confine themselves merely to interfering in enzyme synthesis. On p. 116 we dealt with 'appropriate' (controlled) phages, which did not multiply freely in a lysogenous cell, but only reduplicated themselves in association with the bacterial genome. A repressor acts as a restraint on unrestricted multiplication. The same repressor also prevents the activity and multiplication of any additional phages with which the cell has become infected, and the lysogenous cell is even immune against further infection with the same phage. It can be demonstrated that in this case the repressor is produced from a gene, and even from a *phage gene*.

What is true of phages, however, is also true of all episomes, including the RTF. A precise knowledge of everything that goes on in relation to repressors may thus, unexpectedly, be of the highest practical significance, and so it is to be hoped that the reader will forgive us for having explained the operator-regulator theory to him.

A final remark may be made on the subject of the immunity of lysogenous cells. This is almost always a *limited* immunity. The presence of prophage P_2 indeed makes the cell immune to three or four more phages of the type P_2, but if the cell is infected with a 'surfeit' of, say, twenty or thirty P_2 phages, the immunity breaks down altogether. This agrees well with the suggestion, put forward above, that only a very few molecules of any particular kind of repressor are present in one cell.

4.05 A Timetable for the Release of Genetic Information

A Grub Has No Need for a Wing Pattern

Up to now we have concerned ourselves almost exclusively with single-celled bacteria. Can what is true for them really be applied without modification to higher organisms with nucleated cells? There are several reasons why not:

1. The bacteria have developed their regulatory mechanism in order to be as flexible as possible. It always permits an individual cell to carry out further induction, repression, and de-repression to suit widely changing environmental conditions, with the object of guaranteeing *the survival of the individual cell* against the selection pressure exerted upon it by its companions. In multicellular organisms, on the other hand, the individual cells find themselves in a completely uniform environment: they work together, and among them retain various parts – and varying amounts – of the genetic information, in permanent storage. Regeneration and budding are, of course, exceptions to this situation.

2. The following observations are more detailed, yet in spite of that, more obvious. In *Salmonella*, which we met (on p.134)in connection with intestinal bacteria, the structure genes for the ten enzymes that are needed to synthesize the amino acid histidine are set in an orderly arrangement, in a way that is already familiar to us, all together in a single operon. This is not so in the mould *Neurospora*, or in yeast cells. In these organisms the genes for a group of enzymes that work

together may be divided between several chromosomes – hence the three systems of metabolic processes must be identical here.

We know that in hemoglobin the information units for the four chains (see p. 30) do *not* lie close together, yet in spite of this, the different enzymes do not work in any way independently, but are well co-ordinated.

Operons, in the sense of intimate relationship between structure gene and operator gene, can obviously not be dealt with here. Nothing is yet known about the precise mechanism by which genes from different chromosomes can be reprieved at the same time, and uniformly.

3. Even if the regulatory mechanisms which bacteria have developed for synthesis and for breakdown of substances can operate in the cells of nucleated organisms with certain modifications, there must be additional arrangements specially suited to the development of a higher organism. A bacterial cell grows, and divides itself (a process which sometimes takes only twenty minutes); then the daughter cells grow, and divide in their turn; and so on. With multicellular organisms it is different. The agave blooms for the first time after many years; the insect larva must pupate before it can change into an adult insect. Why should the seed produce colouring matter for the blossom, which may not be needed for twenty years? What would the grub do with wings, or wing pigments?

Here we have a *succession in time* appearing in the release of information, a timetable, which determines how long the gene for blossom colour shall be repressed, and when it shall be 'de-repressed.' Is this a case of yet another *genetic* timetable? And is there a 'timetable gene'?

We still do not know, but there are many indications that the release of information can be directed not through the nucleus itself, at least not directly, but through 'external environmental factors' – among which must be included the immediate surroundings of the nucleus, i.e., the cytoplasm. In many cases it is certain that *hormones* play a part in formulating the timetable: sexual hormones, 'juvenile hormone' in pupation, and such like (cf. p. 223). This is a very wide field, which is only just beginning to be investigated.

It is even more astonishing to hear that in a few cases the de-repression of the release of genes can be seen under the microscope.

4.06 Giant Chromosomes Show the Timetable of the Release of Genes

Active Genes under the Microscope

We know from study of mitosis that no chromosomes can be seen in a nucleus when it is not in process of division. Chromosomes first become visible during prophase, as elongate threads, not tangled together, beset

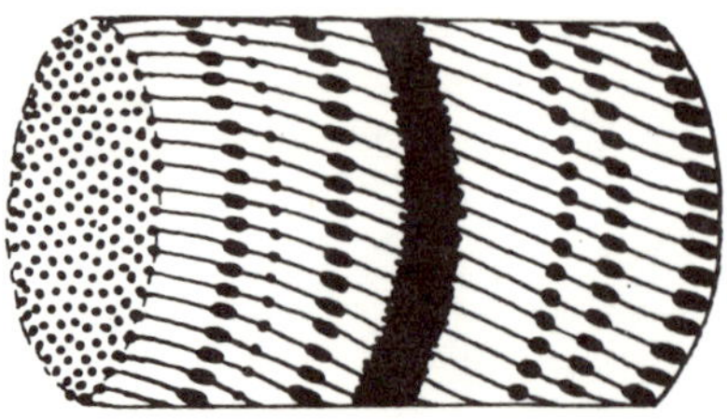

132. Above, diagram of the banding of giant chromosomes. Below, a piece cut out of a giant chromosome.

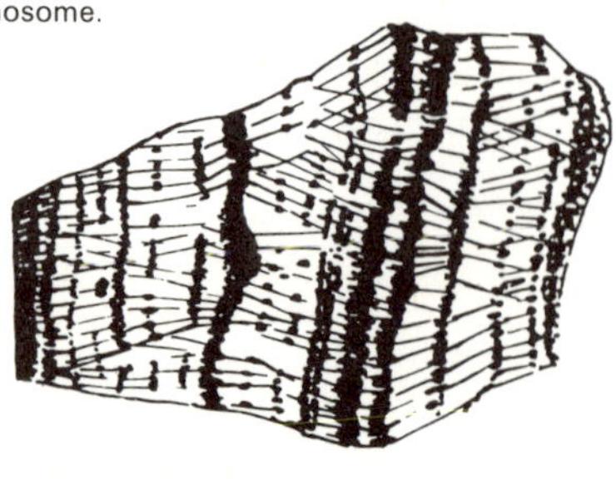

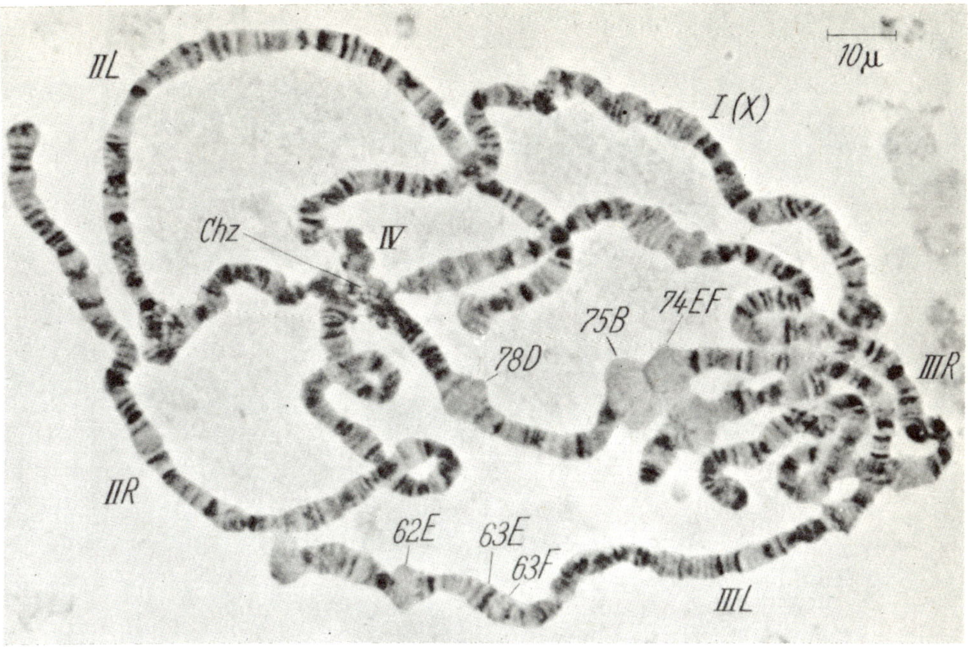

133. Giant chromosomes in *Drosophila*. Roman figures as in Figure 46; Chz . . chiasmata; arabic numerals and capital letters indicate the puffs, which include some of those detailed in Figure 138.

with grains ('chromomeres'), and usually already split into chromatidia. It is rather different in many insects, especially in two-winged flies and midges. A few of their cells, and even whole tissues, have nuclei which remain, as it were, in permanent prophase. Their chromosomes are always visible, always elongate, and beset with chromomeres. In addition, the chromosomes have become elongate *without splitting*, and this has occurred not just once or twice, but much more often. Thus there are close-lying bundles of (sub) chromatidia; up to 4,096 (2^{12}) have been estimated. Such bundles are called polytenia (Greek *poly*, many, and *tainia*, a band). Lastly, they still lie in 'somatic pairing'; the cells are still diploid, and the homologous chromosomes of their nuclei lie alongside each other, just as if a meiosis was in progress. These are actually giant chromosomes, and under the microscope they show many more details than do the pachytene chromosomes of p. 88.

The most remarkable details are the so-called cross-bands. Since the chromomeres of the innumerable individual strands have come to lie very exactly alongside each other, the cross-bands become so narrow that they appear to merge together. They become more distinct if they are stained, and particularly with a stain that will bond with DNA. The thickness of the cross-bands and their distances apart are very characteristic. If cross-bands from cells of different tissues of the same species or genus are compared, the numbers generally agree (so long as the nuclei are in the same stage of development – we shall talk about this shortly).

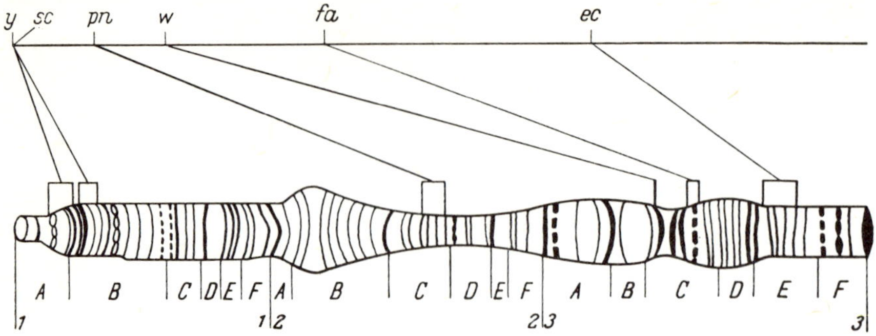

134. Giant chromosome of *Drosophila,* showing how the pattern of rings (below) compares with the chromosome map (above).

It is a great stroke of luck that such giant chromosomes are present especially in the salivary glands of the fruit fly *Drosophila,* which we have already mentioned several times. *Drosophila* is by far the best-known animal genetically, and was the first for which a chromosome map was constructed. If the chromosome map is now compared with the muster of cross-bands, there is such widespread agreement that it may be said that the cross-bands correspond to genes in the way that they become interchanged during experimental crossing over. It is a remarkable feature of giant chromosomes that they allow us actually to photograph genes, but they have other special characteristics as well. We have previously said that the cross-band count of different cells is generally similar, *so long as their nuclei are in the same stage of development*. This implies that when different stages are compared deviations appear. At one stage a broad band surrounding the chromosome, perhaps comprising several of the cross-bands, is disintegrated, splitting the chromosome into two parts. The separation takes place as if the parts were 'puffed' away from each other, and such a separated piece is called a 'puff.'

Balbiani discovered the same thing in 1881, and described it, whence it gets the name of a balbianic ring; but this fell into oblivion for fifty years.

Such puffs are not a lasting arrangement. After several hours the coils close up again, and the original cross-bands appear as before. If a comparison is made between different

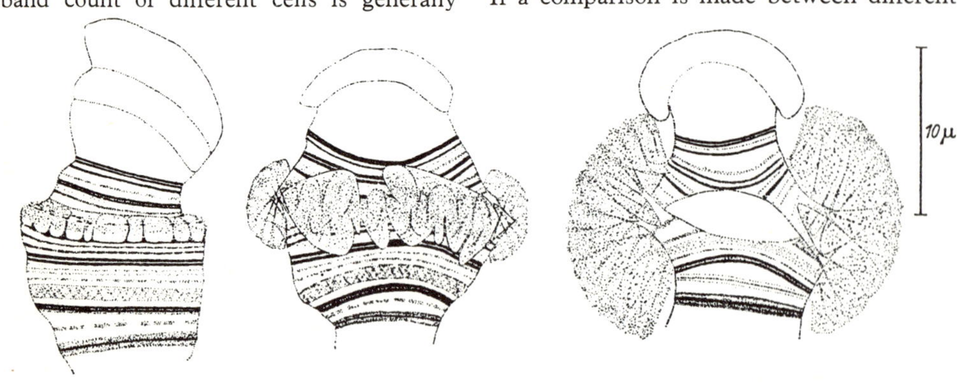

135. How a puff arises in a piece of a giant chromosome.

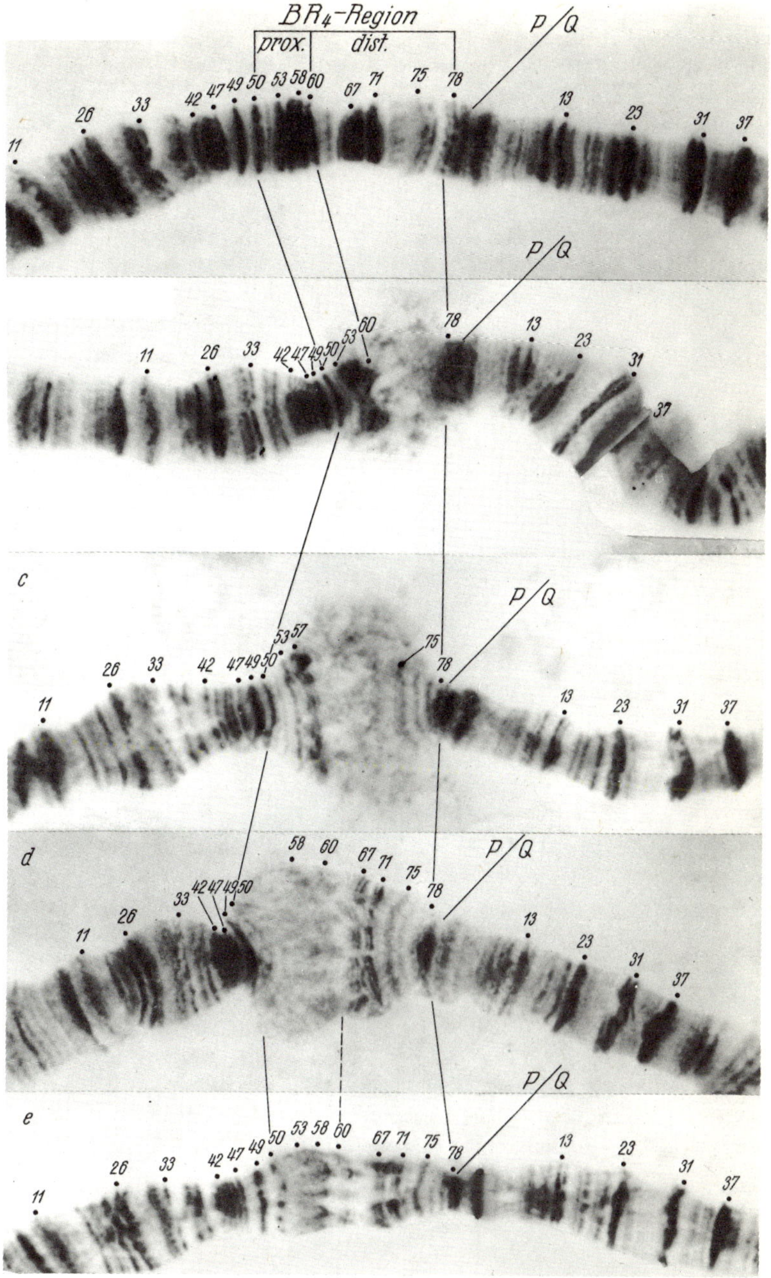

136. A puff comes into being (at 75 in a), spreads (57 in c), and dies away again (53 in e).

cells – and not merely between different stages of development – it appears that their 'puff musters' are different and even specific. Finally, it was discovered from this that there is a time sequence in the formation of puffs.

This phenomenon is comparatively simple in the non-biting midge *Acricotopus*. Like most insects, this fly passes through a metamorphosis, i.e., it changes its form, going from larva to pupa, then to the winged insect. During this metamorphosis about fifteen cross-bands of the chromosomes are puffed, in the sequence in which they follow one another on the chromosome. Thus it appears as if the puffs migrate slowly along the cross-bands from 1 to 15, making one suspect that the genes (plus the cross-bands?) were successively activated. Is that so?

Before we answer this question we must look a little more closely at the links that close up such a balbianic ring. Figure 137 shows this diagrammatically. Only four strands of the polytene chromosomes are pictured, and these are 'inverted' towards the outside. It is to be understood that from a thousand up to as many as 4,096 such loops of one chromosome may lie closely together, becoming so dense that the balbianic ring becomes visible.

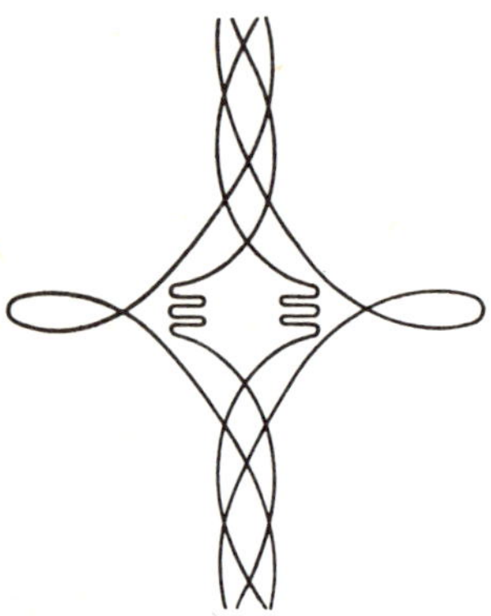

137. How loop formation leads to a puff in a giant chromosome.

But where do the loops come from? Do they arise from DNA, RNA, or protein? After staining, it becomes evident that the mass of the DNA in a puff is no greater than it is in an unpuffed cross-band, so that if the DNA has been changed, it can only be in its *form*. RNA is normally not to be found in the chromosome, but immediately after a puff has formed, there it is, and it can be demonstrated with an RNA stain or by using radioactive markers. We have previously (p. 69) talked about marking the DNA of bacteria with tritium-marked thymine. Now let us take suitably marked uracil (U is used instead of thymine because it is incorporated into RNA and not into DNA; see p. 42). We find that the puffs are marked, but the rest of the chromosome hardly, or not at all. Finally, the presence of protein can be demonstrated in the puffs, or more correctly, the presence of complete (ribo) nucleoproteids, compounds of (ribo) nucleic acids with proteins. It is undoubtedly these which make it possible for us to see the bands, because individual strands of DNA or RNA are much too thin to be observed microscopically.

After all this, we have to admit that in the unpuffed cross-bands the DNA is very strongly entangled, possibly still held off by a basic protein of the histone type (see p. 216). The genetic information is not available. In the puffs the first step is the untwining of the dense packages, and the 'outward inversion.' This exposes the information and makes the gene active. To this extent the timetable of puffing is thus the timetable of information release.

The co-ordination of the puffing with the spatial relationships of the cross-bands leads one to suppose that here we might now be dealing with the real operon, that the cross-

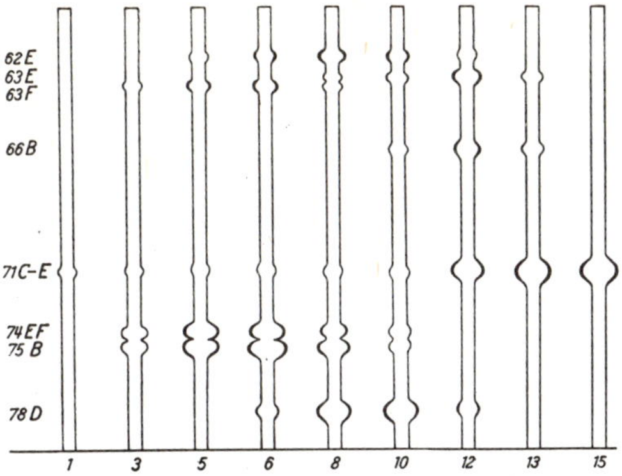

138. Puff formation in a giant chromosome, shown as a sequence in time, from left to right. Figures at foot indicate hours; letters and figures on left should be compared with those in Figure 133.

bands (or the puffs) correspond to the structure genes, and that the first puff can be compared to the operator gene. But one should be cautious about assuming this. In the first place, quite different dimensions are involved: with the operon we are in the *molecular* field, whereas with the polytene chromosome we move into the *microscopic* field.

An experiment with *Drosophila* may be described, where a particular piece of chromosome underwent about a hundred puffs during pupation. Figure 138 shows several of these in chronological order, and distinctly shows that *here* the chronological order of the puffs is *not* the same as the spatial sequence of the cross-bands. So the timetable of puffing is by no means always linked to the sequence of cross-bands.

The timetable itself is probably anchored to some genetic factor, but *when it is to be put into operation* is frequently determined by something in the environment. Such 'releasers,' which, so to speak, fire the starting pistol, are sometimes hormones, as, for example, in the case of pupation in insects. The hormone here is called ecdysone, and it is certainly produced with the co-operation of one or several genes. When it shall be cut off, however, is directed by environmental factors and by the state of development attained. The ecdysone activates several genes (causing puffing); these produce substances which in their turn activate other genes; and so on, till step by step everything is prepared that is necessary for pupation to take place.

4.07 The Fine Structure of Chromosomes

Coupling and De-coupling

Several times already we have nearly been driven to considering the fine structure of the chromosomes (see, for example, p. 78); now we can avoid it no longer. So this is the time for a summary of the most important discoveries which enable us to construct a 'chromosome model.'

1. Usually no chromosomes can be seen in the nucleus so long as it is not dividing, and the electron microscope does not give any clear, unequivocal picture. In contrast to this, at times of mitosis and meiosis the nucleus shows strongly contrasting chromosomes, in an orderly arrangement.

2. Experiments in crossing and recombination have already shown that all the genes in a chromosome follow closely after each other in a linear arrangement; this means that there must be a strand running through them, a double helix of DNA. Admittedly, this strand ought to be strongly contrasting, and much folded, even more than it is in the genomes of phages and bacteria, otherwise the compactness of the chromosomes, which makes them visible under the light microscope, could not be explained.

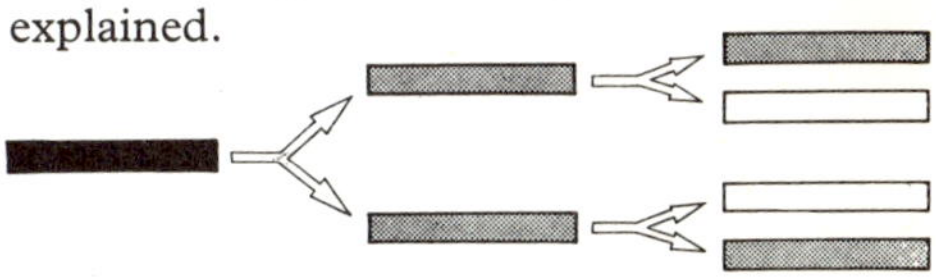

139. Semiconservative transmission also occurs in chromosomes from the root tip of a broad bean.

The length of the genome in phages and bacteria compares with the length of the genome in chromosomes of nucleated organisms in the following proportions: 1 (phage): 100 (bacterium): 10,000–100,000 (nucleated organism). The DNA helix from *one* chromosome from *one* cell of the crested newt, if stretched out, would extend to one metre, whereas the chromosome itself is only one micron long!

3. Reduplication experiments lead to the conclusion that a chromosome is composed of two sub-units, which separate, after which each generates its missing half. Just as was done for bacteria, experiments with radioactive thymine were carried out with root tips from the broad bean (*Vicia faba*), with the same result. These chromosomes, too, were passed on in a 'semiconservative' way (cf. p. 69); at any one time one half is 'old', and the other is a new construction. From this, the further deduction must be made that:

4. For the chromosomes to be duplicated, the *entire* double helix of DNA must be opened by means of the 'zip-fastener mechan-

140. A double helix of DNA with two rotatable couplings, is shown according to two different 'models.'

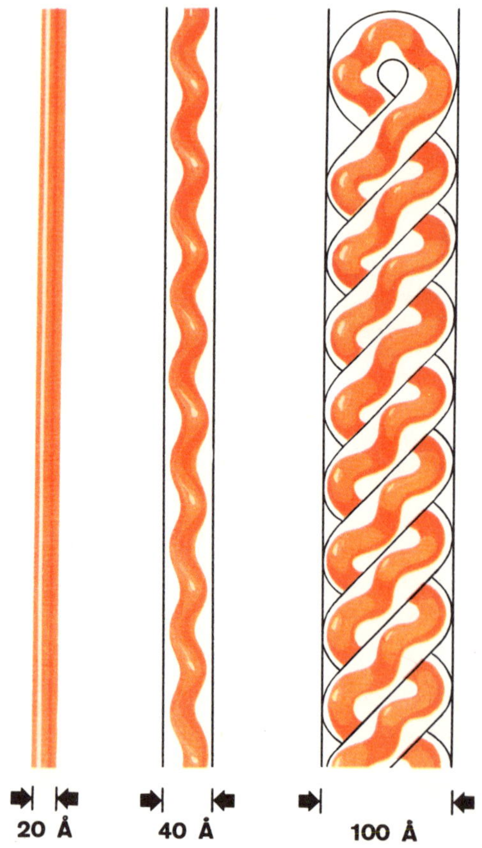

141. By the attachment of histones, and additional twisting, a double helix of DNA may attain a thickness of about 100A.

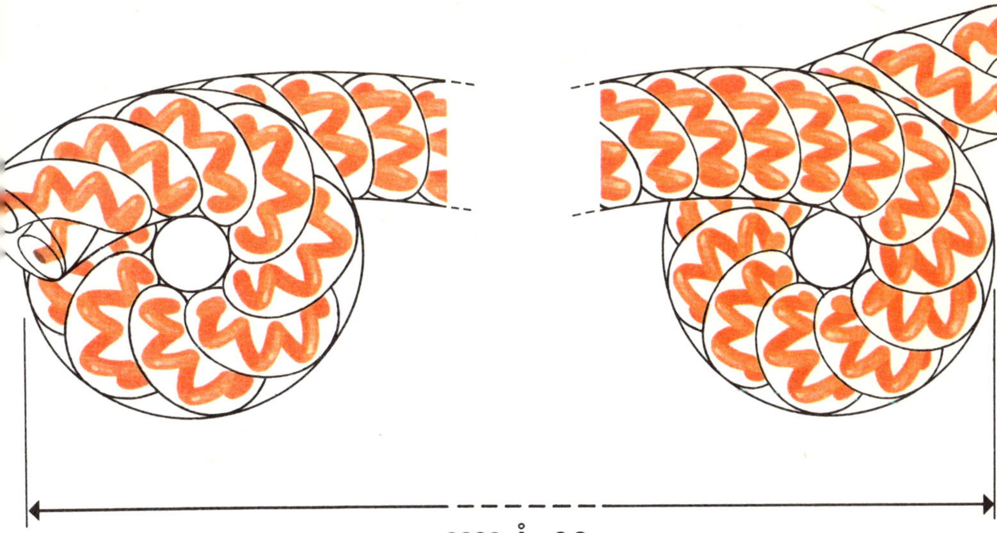

142. The thick strands shown in the previous figure may be coiled round themselves.

ism.' This can come about only by *rotation*, but how great must the speed of rotation be to embrace the complete chromosome (with a DNA helix one metre long!) if as many as 10,000 r.p.m. have already been attributed to the bacterial genome?

5. The giant chromosomes have thin, inverted loops; are these modified double helixes, or single strands stretched out?

At the risk of repeating, it is not possible at the present time to draw up a chromosome model that will show all these facts clearly and satisfactorily. Of course, many models have been made, but each of them has at least one weakness.

Either we postulate a double helix of DNA, and then we can explain our experiments in crossing over and duplication, but cannot say how the zip-fastener mechanism is able to function at such a speed.

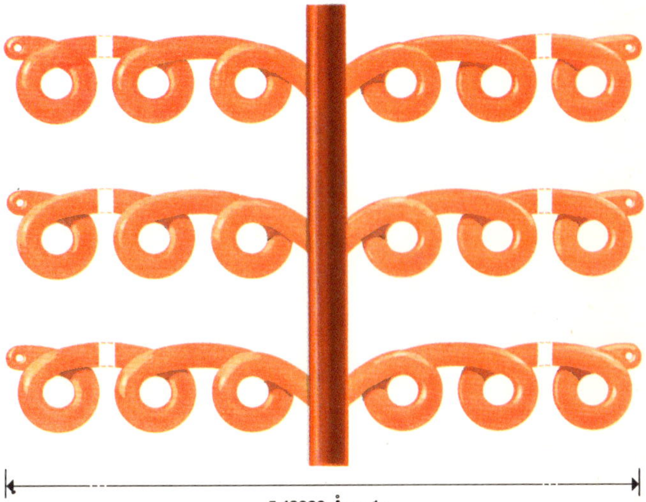

143. The 'super-coils' of Figure 142 may be arranged round a central stem, like that of a rotary brush, and then the cross-section of the whole structure begins to approach that of a chromosome.

Or we start by stipulating that the double helix of DNA consists of more or less independent sub-units. These sub-units might be like the teeth of a comb, and stand up from a continuous shaft, which itself need not be made of DNA. Since, however, the sub-units must be linked together – the results of experiments in crossing over make this essential – there is no choice but to introduce rotatable couplings. At any given time the rotation may be taking place between any two couplings.

The short sections that lie between couplings can be inverted at the same time, but they must be put together again correctly by very precise reduplication. This proceeding is obviously not without its dangers: mis-couplings may occur during the exchanges.

What is still wanted most is a representation of the double helix that runs right through the chromosome, and which measures 20 ångstroms across. Probably it is itself coiled through the adhesion of the histones, and thickened up to a strand about 40Å across. Such a 40Å strand can be 'inverted upon itself.' It then runs firstly into a loop, then reverses, and coils up with its second half lying alongside the first. If this 'strand with loops' is then itself coiled, there arise 'super-coils,' up to 0·3μ long, and therefore visible under the microscope.

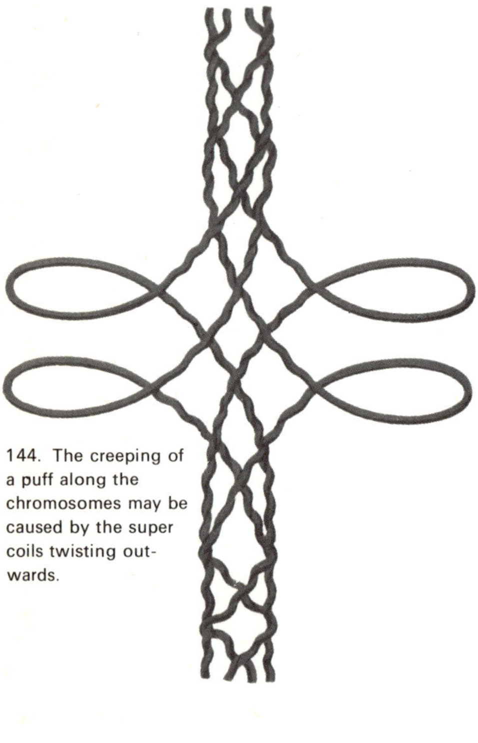

144. The creeping of a puff along the chromosomes may be caused by the super coils twisting outwards.

It must further be made clear that super-coils of this kind (each of which possibly correspond to a 'chromomere') are arranged around a central rod, from which they stand out in all directions, more or less at right angles, somewhat like a roller brush. Thus we arrive at a body that represents a chromosome very well. This compact, contrasting structure must naturally disentangle itself considerably at interphase, in order that the DNA helix may reduplicate itself.

There also remain to be explained the high, and sometimes very high, speeds of rotation at which the inversions are carried to a conclusion. That is the weakest point of the model, but we can at least imagine that during the puffs individual super-coils invert themselves, and thereby expand and expose the actual double helix, with its genetic information. At any rate, this is an advantage in comparison with other models, and so we must make the most of it; the theory is in many ways uncertain, and in many ways needs to be improved, but it is useful in this one respect. Moreover, we have nothing better as yet.

Perhaps it will be in chromosomes that it will become particularly clear how closely structure and function fit together, and hence how careful one must be not to make models *just* of structure or *just* of function – we generally come to grief when the other aspect is revealed. On the other hand, it is consoling for those who make chromosome models to take an actual photograph of one

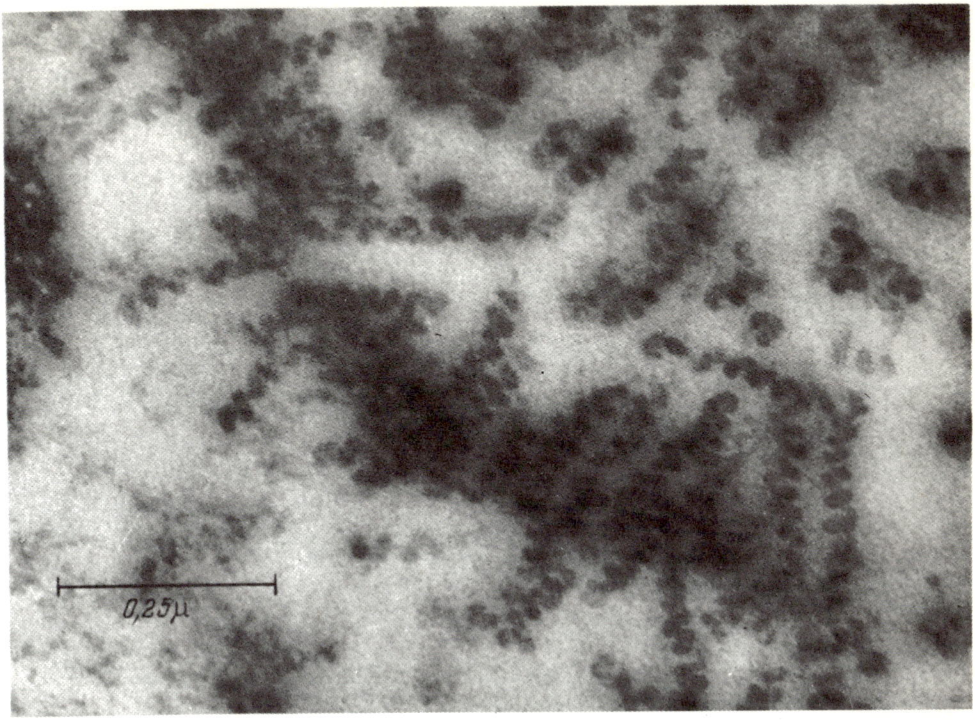

145. Electron-microscope picture of the nucleus of an amoeba. The coils that are visible are probably super-coils.

with the electron microscope. This shows just such 'curly' super-coils, though whether they are really individual genes or complete operons (if there are such things in the higher organisms) or still bigger units must await further investigation.

This brings us back to the starting point of this chapter, the regulation of the release of information.

There is no need to point out that we are still a long way from being able to survey the complete regulatory system of all organisms. Still, what is known so far is impressive, and permits us to understand much that a very short time ago was still obscure: the great adaptability of bacteria, their economy in the use of materials and of energy, their differences from the multicellular animals, and the timetable according to which puffs are generated and information is released. We already know the essential basic paths by which regulation goes on in the cell, and it is not to be doubted that soon further principles, and fuller details, will be discovered.

It is probably even more important to have an overall understanding of the process. The output of information does not just follow rigidly a preset punched card, but is flexible, and above all *has a widespread dependence upon external factors*. The nucleus of the cell is not unconditionally the centre from which each and every function of the cell proceeds; it is rather to be compared with the archives, from which information is abstracted – or the reverse – according to need, via the cytoplasm, and by means of inductors and co-repressors. The nucleus is not the dictator of the cell!

5. MOLECULAR MEMORY

5.01 From the Greek Philosophers to the Biochemists

The Search for 'Wax Tablets' and 'Inscriptions'

The first four chapters of this book were about physiological topics, as follows:

1. Genetic information written into the molecules of the nucleic acids by means of four symbols.
2. Passing on of the 'molecule of information' to the offspring.
3. Extraction of information and converting it into particular enzymes, into regulatory mechanisms, and eventually into complete phenomena.

All this is undoubtedly molecular biology, or within narrower limits, molecular genetics, and so it would seem that molecular biology was nothing else but the biology of nucleic acids. This is not the whole truth, but one thing at least cannot be denied: no sooner had nucleic acids been envisaged as arbiters of control, or at least as archives for the storage of information, than the whole of biological research started on an explosive expansion. It became possible to explain occurrences that were among the very basic phenomena of all living things. Until twenty years ago no one would have dared to hope for such a thing.

Do these three points, however, mark the limits of potentiality of the nucleic acids? The assumption that the nucleic acids might also be the omnipotent driving and regulating mechanism for other phenomena of life is not entirely empty speculation. We can follow the example of distinguished biologists and look upon the genetic information as a kind of collective 'species memory.' The fertilized ovum, the starting point for each individual, on account of its genetic information, 'recollects' how its ancestors were constituted and develops itself accordingly; and at once it occurs to us that we ourselves have an *individual*, personal memory, that we can recall incidents, feelings, sensory impressions from far back in our own lives. Could it be that nucleic acids play a part in these too?

If we set up such headlines as 'Memory by Syringe,' 'RNA and Learning,' our reflections will be neither original nor misleading; on the contrary, a large number of research workers before us have suggested the same idea, and many laboratories have concerned themselves with the relationship between memory and nucleic acids. Can we find out what has come of all this?

What actually happens when one remembers a fact or an incident, or 'keeps in mind' a name? That we should remember things is a familiar occurrence to us, so familiar that we are no longer surprised by it and do not often stop to think about it. If a name 'escapes' us – if there is someone whom we are absolutely certain that we know, only we cannot bring out his name on the spur of the moment, then we rack our brains for what may be in our memory, until eventually the name strikes us again.

This happened to people in all periods, and if they were philosophers, they still racked their brains until the forgotten name again came to mind. The early Greek philosophers thought the spirit of man to be a kind of wax tablet. At the time of his birth it is empty, and then the 'hand of experience' engraves incidents upon it; hence the term *engram*, from the Greek, for a single trace of memory which is 'engraved.' The tablets referred to, of which perhaps man had several, were stored away in our brain, and held or released according to need.

This is strictly correct, but it tells us nothing about either the wax tablet or the inscriptions on it. The questions start at this point, and as Professor McConnell says memory research is nothing more than questions about and the investigation of these 'wax tablets' and 'inscriptions.'

It might be quite amusing to trace out how the study of memory has changed since ancient times, how first of all, criticism of the accepted opinion was entirely formal, until eventually the time was ripe and the researchers were ready to put the study of memory on a firm basis. But our task here is not to present a history of biological research, and so we must pass over the ingenious, painstaking, and most praiseworthy researches of the past, and jump straight from the Greek philosophers to modern biochemistry.

The approach of modern biochemistry is more likely to draw a comparison with shorthand notes stored away and waiting for an order to release them. May nucleic acids play an active part in this? Do they correspond to the marks on the wax tablet, and do the changes of form of the nucleic acids constitute the inscription, memory's trace?

Here is an observation that suggests this. The genetic information, species memory, is transmitted from cell to cell, and from organism to organism, and is inherited, by means of the double helix of DNA; but individual memory, alas, is not hereditary. Our store of skill and experience, everything that we have learned, often unique, cannot be passed on to our children. They have to learn it all again. If nucleic acids are concerned in this process, then undoubtedly it will be RNA only, and not DNA.

Hence it at once follows that we cannot employ genetic methods – as reliable, as precise, and often as elegant as they may be – to the experimental investigation of these problems. Nothing can be done here with studies of recombination and mutation, and we must look for some chemical or physiological methods to take their place. What, then, should we actually be studying?

5.02 First Experiment
Nucleic Acids Once Again

If we think it likely that memory has a material basis, and that this material basis can be represented by RNA (as the wax tablet) and by the assembling or arrangement of the RNA molecules (as the inscription), then qualitative and quantitative measurements ought to give some indication of this. It is known that in the human brain, for example, the amount of RNA increases up to the age of forty, remains high and constant from forty to sixty, and thereafter progressively diminishes.

It is remarkable to find that the curve of ability to learn follows this quite closely, but the parallel should not be overrated. In reality, the RNA curve probably gives an expression of the rising and falling of general metabolic activity, which, on its part, also follows spells of active and less active stirrings of memory. Moreover, it is certain that a considerable part of the genetic RNA is messenger-RNA, transfer-RNA, and ribosome-RNA, and will therefore be necessary for enzyme synthesis and the like, and so there is no question of its being used for memory recording.

New discoveries are to be expected from the analysis of specimens during and shortly after an exercise in learning. Plants are unsuitable as experimental subjects, since they seem to have no individual memory in the accepted sense. On the other hand a variety of different kinds of animals suggest themselves, the docile and intelligent rats, but also goldfish – often decried as stupid – and even primitive flatworms (Planaria).

146. A flatworm, *Dugesia doratocephala,* a species that is used in research.

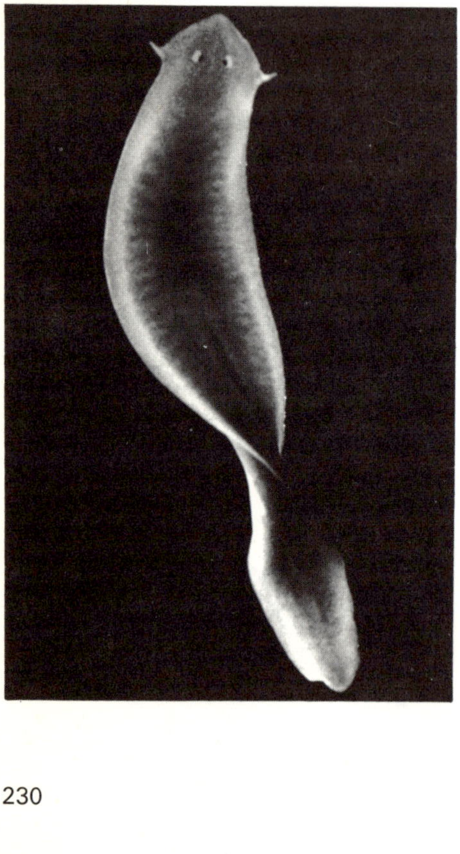

It is very easy to teach a drill to rats. They can be made to balance over a tightrope to reach their feeding trough, always to take, say, the left fork at a junction, and so on. They learn this in the course of a few days, depending on how much practice they are given. During the experiments in learning, the RNA content of the brain remains steady at about 50 per cent, but there are additional *qualitative* differences. During the first day of learning the brain contains RNA that is rich in adenine and uracil (which are complementary to each other, see p. 40, and so are reckoned together). On the ninth and tenth days, on the other hand, the so-called slow-learning stage, there is a new construction of RNA that is rich in guanine and cytosine (also complementary to each other).

This 'late' kind of RNA has a general resemblance to ribosome-RNA, but it is not quite known what is the significance of this. At any rate, it seems as if the change of base grouping – from A + U in the phase when the rat is capable of learning, to G + C in the phase when it is memorizing what it has learned – is important.

It might be a good idea at this point to try to be more precise. So let us begin with the following consideration. If an animal has learned a quite definite piece of drill, then it has at its disposal a corresponding inscription (engram), of a non-interchangeable kind of RNA. Of course this amounts to a very little out of the total RNA present in the entire organism, and even very little out of the RNA of the brain, where possibly all the other possible engrams are deposited. It was at first thought that this would be too little to be extracted, isolated, and analyzed. How would one ever know that the right molecule of RNA had been isolated?

A possible way of testing this would be available if the total RNA, and at the same time the RNA specific to a particular piece of drill that had been learned by one animal, could somehow be introduced into another animal of the same species which had a life history as similar as possible to that of the first animal, but which had not learned that bit of drill. Then we should hope that the second animal would either know this drill without learning it, or at least would learn it more quickly than a third animal which had neither learned it nor been indoctrinated.

5.03 Flatworms can be Taught

Success and Vexation

One group of research workers chose Planaria as subjects for their investigations. These are flatworms which are found in brooks and streams, in ponds and lakes, and which are not infrequently introduced into aquariums. Planaria are hardly ever more than two centimetres long, they have a head, and a tail piece, and in general are exceptionally simple animals. Their nervous system and brain are very primitive, and so is their digestive system, since they have no stomach. Most of all, their production of acids and enzymes is extremely small – yet even so, these are exactly the same substances which in higher animals break down any RNA of external origin before assimilating it into the cells. To this extent the Planaria are actually especially suitable for experiments on the transfer of RNA.

Planaria are best known, of course, for their quite remarkable powers of regeneration. If a flatworm is cut in two, the head end grows a new tail, and the tail piece a new head. This requires about one month, but by that time there are two flatworms, each of normal size and functioning normally. Since this regeneration is based upon mitosis, (see p. 73), the two animals are genetically completely alike.

This power of regeneration has been made use of for the first experiment in the transmission of memory. An individual planarian is allowed to learn something, cut in two, and then after successful regeneration there are three possibilities: both regenerates may have learned their lesson; or only the head end; or neither. Yet can such a primitive animal as a flatworm learn anything? It is indeed rather difficult to get a flatworm to take in anything, but it is actually possible.

Planaria are unhappy if they are suddenly exposed to a strong light. They draw themselves together, contracting their body. Yet, since the light obviously does them no harm, they become accustomed to the level of illumination, and become increasingly indifferent to it. If the experiment is repeated often enough, they finally show no reaction at all.

It is different if the light stimulus is combined with a second stimulus that is certainly unpleasant to them, such as a slight electric shock. In order to deliver this shock, the planarian is allowed to crawl in a plastic trough about a foot long and a half-inch wide, filled with water. Over the channel hang two lamps to provide the light stimuli, and at each end is an electrode, which can be connected to a source of electric current. When the switch is pressed the circuit is completed, and a current flows through the water; the planarian normally reacts by violent contraction.

And now for the association of stimuli. The lamps are turned on for two seconds, and then *in addition* to the light, the current is applied for a further second. After the third second the light and the electric current are simultaneously cut off. This treatment is repeated at convenient intervals, which may be about two hours, and the worms begin to react by contracting during the time the light is

147. A simple apparatus for experimental training of flatworms. In front is a plastic trough with two electrodes; the two lamps provide illumination that can be switched on and off.

shining – though in the long run they become accustomed to it; they have learned to interpret the light as a signal that an electric shock is imminent. If the lesson is prolonged over several days (totally about 150 experiments), the animal reaches the point at which it contracts for almost the whole of the two-second illumination (90 per cent of the time); whereas if the experiments last only a short time, the worm learns to contract for only about 50 per cent of the time it is illuminated.

Up to this point we have been dealing with isolated phenomena of learning, but the animals that have been taught are cut in two and allowed to regenerate. One month later they are again trained (on the assumption that during this month they have forgotten everything), and it is noted whether they learn more quickly – i.e., whether they reach a 90 per cent success in a shorter period of time. The result is astonishing. The head ends, which regenerated the missing tails, need hardly any new practice at all before they are fully trained. They have retained what they learned as well as undivided animals.

This is not, perhaps, so surprising after all, since the primitive brain is located in the head end; but the tail pieces, which have had to regenerate complete new heads, differ hardly at all from the regenerated head ends, and they still retain almost everything that the complete animal had learned. *The newly built brain contains the old engram.*

Apart from feeling astonished at this result, we cannot avoid the conclusion that an engram has a material basis, and apparently can be stored not only in the brain but in the entire body of a planarian. Of course these experiments contribute nothing to the hypothesis that RNA is responsible for transmitting memory, and so we must bring into the picture some things called *neoblasts.* Neoblasts (new cells) are undifferentiated – i.e., not yet specialized – cells which migrate around the planarian body in great numbers. After a worm has been cut in two the neoblasts migrate to the cut surface and there form themselves into a complex of cells from which the missing part of the body is regenerated. The neoblasts, however, are very rich in RNA, and so they may carry the engrams about with them.

The following consideration takes us a step further. If the RNA carries the engrams, or represents them, then the engrams can be destroyed in the same way that RNA can be destroyed. A destructive agent for this purpose is *ribonuclease*, shortened to RN-ase, an enzyme that breaks up RNA into its component parts, but which leaves DNA and all other substances untouched. (On p. 121 we learned about its opposite number, DN-ase; it was the fact that DN-ase interferes with transformation, and with the transfer of genetic information that gave the first evidence that only DNA could be considered in the search for the hereditary substance.)

If sliced worms are allowed to regenerate in pond water to which RN-ase has been added, the head ends (which have only the tail sections to grow) remain completely unaffected; they have learned their lesson, and their engrams are not vulnerable to the enzyme. It is different for the tail sections, which have to reconstruct a head and brain with the help of neoblasts. These regenerated animals have forgotten their lesson; they have *not* retained the old engram, presumably because it has been destroyed by the RN-ase. Forgetfulness can be induced in the same way by injecting a solution of RN-ase into the body cavity of the planarian that has been trained.

This work has given a hint, but no more than a hint, of the part played by RNA in the phenomena of learning and retaining what has been learned. Of course, RN-ase attacks not only the (hypothetical) memory RNA but also the part of the RNA that is necessary for general synthesis of proteins. Perhaps the RN-ase is effective only when the animal concerned is in a generally enfeebled state? Since the experiments on regeneration and RN-ase were not entirely satisfactory, a search was made for other possible ways of introducing the engrams from 'trained' animals into 'untrained' ones.

One idea soon attracted considerable attention, as being very effective, but extremely drastic, and insensitive, *the transmission of memory by cannibalism*. When it is very hungry the planarian *Dugesia doratocephala* does not hesitate to eat other individuals of the same species. Hence it is easy to arrange that hungry, untrained planarians should take up the memory substance of trained animals, and then to test whether they have assimilated the engrams of their victims. Of course, as a control, other untrained flatworms must be supplied with *untrained* victims. The result: those experimental animals that had been given bits of trained worms to eat showed up their colleagues that had been fed only on untrained worms as distinctly inferior, *from the first stage of training onwards*. They learned their lesson easily, much more quickly than the others.

This spectacular result remained in doubt for a long time, because at first it was not possible to repeat it in other laboratories. Dr Morey, an English psychologist, expressed his annoyance with the worms more or less in the following words: 'Obviously British flatworms, in contrast to American ones, cannot, try as one will, be induced to eat their own kind, except when they are quite ready to do so. This proves that in our country gracious living has spread downwards even to the worms, but I am very concerned about this, because I had hoped to carry out further experiments on these worms on the lines of the American experience.'

Since then these difficulties have been overcome, the doubtful result has been verified, and it is today accepted, as confirmed. The biggest contribution to this has been a new series of experiments, less reprehensible ones: from hundreds of trained planarians (and, as controls, from untrained ones) so much RNA was extracted that it could be injected with a hypodermic syringe. The result was exactly the same as before. Those planarians that had been treated with RNA from *trained* planarians learned distinctly more quickly than their companions that had received RNA only from untrained animals.

So it must be taken as fully proved that the RNA is involved in the phenomena of learning and of retaining what has been learned in the form of memory. But how? Is the RNA only the wax tablet, or only the engram, or both equally? During these injecting and cannibalistic experiments, has the fragment of learning or the engram specific to it actually been transferred, or has the shot of RNA merely raised the general level of the animals' ability to react to stimuli, and to learn? Experiments with planarians can give only inexact solutions to these problems, since the final reaction – contraction in response to a light signal – is not a particularly specific one. It could be a general shrinking from all stimuli. Perhaps the animal has just become oversensitive?

5.04 Rats Are Better Pupils

Specific Memory Made Visible

Planarians cannot be taught any specific, unchangeable reaction, because they are of too primitive a construction. In the search for intelligent subjects upon which to experiment it was remembered that rats were docile animals. Perhaps a successful transfer of memory could be achieved in them, but this time with a piece of trained behaviour that was less ambiguous?

Out of a multiplicity of experiments, carried out in different laboratories, we shall single out one and describe it in detail. It will serve as a prototype of many others, and will also show how carefully such experiments must be planned and carried out if the results are to stand the test of time.

Eight rats of the same origin, age, and weight were introduced, each by itself, into a small cage; they had been starved for two days beforehand. In the cage there was a feeding trough with two small food pellets (weighing only 4·5 milligrams). After the rat had become accustomed to its new surroundings, it began to run around, and of course, to inspect the trough. The two pellets were at once eaten, and at this instant a container high above the trough opened with an audible click, and another pellet of food fell into the trough. This was repeated several times, each time with a click, at first in rapid succession, and later at longer intervals.

A further stage in training was to allow the click to be made when the rat was away from the trough, at first a short distance, then later, when it was farther away. During the intervals the rat could inspect the trough as much as it liked, but without receiving any reward in the form of an extra pellet.

The click experiment was repeated two-hundred times on each of the first four days, and one-hundred times on the fifth, and last, day. At the end of that time each rat had learned to make for the trough as quickly as possible, immediately the click sounded, from wherever it found itself inside the box at the time. Without any click, in contrast, it left the trough almost entirely untouched.

Nine equivalent control rats received their food during the same period, and in the same quantity, without any training.

After their training was finished, and they were 'conditioned,' the rats were anaesthetized with ether, and killed, and then, working as quickly as possible, one gram at a time was taken from particular areas of the brain. After it had been extracted and purified with great difficulty, between 0·7 and 1·1 milligrams of RNA remained. The same thing was done, of course, to the nine control rats. The RNA that had been extracted could be injected any time within eight hours, and was injected straight into the body cavity of new experimental rats, so that the RNA could reach the body fluid as quickly as possible – one probe only in each rat.

The 'new rats,' completely untrained, had been introduced into the cage for fifteen minutes at a time on each of the five previous days, so as to accustom them to the cage. During each of these quarter-hour periods the rats were given two clicks, though without these being followed up by a food pellet. On the contrary, powdered food material was scattered around on the floor of the cage, so that the rats remained lively and responsive. The most important thing was to ensure that the rats did not become interested in the food trough, perhaps attracted to it by smell.

Now, at last, the actual test could begin. One rat was introduced into the cage, in one corner of which the food trough was placed. After a period of two minutes for familiarization, clicks were given five times in succession at intervals of at least one minute, with the trough empty. The series of five clicks was repeated five times in all; four, six, eight, twenty-two, and twenty-four hours after the injection, giving a total of twenty-five clicks.

A *positive* reaction was recorded if within five seconds after a click, a rat poked its nose into the food trough, or within a narrow 'demarkation zone' round about. Two testers, independently of each other, determined whether the reaction was positive or not, and only those experiments were accepted in which the two testers agreed. In order to avoid any prejudice, the two testers did not know whether the extract came from trained or untrained rats; the rats carried numbers, and the interpretation of these was not revealed until after the tests had been completed.

And now for the results. The control rats, which had received RNA from untrained donors, reacted to the clicks either not at all, or only occasionally, as if by chance. From eight rats (one was excluded, because it remained immobile, sitting in a corner of the cage) each reacted only one single time out of a series of twenty-five clicks. Of the seven rats which had received RNA from trained donors (one was excluded here too) each came to the feeding trough seven times during the series. Thus these rats reacted *seven times as frequently* as did the controls.

This cannot be attributed to chance, nor can it be explained in terms of a uniform escalation of activity or sensitivity. Why should the rats on the stimulus of a single click (which up to now had meant nothing to them) run to a trough to which previously they were quite indifferent? We must assume at this point that the injection brought about a transfer of some specific memory content, since the rats 'remembered' an experience which was not their own, but belonged to other rats first trained and then killed.

Of course, this would have been much more impressive if each of the experimental rats had shown a positive reaction to *every one* of the twenty-five clicks – but that would really be too much to ask. Much material is undoubtedly lost during the processes of extraction and purifying necessary to obtain the RNA; what remains may have undergone certain changes (e.g., changes of form of the molecule of RNA, which most likely may have partially detached the engram) during the process. In addition, it certainly cannot be expected that after injection *all* the RNA that has been released into the body cavity will find its way into the nervous system and the brain; here, too, some loss must be expected.

On top of this, as McConnel showed very forcibly, another significant fact must be taken into account: 'An animal that has learned an exercise, no matter how simple, has of course assimilated thousands and thousands of impressions in the process. A rat in a simple maze learns not only whether to turn to the right or to the left in order to reach the food, but at the same time, and in addition, it learns that the maze has a particular colour, condition, height, breadth, length, and smell, that the food is to be obtained in such and such a way, and other things too numerous to mention. All these innumerable "micro-memories" go to build up what the research worker dismisses as a "simple experience in learning."'

After all this, one really cannot but wonder that the injection experiment should give such decisive results, which make it quite certain that RNA takes part in the fixation of specific units of memory, of specific engrams.

The latest experiments make this still more distinct. This time in addition to rats that were 'click trained,' some were used that had learned to run to the food trough after a light signal. Once again brain extract was injected, and of course from different untrained animals, but these now received light and click stimuli irregularly mixed, in the same experiment. Animals with 'light RNA' ran for food more often after a light signal than after a click; rats with 'click RNA' did the reverse, and 'preferred' the click to the light.

For many specialists this put the matter beyond doubt; here we have reached the material basis of memory, the chemical nature of 'wax tablets' and engrams. Yet the chemistry of memory is still in its infancy all the same, and we are still a long way from a molecular-biological view of it. Of course, the way towards it is to a certain extent mapped out; one day, using better methods of extraction and purification, and ever more rigorous conditions of experiment, we shall be able to transfer individual molecules, and individual units of memory, and to trace exactly their effects in the recipient organisms.

5.05 Surprises from Goldfish

Doubts Whether the 'Mathematics Pill' Will Ever Materialize

We are never proof against surprises, as an experiment with goldfish will show. By using mild electric shocks, such as were used for the flatworms, we can make them swim in complicated patterns, and overcome obstacles in order to reach food. Such experiments can go on for months. It is different, however, if immediately after the lesson the goldfish are given a squirt of puromycin, an antibiotic, which restricts the synthesis of proteins. Then the fish forget what they have learned. This does not take place if the injection is given after one or two hours delay; it then has no effect on what has been learned or remembered.

This suggests that what the puromycin inhibits is only the *transformation* of recent sensory experiences into a lasting memory. Can it be that the engram is located, not in the RNA itself, but in special proteins? This is not unlikely, since we already know (from Chapter 1) that specific enzyme proteins are coded onto the RNA. Why should not some of these be specific memory proteins?

The reader will find that there is much more that is obscure about the study of memory than there is about the study of enzyme synthesis, or the regulation of metabolism. On the other hand, the experiments and the deductions made from them should be clear to him, and so he will certainly ask whether one day it will be possible to indoctrinate animals with particular items of knowledge, without actually training or teaching them, just by spraying them or dosing them with a pill.

The possibility of acquiring ideas, foreign languages, or mathematics just by taking a pill is certainly fascinating, and its consequences are almost limitless. There are people who really expect to read newspaper advertisements offering such things as 'Tablets for Driving License at Bargain Prices' or the opportunity to 'Become a Seasoned Globetrotter with Our Memory Spirit B, Extended to Cover the Following Itineraries . . .'; or to see the headline 'Chief Prosecution Witness in Murder Trial Given Memory Drug.'

Now it is quite reasonable to anticipate new and even more startling sensations, and these should arouse neither uncritical enthusiasm nor flat rejection. Each of us, according to his temperament, should be either prudently optimistic or benevolently skeptical.

Undoubtedly, one day 'memory molecules' will be isolated and it will be possible to regulate their transmission. The question is whether much will be gained by that. Of course, it is attractive to think of buying one's experience at the drugstore and swallowing it then and there, thus getting experience quickly and without any of the (sometimes fatal) risks that are normally involved. But will the body retain 'strange' experiences and assimilate them with its own? We shall have more to say in the next chapter about the reactions of immuno-biology. In any case, up to the present, all injections must be repeated if the effects of learning are to be retained for more than one day.

Furthermore, is there not a danger that man will forget how to acquire experience for himself, and so become completely dependent on pills? No matter how rewarding this process may become, a man who acquires a completely programmed memory (very likely suited to his station in life) will never get the better of one who is more adaptable because he has learned for himself. It is surely no accident that a parent's individual memory is *never* inherited by offspring. If it were advantageous to be able to pass on what one had learned during one's life, then living organisms would have acquired this power long ago in their evolution.

6. IMMUNITY AND MICROBIOLOGY

6.01 Attack and Defence

What It Means To 'Be Immune'

One often reads that 'life' is threatened and put in peril: being threatened is indeed one of the facts of life, which has resulted in a continual improvement in living creatures. The reader will be aware of this, though he will remember that the first chapter of this book stressed that one should not talk about 'life' as if it were a thing in itself. Moreover, it is amazing to realize that this 'life,' though threatened unremittingly, has existed, probably without interruption, throughout billions of years, and that it manifests itself in untold numbers of organisms and individual living things.

It would be much more correct to say that the organisms, the individuals, are continually threatened: catastrophes of every kind, aggressive rivals, various other animals, and most of all, bacteria and viruses without number threaten organisms continually with illness, injury, and death. Nevertheless, there exist millions of species, comprising thousands of millions of individuals; they populate the entire earth, and evolve continually and energetically – hence they must be well protected against disease and injury (not against *death*, though insofar as they beget offspring, they may be said to 'live again' in them).

In fact, the organisms themselves, which range from single cells, through plants and animals, to man, have defensive and protective arrangements equally diversified. Obviously it is outside the scope of this book to catalogue all these devices, and we must necessarily confine ourselves within limits. So we shall deal at this point with biological devices protecting against *infection*, and in particular infection of the human body – though most of what follows is true for all vertebrate animals.

The most important protective devices are (a) the *skin*, which is impervious to bacteria and viruses; (2) the *hairs* with which the bodily orifices are surrounded; and certain secretions, such as (3) the secretion of the *sebaceous* glands of the skin, (4) the secretion of the *tear glands*, (5) the *spittle*, and above all, (6) the secretions of small intestine, which break down a large number of 'foreign' substances into simpler compounds, disintegrate these, and so render them harmless.

These defensive arrangements are obviously of universal application, effective against every possible foreign substance or foreign body not just against one special kind – i.e., they are effective not *only* against tubercular bacilli, or *only* against infantile paralysis. They exist, therefore, from the outset; they are inherited, and in the course of development of a living organism they are produced, so to speak, automatically, and do not come into existence for the first time when the body is being immediately threatened.

All these things actually form part of the immuno-biology that is the subject of this chapter; but what is understood in biology by immunity, immunity protection, and immunity defense is what comes into being if the above-mentioned general barriers against infection by bacteria and viruses should be breached. Then everything is altered. The person is – unfortunately – not immune in advance to typhus or diphtheria, to measles or tetanus, though he can become immune to them. As a rule, immunity is acquired during the lifetime of an individual, and in addition, a person can be made immune in various ways, so that one can distinguish between active immunity and passive immunization.

Passive immunization is usually brought about artificially. If an infection is a danger to life – if, for example, because of contamination of a wound there is a risk of tetanus – then a 'serum,' more precisely an 'anti-tetanus serum,' is injected into the bloodstream of the injured person. This makes the pathogenic bacilli harmless, and so removes the risk.

An organism is actively immunized when it has overcome an infection *by its own efforts*. Of course, it is sometimes risky to rely upon the power of the organism itself, since infections, such as that caused by the tubercular bacillus, can overpower the organism, and having done so, be fatal. The device of protective inoculation has been invented to minimize this risk. The principle is that the affected person is given a weak dose of bacilli, or of the poisonous product of the bacilli, preferably during childhood when the danger of inoculation is greatly reduced. Such an 'artificial infection' is thus guaranteed harmless (cf. pp. 273–4); in most cases it does not even produce any visible symptoms.

The body itself makes its own anti-serum, and is thus equipped to deal with the germs, which are few and weak. And once the body has 'learned' how to produce anti-serum – regardless, incidentally, of whether by inoculation or by its own efforts – it retains the knowledge, and is now in a position to concentrate immediately and with still larger quantities of anti-serum against any new infection of the same kind. Of course, this acquired character is not inherited any more than is the individual memory.

A further and very important difference between true immunity and the unspecific protective devices of page 238 lies in the following: a person who has once had diphtheria or typhus, measles or cholera, has generally become immune for the rest of his life – but unfortunately only against the germ for the disease concerned, and never against all the others. A person who has recovered from typhus is just as vulnerable to diphtheria, measles, or cholera as he was before. The immunity thus acquired is not only postdated, but it is also strongly specific, not universally applicable to any alien or hostile infection. Just for the record, we might mention briefly here something that will be gone into in more detail later: immune reactions are somewhat 'capricious.' They do not work every time, and for every person, or every horse, or every rabbit, as reliably as might have been inferred from the description above.

After we have defined immunity to some extent, the points to be investigated in this chapter arise almost spontaneously:

1. What are the consequences of the specificity of the anti-serum or protective substance?

2. By what means does an organism acquire its immunity?

Naturally the two questions are interconnected, since it is through the presence of a protective substance that an organism becomes immune; and yet the division of the problem into two stages is justified. Stage one sees the production of a substance which makes toxins harmless, and this undoubtedly is a matter of chemical (or biochemical) processes. On the other hand, once immunity has been developed, the complete 'apparatus' for immunity is shared between the cells, or the organs, and this is a biological problem. We shall see later that the locations of the two phenomena are different.

Let us deal first with the problem that is possibly the simpler, and for which we can make use of some things from the first two chapters of this book. This is the first point: what is the importance of the protective substance?

6.02 Antigens and Antibodies
Two Hostile Brothers

We must become acquainted with the two basic ideas without which a discussion of immuno-chemistry and immuno-biology is not possible. They are mutually dependent on each other, and neither is conceivable by itself.

The one is that of the 'foreigner,' a bacterium or a virus, or some other poisonous substance coming from outside (being exogenous) and threatening the individual: this is the *antigen*. 'Anti-gen' is actually a contraction of anti-somato-gen (*anti*, against; *soma*, body; *gennan*, to bring about, to actuate). In other words, it is that which evokes the antibody; so here we are already with the second concept, the 'opposite number,' the antibody. Antibodies are substances which the threatened organism, whether human or animal, generates against the threat of the antigen; they are therefore peculiar to the body, which generates them to trap the antigen and make it harmless.

Any foreign substance that does not excite the production of antibodies is for that reason not an antigen, however dangerous it may be in other respects; on the other hand, an antibody never exists by itself, but only when an antigen is present and has been released. Consequently, not every foreign substance is an antigen, but every antigen is a foreign substance, or foreign body. The antibody, the protective substance, is matched only against the antigen that is in action at that time, and not against anything else. In other words, the antigen, which is always foreign, stimulates the body to produce antibodies that are quite specific, and the presence of such antibodies makes the organism in question immune against that particular antigen, sometimes for its whole life, but sometimes only for a short time.

It must be understood that the two basic terms 'antigen' and 'antibody' were originally mere formal concepts, and they still *explain* nothing; but the concept was so ingeniously simple, and so logical, that it quickly proved to be extraordinarily fruitful. In the first place, it provoked a profusion of questions: What substances, in general, can function as antigens? What are antibodies? How and where do they arise? What goes on between antigen and antibody? How does the organism know that the antigen is 'foreign'?

However, the concept brings not only questions, but also the possibility of being able to study these matters experimentally, and to find answers. We shall try to answer all these questions faithfully, though admittedly in a rather different order than that in which we have posed them.

We shall begin, with the fact that antigens and antibodies are highly specific to each other. Thus an antigen is matched to one very special kind of antibody – an antibody to one kind of antigen – and no other (though presently we shall mention some exceptions), just as a key is matched to a lock, positive to negative, template (matrix) to impression. The reader will be reminded of the specific enzyme proteins, of codon and anticodon, of repressor and regulator gene, and the resemblance is not misleading; in fact, the phenomena are essentially similar, and that is why we have dealt with specificity at the beginning.

The cause of the specificity is usually a quite definite, invariable molecular structure: in the sequence of the amino acids, or of nucleotides, in the shape of folds, or similar things. For the chemist these are welcome starting points for experimental research.

6.03 Stage One: The Serum
Here Antigen and Antibodies Meet in Conflict

Before we consider the chemical reaction between specific antigen and specific antibody, we must set the scene of action, which takes place in the blood, or rather, in the system of blood vessels. Blood is a fluid, its red colour arising from millions of red blood corpuscles (erythrocytes) which at certain times contain the red pigment hemoglobin (as we have seen on p. 30). Along with the red blood corpuscles, but in significantly smaller numbers, there are white corpuscles (leucocytes), as well as a whole mass of other cells, which are known as macrophages, or 'devouring cells' – we shall have more to do with these later.

All these cells swim in a surrounding fluid, the *serum* (Latin, skimmed milk). The cells can be separated by allowing the blood to coagulate, and then a pure, cell-free serum remains. This is not, however, by any means just water, but contains, among other substances, a great number of albuminoid bodies known as globulins (from the Latin word *globus*, spherical, since their molecules often have a spherical shape). Globulins appear in several forms, which are distinguished by the

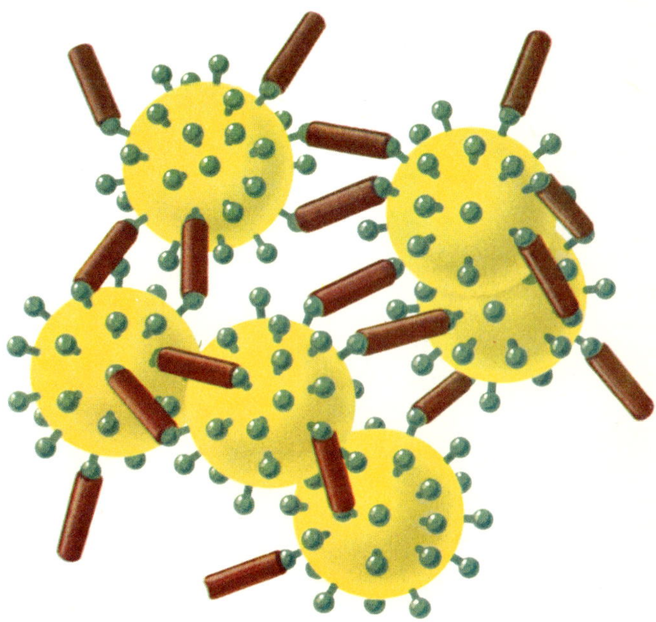

148. Antibodies (brown) destroy antigens (yellow with green knobs) by binding them loosely together into a *precipitate*.

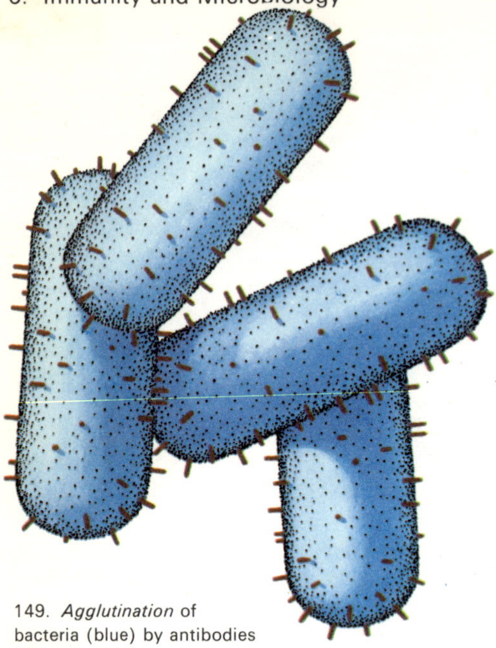

149. *Agglutination* of bacteria (blue) by antibodies (brown) which cause them to stick closely together.

first few letters of the Greek alphabet as alpha, beta, and gamma globulin. The two most important for our problem are gamma globulin and a sub-form of beta globulin, β_2-globulin

Gamma globulin and β_2-globulin at first have nothing to do with antibodies, as long as no antigens are to be found in the bloodstream, and also as long as the 'unspecific' barriers to infection, from the skin to the small intestine, are still functioning. If these defenses are overcome, then foreign substances enter the blood, and this acts as a signal to certain of the body cells to generate antibodies which are specific and matched to the antigen, and to release them into the bloodstream. Thus cells also are involved, and now we have reached stage two mentioned on p. 239. But we must defer all that for the time being: what these cells may be, and how they are capable of giving rise to the appropriate antibodies, must be worked out later. At the moment we are interested only in *the finished product, the antibody*.

It happens that all antibodies are both chemically and physically indistinguishable from gamma globulins, yet in one respect they are very easily distinguished from those particular gamma globulins that were present in the serum before the infection: they attach themselves to antigens, and react 'serologically' with them, and are thus *serologically different* from the original globulins. For this reason, the serum when it contains antibodies is known as an *anti-serum*.

Thus in the serum after an infection we have, in addition to the original globulins, antigens and antibodies together, and these two react serologically with one another. This reaction appears completely different, according to the kind of antigen that has invaded the bloodstream.

Things are comparatively simple if, in an experiment, we use as antigen a 'foreign' globulin or some poisonous substance that we have extracted from a dangerous bacterium, e.g., the toxin (Greek *toxikon*, an arrow poison) of tetanus, from the bacterium *Clostridium tetani*, the causative agent of lockjaw. If we administer at the same time the toxin and an anti-serum from another animal suffering from the same disease (or from a culture in a test tube), then a fluffy substance separates out, this is called a *precipitate* (Latin *praecipitare*, to fall down). This phenomenon is known, not unreasonably, as *precipitation*.

If, instead of extracting the antigen, we infect our experimental animal with the bacterium itself, then things happen differently. The bacterial cells, which pre-

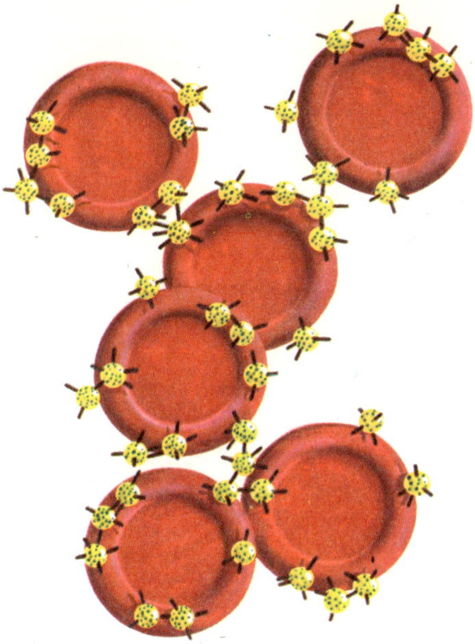

150. *Hemagglutination*. The antigens are allowed to attach themselves to the surface of red blood corpuscles, and then are themselves tied together by antibodies (brown).

viously were separate cells surrounded by fluid, now become aggregated into clumps; they agglutinate (Latin *agglutinare*, to stick to). The reaction product, the agglutinate, sinks to the bottom at once.

A method that is a special variant of this, which has been of great service in serology because of its precision, is the so-called *hemagglutination* (see table).

Table 5

Reactions between Antigen and Antibodies

1. Free antigen + antibodies (anti-serum) → precipitation
2. Cell-bound antigen + antibodies (anti-serum) → agglutination
3. Antigen bound to blood corpuscles + antibodies (anti-serum) → hemagglutination.

As in the first experiment, the toxin (the antigen) is extracted, but it is then allowed to attach itself to the surface of red blood corpuscles. If the corresponding anti-serum is now applied, then this time there is formed an agglutinate of the red blood corpuscles.

It is a common feature of all three systems that the dangerous antigen is no longer able to swim around freely by itself, but through the agency of the antibodies, it has been converted into big, insoluble complexes. In this form the precipitate or agglutinate can be devoured and digested by the macrophages of the blood. It has been rendered harmless – but more details about this later.

The ability to cause precipitation or agglutination originates in a special property of antibodies, that they are *bivalent*. One molecule of the antibody can attach itself to two molecules of antigen at the same time. Antigens, on the other hand, are mostly *polyvalent*, i.e., they can attach themselves simultaneously to several molecules of antibody. At first glance this does not tell us much about specificity, but it goes a little way towards explaining how a precipitate forms. The bivalent molecules of antibody bind the antigen molecules into giant, net-like complexes, which, because of their size, can no longer be held in solution and are deposited as an insoluble precipitate.

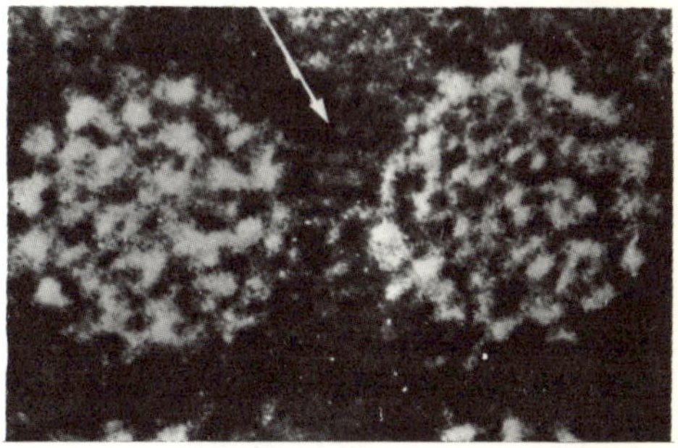

151. An antibody connecting two spherical virus particles – photographed for the first time under the electron microscope!

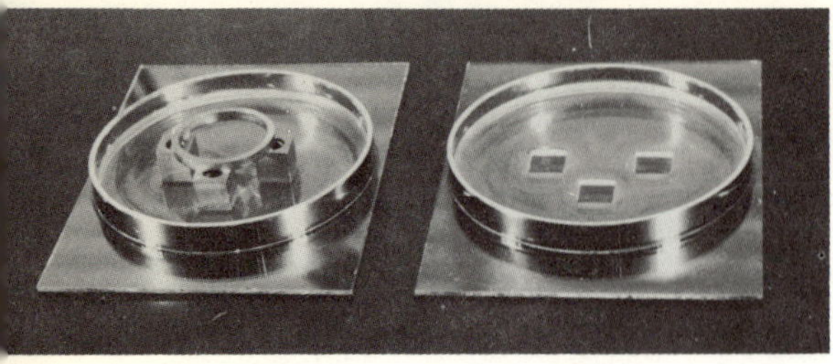

152. Making agar plates with holes, which can be filled with either antigens or antiserum.

What appears in the test tube as a more or less uniform cloudiness (caused by the precipitate or the agglutinate) can be made very much more sharply defined if the reaction is allowed to take place in a semi-solid, gelatinous medium, e.g., on an agar plate, such as we used previously for the culture of bacteria. Such plates can easily be poured in such a way that two holes are left. Into one of these is poured the solution of antigen, and into the other the corresponding anti-serum (any other would not react). From both centres the enclosed substances slowly migrate into the agar in all direction ('diffusion'), and somewhere between the two centres they meet, react with each other, and are precipitated. The precipitate appears as a fine, sharp line.

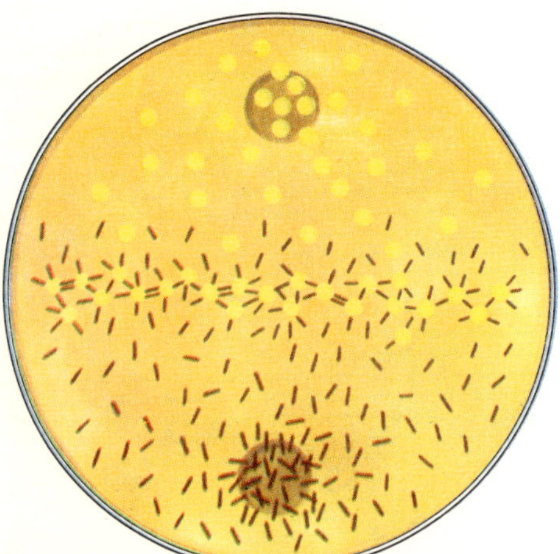

153. An agar plate with two holes. In the upper are placed antigens (yellow), and in the lower antibodies (brown). All move freely, but where they meet each other they come into combat.

There are two important qualifications to be made here:

1. The zone of precipitation is not exactly identical with the zone of contact. If many antibodies fall upon a few antigens (as happens if an anti-serum is used), then there is little or no precipitation. The same thing happens if there are many antigens and very little anti-serum. It is only in an intermediate situation, when both partners are present in equivalent concentrations, that precipitation is clear and at its maximum. The term 'equivalent' means having the same value, or exactly corresponding; the concentrations are not *the same*, but are of equal value. What does this mean?

If the amounts of the two reagents are measured in the 'zone of equivalence,' it is found that the two are related to each other in the proportions of small whole numbers such as 1:1 or 1:2, or 1:3 or 2:3, or the like. The chemist calls these *stoichiometric* proportions, and he interprets them as a certain indication that the two partners are linked together by chemical bonds, and that they can be investigated by well-tried chemical methods. Precipitation and agglutination thus originate as chemical reactions.

2. And now the second qualification. On the agar plate, the anti-serum (antibodies) and antigen are not always separated by a single line; often they are separated by several lines. This is the case, for example, in Figure 154, where the anti-serum is an anti-diphtheria serum from a horse, and two antigens (I and II) are both diphtheria toxins. It can be seen clearly that there are several lines of precipitation between the anti-

serum and antigen II. This is explained when we discover that antigen II is a 'crude' toxin, which has not been purified, and so still consists of a number of components, whereas antigen I, in contrast, is a highly purified fraction, which obviously retains only one component.

This problem of specificity takes us into the realm of the chemistry of immunology. Is the above experiment evidence for or against a high specificity of serological reactions? This question can be answered either yes or no, according to whether one assumes that the anti-serum contains several kinds of antibody, or only one. If only one kind of antibody is present, then this antibody must be able to react with all the different fractions of antigen II – so the reactions themselves must be *non-specific*. On the other hand, if several kinds of antibodies are present, then each one can react with a corresponding fraction of antigen II, producing several lines of precipitation, whereas with antigen I, which is homogeneous, only one of the antigens can react, and the others give no precipitate. In this case the antibodies must be *highly specific*.

This question must therefore remain unanswered until we know more about the antibodies.

6.04 The Antigen and Phagocytosis

The Structure and Fate of the Antigen

With regret, we must go back from antibodies to the antigen; this is tiresome, but it cannot be avoided. We have already noted that antigens and antibodies are mutually necessary to each other and that the one is defined in terms of the other. So if we want to know something about the specificity of antibodies, we must check back to the antigens, like this:

a. What property is it that gives a substance, a virus particle, or a bacterium, the ability to stimulate the formation of anti-

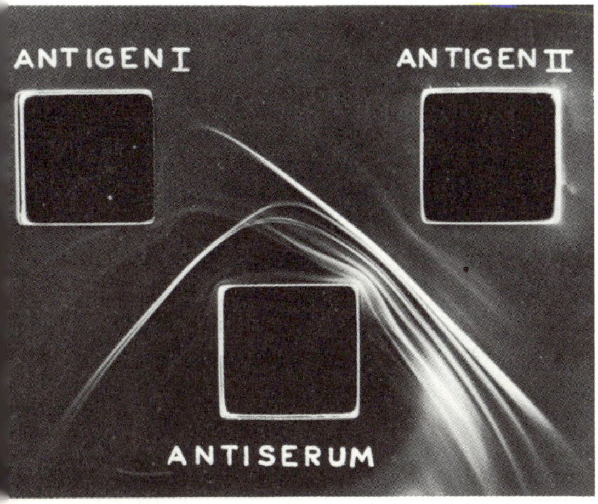

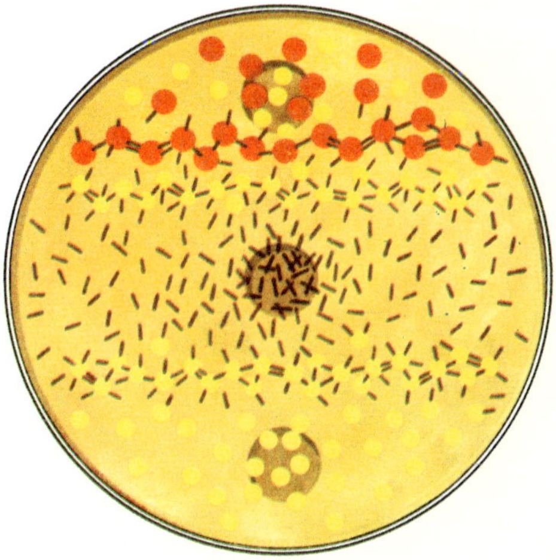

154. Pure and impure antigens. Antigen I has been purified, giving only one line of action with the anti-serum. Antigen II, on the other hand, has not and so includes several different antigens, each of which is attacked by the anti-serum along a different line. 155. Case II of Figure 154 shown schematically. The anti-serum is now in the middle of the plate, and operates against: (a) a single type of antigen (yellow) in the lower half of the plate, with which it forms one line of action, and (b) a mixture of two types of antigen (yellow, red) in the upper half of the plate, where the antigens move at different speeds, and so two lines of action develop.

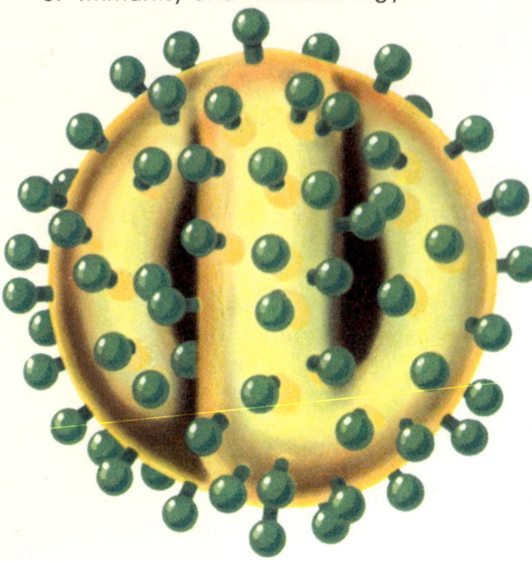

156. A protein molecule (yellow) with many side chains (green), of type S_1.

157. If surface of protein bears only side chains of the type S_1, shown in Figure 156, then only antibodies against these (anti-S_1) will be produced.

bodies in a particular organism – i.e., what property gives the substance the *nature of an antigen*?

b. How far can, or must, an antigen be chemically changed before it stimulates the production of correspondingly different antibodies? In other words, how far-reaching is the specificity of antigen/antibody?

In answer to a, it can be said with certainty that *it is not the chemical composition of a substance that makes it an antigen*. Initially it was thought that only proteins could act as antigens, but this view has not been confirmed; instead, it has been found that carbohydrates, fats, and nearly all classes of chemical compounds can be antigens. This was very important for research, since proteins, not least because they usually have a high molecular weight, are substantially more difficult to produce than, for example, carbohydrates, especially sugars (polysaccharides). Today, in fact, the latter are among the antigens that have been most widely studied.

Nevertheless, it is remarkable that nearly all antigens do have a high molecular weight. Small molecules, by themselves, remain ineffective, possibly because the body can easily get rid of them, until they attach themselves to a big molecule, as we shall see presently. Nevertheless, the mere size of molecules or of complexes of molecules is not enough to make them into antigens, i.e., to endow them with the functions of antigens. One requirement is that the substance must be capable of being digested. What have we to say about that?

In general, small molecules such as those of urea, lactic acid, or glucose are dissolved in the body fluid, or in the blood and are also taken up by the cells in a dissolved state. That is to say, they are carried more or less passively through the outer boundary of the cell, probably through fine canals (see p. 195), and thus arrive at the interior of the cell. This route is barred to bigger molecules,

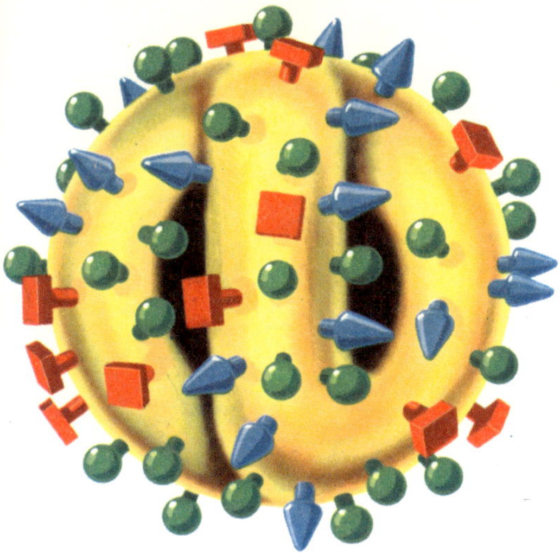

158. A protein molecule, this time with side chains of three different kinds.

whether they are dissolved or not, and to bigger particles. These are 'devoured' by phagocytes, a phenomenon that we have encountered before, on the occasion of our glance into the cell on p. 177. *Before a substance can function as an antigen it must be vulnerable to phagocytosis* and it must penetrate into the interior of certain cells – and obviously only there, and *not in the serum,* can antibodies be formed.

These cells are in fact the macrophages of p. 241, belonging to stage two, which we now meet for the first time; but we must defer consideration of them for a little while longer, while we go further into the fate of antigens after phagocytosis.

We must bear in mind that antigens are substances of high molecular weight, which are also vulnerable to phagocytosis. If that is so, the big molecule must be actually tested by changing it a little, or adding new, small groups to it, to see whether this produces *other, different antibodies* – and in this way we should arrive at the second of our checks, check b.

These basic experiments were carried out by Karl Landsteiner, who worked with *protein* antigens. If these were injected into the bloodstream of an organism, there arose antibodies which were directed against the foreign protein, 'antiprotein antibodies,' so to speak. We shall refer to these as simply 'anti-proteins.'

We know that the proteins are substances that have a reaction potential resulting from their side chains, the residual chains R_1 to R_{20} of the first chapter of this book, onto which may easily attach itself an additional side chain, which we will call S_1. If an antigen that is provided with S_1 in this way is injected into an experimental animal, then in

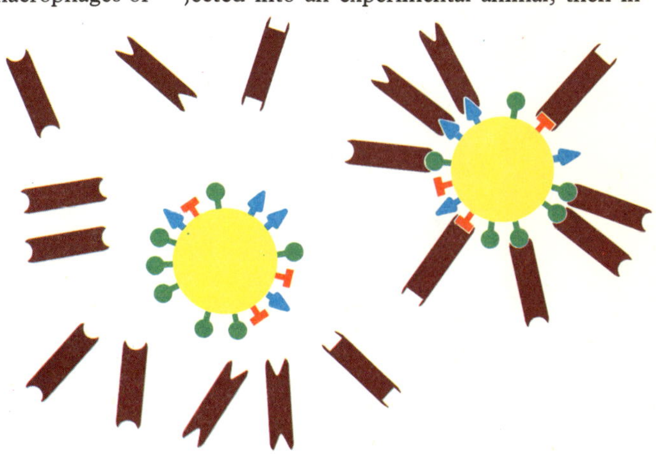

159. The protein molecule of Figure 158 stimulates the production of three different kinds of antibodies, anti-S_1, anti-S_2, and anti-S_3.

the vicinity of the anti-protein there will arise an antibody anti-S_1, and this will react only with S_1, for which it is thus highly specific. If, now, many such S_1 side chains are attached, so many that almost the entire outer surface of the antigen-protein molecule is covered with them, then the original antigen is completely overlaid, and it no longer constitutes an anti-P any more, but it is still anti-S_1 (see Fig. 157).

Now, of course, the outer surface of the original antigen-protein can be decked with *different* side chains – besides S_1, with S_2, S_3, and so on – and immediately the corresponding antibodies, anti-S_1, anti-S_2, anti-S_3, and so on, can be detected. This means, however, that an antigen consists of several parts:

1. A ground-substance of high molecular weight, which we might label the 'bearer of the antigenity' (in this case the antigenity for anti-P).

2. One or more groups of small molecules. These bring about, and determine, the specificity, and so they are called the *determinant groups*.

Obviously if an antigen can bear different determinant groups, then we can never make a *universal antibody*, but we must have at least as many different kinds of antibodies as there are different determinant groups present. This often gives rise to a whole 'family' of antibodies. Every member of the family is adapted to be strongly specific to one particular determinant group, and enters into a chemical bond *with this group, and with no other*.

The 'family' of antibodies can grow bigger. If one of the side chains, for example, S_2, is very long, then the appropriate group of antibodies may be completely specific, i.e., *matched to the complete length of* S_2.

Intermediate stages are possible, however, to the extent that the specificity of the antibodies may extend only to three-quarters, or half, or only one-quarter of S_2. We need not concern ourselves with these matters here.

Even more detailed and comprehensive are the results of experiments with antigens that are not proteins, or in which the protein component plays only a subordinate part. Here again we are able to refer back to something we already know, namely to the Enterbacteriaceae – *Escherichia* coli, the 'work horse' of molecular geneticists, the dreaded *Salmonella*, the pneumococci – which were used in the earliest transformation experiments. We have heard about the S-form and

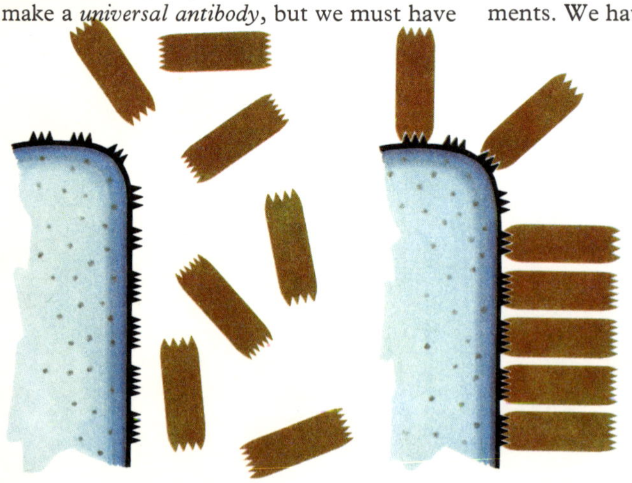

160. The substance forming the capsule of a bacterium bears its determinant groups – which are polysaccharides – directed outwards like the teeth of a comb. In this position they react with antibodies (brown).

the R-form of the pneumococci; the S-forms (smooth) have a capsule and are infectious; the R-forms (rough), on the other hand, have no capsule and are harmless. In transformation experiments it was the procedure to inject such considerable quantities of S-bacteria into non-immunized mice that they were overcome, and succumbed (see Fig. 69).

In experimental immunology, however, use is made of very small amounts of antigen, so that the animals that are being experimented upon may survive and produce antibodies. Very soon it comes about that the *capsule itself,* or any substantial part of it, actually has the functions of an antigen, and will initiate the formation of antibodies – and the R-form again (having no capsule!) is harmless. Now the capsule substance is, of course, rather complex in constitution: proteins, sugars (polysaccharides), and fat-like substances (lipoids) all play their part in it. It should be known correctly, if formally, as a polysaccharide-lipo-protein. The only parts of the complex to function as a determinant group, however, are the polysaccharide constituents; not the lipoids, or the proteins. This may be consistent with the fact that, as Figure 161 shows, all the sugars lie on the upper surface; they project like the teeth of a comb, towards the 'outside,' and thus are the first groups to react with other substances. The lipoids and proteins, on the other hand, lie in deeper layers of the comb, and are normally 'concealed.'

More than a dozen different sugars may act as building units for the polysaccharide part, though actually only three to five of these are present at any one time. They may even be gathered together into units, which are then regularly repeated. For every combination of three or five sugars there is usually formed at any one time one serologically reacting antibody, a 'serotype.'

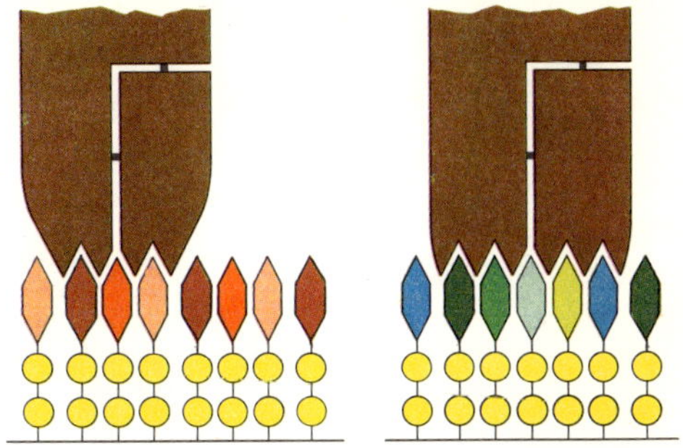

161. Schematic reconstruction of a capsule of a bacterium. Below (yellow, linked together) are the lipoid and protein constituents, and attached to these are the polysaccharide constituents (various shades of red and of blue and green). The polysaccharides operate in groups of either three (left) or five (right), and build up the antibodies (brown) out of smaller units.

And since the replacement of only one single sugar molecule by another of slightly different structure is enough to change the determinant group, and thus to evoke a different antibody, hundreds of 'serotypes' exist. It would require a whole book just to describe their relationships with one another and to explain their genetic and phylogenetic connections – here we must let it rest.

It is important in our context that we find confirmed in the bacterial capsule what is true for the protein-antigens: the antigen is a complex of high molecular weight, and the determinant groups, after which the specific antibodies were constructed, are small pieces of the whole complex, or small molecules built onto it, *which in themselves may be quite innocuous to the threatened organism* – the sugars in the polysaccharide-lipo-proteins are indeed harmless!

It can be shown that this is important by the following experiment. Antigens and their corresponding antibodies are brought together, and so the formation of an antigen-antibody complex is arranged; we thus have an insoluble precipitate, as in Figure 148.

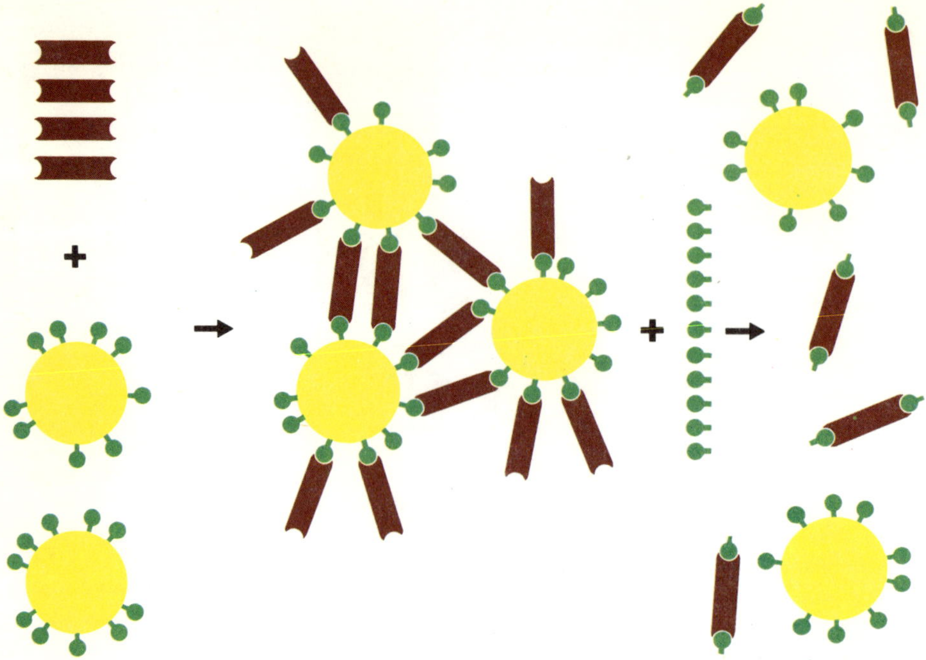

162. If antigens, with their determinant groups, come into contact with antibodies, a precipitate is formed (left); but if the determinant groups are present in excess, they neutralize the antibodies, and the precipitate is taken back into solution again (right).

Now, however, we combine with the precipitate as many as possible isolated (i.e., 'free') determinant groups. (This is no longer difficult if we have attached additional side chains to the protein-antigen, following Landsteiner; we then need only take the unattached surplus. In the capsule antigens the sugar components that sit on the upper surface can be split off by chemical means, isolated, and introduced into the experiment.)

Now, in the precipitate the places of attachment of the (bivalent) antibodies are already occupied by the corresponding determinant groups, which are *attached to the antigen*. The 'free' determinant groups, which have just the same structure, are seeking to attach themselves to the same places at the same time. If the 'free' groups are in excess, then it is they, and not the determinant groups attached to the antigen, that succeed in occupying these places on the antibodies. And the result? Since the antibodies are 'only' bivalent, they can therefore attach only two free determinant groups, and this gives relatively few compounds. The determinant groups that are attached to the antigen have now become free again; the complete net-like complex, the precipitate, is dissolved into free antigens (with unattached determinant groups) and antibodies which have attached to them only two determinant groups, and not whole molecules of antigen. We have taken antigen molecules that were previously harmless, and by adding free determinant groups that were innocuous in themselves, we have set the antigens free to do their evil work.

6.05 Antibodies

Gamma Globulins Are Made Specifically

Now we can turn back again to the antibodies, and understand them with a new insight. We now know that the determinant groups on the antigen molecule decide its specificity and that it attaches itself to the 'appropriate' place of attachment on the molecule of antibody, forming a precipitate or an agglutinate. The determinant groups may be derived from any chemical group; on the other hand, the antibodies, with singular unanimity, are proteins, and indeed gamma globulins. *How is it possible for these globulins to be matched so exactly and specifically to thousands and even hundreds of thousands of different determinant groups?*

This question cannot be exactly answered even yet, but a great many of the essential details are already known. We shall try to put some of these details together in the hope of arriving at a picture that is correct in at least some respects. Starting at the bottom, we must first find out the size of the molecule of antibody.

Strange to say, there are two kinds of these. The one kind has a molecular weight of about one million, exactly like the unspecific gamma globulins in the serum of an organism that has not been infected. In the ultracentrifuge (see pp. 45–6) they turn out to be '19 S-bodies.' These big antibodies appear very soon after the infection, but they are not always very effective.

The other kind is much smaller, since their molecular weight amounts to only 160,000 ('7 S-bodies'). They arise later than the bigger antibodies, and are highly active as antibodies; it is not yet known why this is so, and we shall have to think about this more when we go into the origin of the antibodies.

Now, even 160,000 is still a high molecular weight; many enzyme proteins, for example, have a molecular weight lying well below this. If we now remind ourselves (a) that a molecule of antibody is bivalent, and so displays only two bonds, or places of attachment, and (b) that these two places of attachment are controlled by one and the same determinant group of the antigen, and thus are completely equivalent to each other, then it does not

163. Model of an antibody built up from four smaller units (brown). It has two attachments, each specific for one determinant group (green).

surprise us to hear that over the entire surface of a molecule of antibody only about 1 or 2 per cent of the bonds are 'active,' i.e., capable of forming an attachment. The remaining 98 per cent can be dismissed as insignificant as far as bonding to a determinant group is concerned.

The antibody molecule of weight 160,000 can be successfully broken up by using a protein-splitting enzyme, and produces four fragments: two 'light' polypeptide chains, each with a molecular weight of 20,000, and two 'heavy' ones, of 60,000 each. None of the four units has any longer any activity as an antibody; we have not subjected the antibody to random destruction, but, so to speak, have dismantled it in orderly fashion into four subordinate units. With care these can be reassembled and will give us once again completely functional antibodies.

A model will give an impression of how these subordinate units might be brought together again. Opinion is not yet entirely unanimous on the question of whether the attachment group is temporarily bonded onto a small fragment (subordinate unit) or whether (as in Fig. 63) each attachment group is shared between a big and a small fragment. However that may be, the decisive factor for attachment to a determinant group is the state of the surface of two narrowly defined areas of the molecule as a whole.

Now, we know from our discussion of proteins that a polypeptide chain is only rarely stretched out, but is generally coiled into an α helix, and in addition, may be repeatedly folded, doubled, kinked, and so on. It is easy to show from this that such folding and kinking will produce surfaces of quite different shapes. Exactly the same phenomena take place at the surfaces of 'substrate-specific' enzymes, which we discussed in detail in Chapter 1. Now, this time, it is a matter of *antigen-specific* surfaces of protein molecules; they exactly match, in a complementary way, the structure of the determinant groups, like lock and key, like a signet ring and a seal.

The number of possible variations of pattern is obviously very great, the more so since *one* attachment may be shared between *two* polypeptide chains, one 'light' and one 'heavy.' The two chains may be displaced in relation to each other, giving rise to still more new patterns.

The most difficult question, however, to which we now come, is this: Does the determinant group actually 'stamp' the surface structure at the place of attachment, and if so, then how and where does this take place? This poses the following problem: if we are going to talk about wrinkling and surface structure, then we shall end up with the tertiary or quaternary structures. Moreover, as we found out in Chapter 1, both of these are determined exclusively by the primary structure, through the genetically determined sequence of amino acids. Is there, then, any possibility that the determinant groups of antigens may be further changed in their tertiary or quaternary structure in such a way that they 'fit' to the attachments?

Such a change must be a *lasting* one: one need only consider how anyone who has suffered from diphtheria is thereafter immune against diphtheria for the rest of his life, that is, he has at his disposal the appropriate antibodies. This appears to be a problem that cannot be solved by chemical means alone, and this is even less likely when we consider that the impression which causes matching of the antibodies cannot be carried out except by the intervention of living cells.

And now we must leave the territory of immuno-chemistry and enter that of immunobiology.

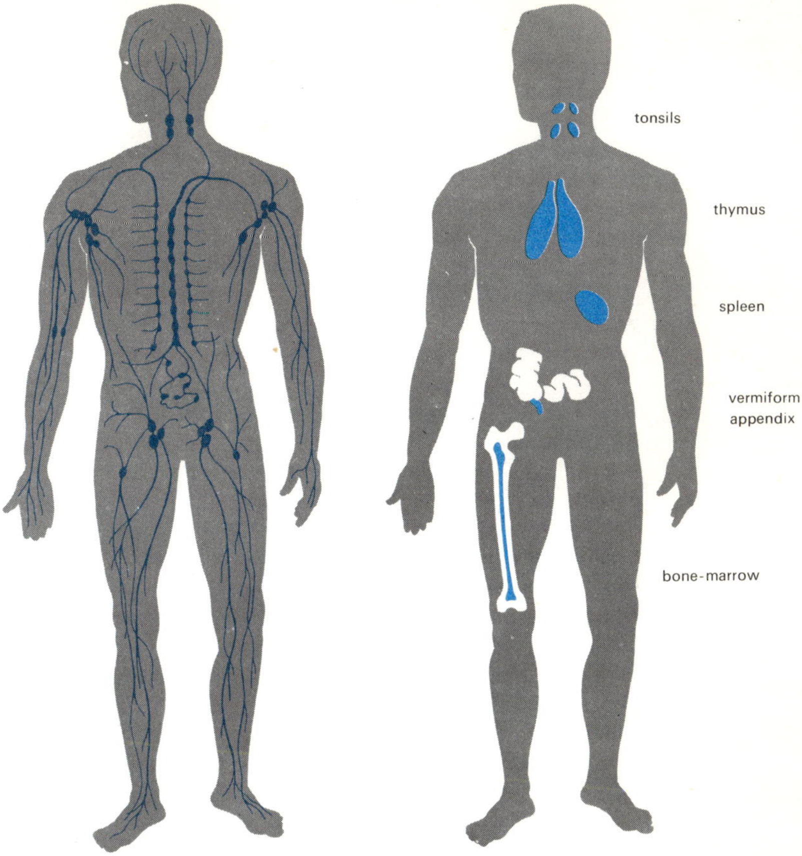

164. Human immunity apparatus: left, the lymphatic system; right, all the other systems, from the thymus gland to the bone marrow.

6.06 Stage Two: The Mechanism of Immunity

Phagocytes Attacking Antigens

Reactions of immunity are undoubtedly more complicated biologically than they are chemically, and from neither aspect are they fully understood in every detail. Their higher level of complexity as well as their very great universal significance is derived from the fact that the cells concerned are *only part of the complete immunity mechanism*, which involves large complexes of cells, and even entire organs. This immunity mechanism is distributed over the entire organism of a vertebrate animal, 'interwoven,' so to speak, like the system of blood vessels or the nervous system.

From this is derived the specialized term *reticuloendothelium*; this network of internal tissues is also called the Reticulo Endothelial System, RES for short. The reader will at once think of the ER, the endoplasmic reticulum (see p. 149), and rightly so; the

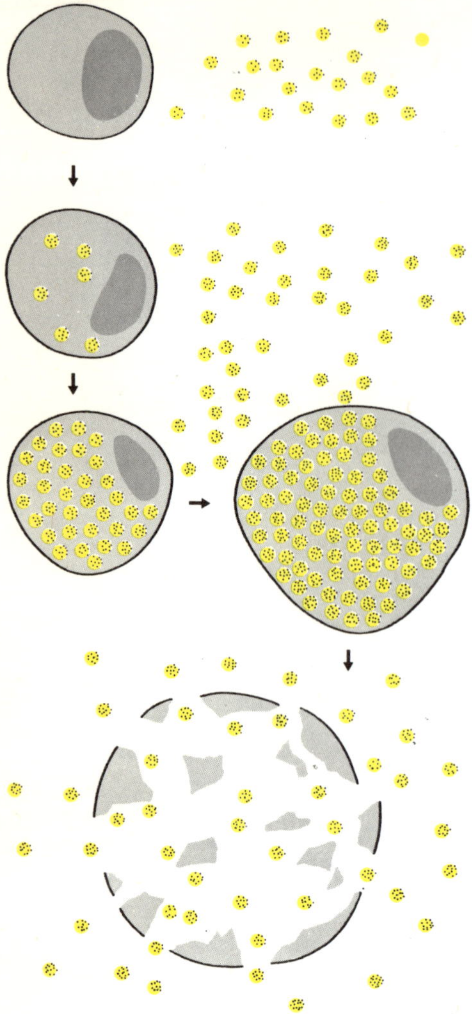

165. Leucocytes devour antigens. If the antigens are present in excess, then the leucocytes 'overeat,' and burst.

ER is basically also an internal network, though one that is found only in single cells and not in multicellular organisms.

To the RES belong, among other organs:

1. The thymus, in the upper chest; this gland when taken from a calf and used for food is called the throat sweetbread.
2. The spleen.
3. The bone marrow.
4. The tonsils.
5. The vermiform appendix of the intestine.
6. The big lymph ganglia.
7. The small follicles of the lymph system.
8. The lymphocytes.

We have now arrived at stage two (see p. 241), and since we have already bypassed this subject several times, we must finally deal with it. Let us retrace what happens when pathogenic bacteria or their toxins arrive in the blood as antigens, having penetrated the unspecific defense barriers of the body through a wound or injury.

Among the millions of red blood corpuscles that are to be found in one cubic centimeter of blood, there are always also a few white corpuscles, or leucocytes. These are part of bigger groups of so-called lymphocytes, and constitute a kind of police force, which immediately engages the intruders and mops them up, making use of the phagocytes that we have already heard about. If there are only a few antigens, then the leucocytes that are already present can deal with them without trouble. The leucocytes digest the antigens (for which purpose they are richly endowed with enzymes), and in this way, the matter is disposed of without any fuss; it never gets as far as outbreaks of disease.

If, however, too many antigens are present, then the leucocytes are no longer in control. They 'overeat,' and swell enormously; they then no longer have the power to break up the antigens. So it comes about that the leucocytes in their turn are destroyed by the antigens and become part of the general debris of the cell. This cell debris, by further decomposition, releases some very disagreeable substances, which – by means that are little understood – provoke a flush of symptoms. One of the most important of these substances is histamine, a near relative of the amino acid histidine, which also plays a part in hay fever.

The appearance of such inflammatory substances mobilizes more lymphocytes, which are dispatched from the bone marrow, the lymph glands, and above all, from the thymus. These 'new' lymphocytes are thrown in as *a second wave* against everything they can find – against the cell debris, against such antigens as still remain, even against the 'old' white corpuscles. They are true phagocytes, and their technical name is *macrophages*, or 'great devourers,' in reference to their very considerable bulk. Frequently this action brings to an end the first act of the drama. If all the antigens have been eaten up and digested, then danger has been averted and the attack repulsed.

This action, however, does not confer *any active immunity*. We speak of immunization only if the body, after it has won the first battle, then goes on to produce additional antibodies which are held 'in readiness.' In the event of a second infection *of the same nature*, enough antibodies are already present to render the new antigens powerless immediately by precipitating or agglutinating them. Which cells produce the antibodies to be held in readiness, and how is this done?

It might be thought that the antibodies must be produced in the macrophages; these have consumed the antigens to which the antibodies are 'matched.' This is not so, however; instead, an entirely new kind of cell comes into action: the so-called *plasma cell*.

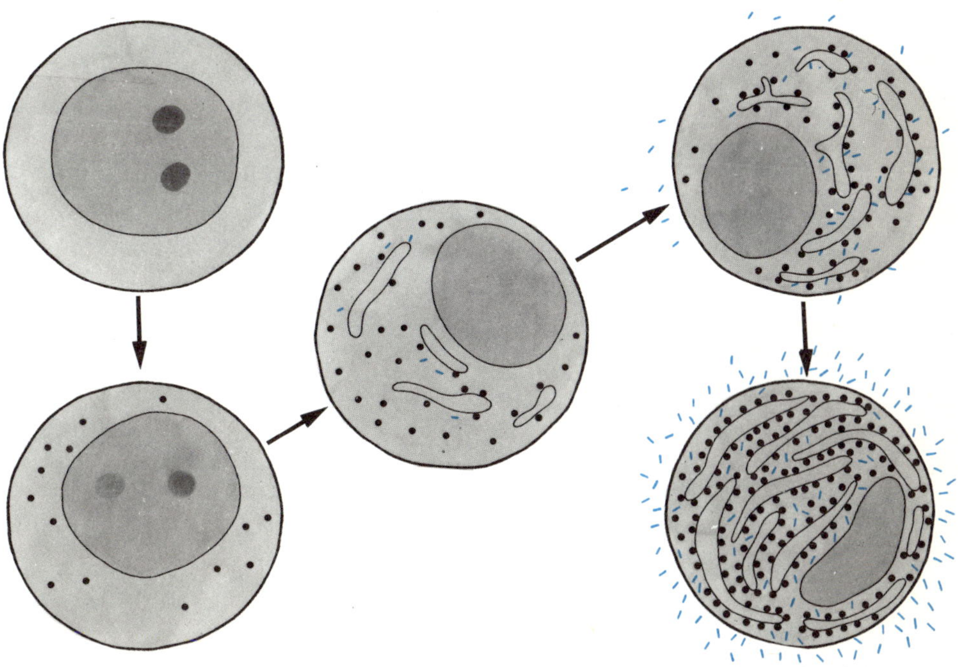

166. Young plasmablasts mature into plasma cells. The endoplasmic reticulum and the number of ribosomes increase steadily until the 'antibody factory' is ready to operate.

6.07 Plasma Cells are Antibody Factories

But How Do They Make the Antibodies Specific?

The origin of plasma cells is fairly clear. They arise complete from undifferentiated cells of the mesenchyme, which is a youthful, loosely woven tissue, from which is developed the connective tissue, in particular. In some way – which for the time being we must express in indefinite and non-commital terms – the mesenchyme is provoked into dividing. The first set of daughter cells are still not greatly specialized (differentiated), and they are called *plasmablasts.* These divide again, about every ten hours, so that eventually, about the fifth day, some five hundred cells have arisen from one plasmablast. At the same time, however, a ripening process can be observed, converting young plasmablasts into mature plasma cells. As can be seen from Figure 166, ribosomes appear in the cytoplasm, a few at first, later on more and more, and these are reinforced by more and more cisterns of an endoplasmic reticulum. In its final state the plasma cell is filled with an extensive endoplasmic reticulum, the membranes of which are thickly beset with ribosomes. *The cell is now fully equipped as a factory for making antibodies.* It can produce so many antibodies that these overflow and are passed into the bloodstream.

Now, it was actually to be expected that the plasma cells would be rich in RNA and in ribosomes; they were to produce antibodies, i.e., specially constructed gamma globulins, and so the cell would naturally need the complete equipment for synthesizing proteins, particularly ribosomes, and these can easily be demonstrated under the electron microscope.

Much more astonishing was the result of an experiment with (attenuated) tetanus toxin, which had been radioactively marked, once again with tritium, the radioactive isotope of hydrogen, which we already know from the experiments with DNA and RNA described on pp. 70 and 222. It turned out that the cells, which were sometimes only plasmablasts, but more often had already become ripe, antibody-producing plasma cells, showed no trace of radioactivity: in any case, therefore, they could contain only minute, insignificant bits of antigen. The phagocytes, on the other hand, were very strongly radioactive, and thus contained significant amounts of antigens.

From this it must be concluded that plasma cells have no absolute need of the antigen for which they are to make the matching antibody. On the other hand, they 'know' exactly which antibody they have to make. They do not make a mistake, and they normally make only one kind of antibody (though exceptions have been observed, where one plasma cell can produce up to four different but related antibodies at the same time). Whence do the plasma cells get their 'knowledge,' or to express it rather more prosaically, *from what source*, and above all, *in what form*, have they received the information according to which they have produced the antibody?

This critically important question cannot be decisively answered at the present time. There have, however, been two different attempts at an explanation. These have been subject to lively discussion in the specialized literature, under the designations 'instructive,' or 'template,' theory, and 'selective,' or 'clone-selection,' theory.

First theory: the antigen 'stamps out' the antibody. The instructive, or template, hypothesis starts off from the idea that the antigen itself remains available as a model, or template. Each determinant group serves to some extent as a template, and the attachment points of the antibody are 'modelled' on this in an exactly matching, or complementary,

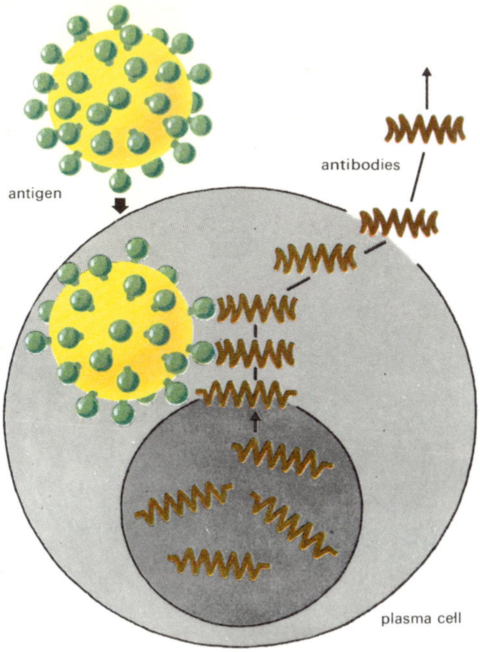

167. Diagram to illustrate the template hypothesis of antibody formation. The antigen penetrates into the plasma cell and serves as a template for the formation of antibodies (cf. Fig. 170).

way. To be precise, two identical attachments must be produced in this way, since every antibody is bivalent.

Two difficulties, in particular, arise from this hypothesis:

a. We have already established that antibody-producing plasma cells exist in which no *antigens* can be demonstrated. What can we make of this? This apparently clear finding is contradicted by many others, according to which antigens can be demonstrated months or years after the first infection, if not always in the plasma cells. Often the concentration is extremely low; nevertheless it may reach about a thousand antigen molecules for every 'immunologically competent cell.'

In other cases, especially in the pneumococci-capsule-polysaccharides, the quantity of antigens hardly diminishes at all over months and years, obviously because in general the destructive enzymes are not present. Of course, these antigens are not effective *as antigens*, otherwise the organism would die. Instead of this, they are in some way or other blocked; it is thought that they have become bonded to molecules of RNA on the macrophages. In this form, doing no harm, but containing all the parts needed for their function as a template, they can be passed on as information to plasmablasts and plasma cells through several generations.

For all that, if there is even a single case where no antigen is present in the plasma cell, this is a serious objection – it must then be true that the antigen has transmitted its information content (it was modelled on the attachment point complementary to the determinant group) to the antigen-binding RNA. And this brings us at once to the second difficulty.

b. Every antibody is a globulin, and thus a protein, and for that reason obeys the laws of protein synthesis, which we went into so fully in Chapter 1. The globulins of the antibodies do not differ either physically or chemically from the other globulins that have no functions as antibodies. Since the method of protein synthesis from genetic information (DNA) is tied up with RNA and ribosomes, and the definitive form (the folding) of a polypeptide chain is determined solely by the genetically directed sequence of amino acids (or will be so determined) – then where is there room left for intervention by the antigen, or by the information that it carries? Certainly not in its primary structure (the sequence of its amino acids), and not in its secondary structure (the helix); perhaps in its tertiary structure?

In this connection there is something that must be said. Many reputable scientists consider that the view that the tertiary structure is exclusively bound up with the primary structure is a dogma for which any convincing proof is completely lacking. Instead of a great deal of argument, two experimental examples may be put forward:

The one we owe to Professor Haurowitz, one of the best-known chemists working on proteins and immunology. He took so-called azoproteins–that is, proteins with antigen properties, which additionally have been equipped with a very special nitrogenous determinant group, an azo-group – and injected them first into a hen and then into a rabbit. Both animals formed antibodies against this particular antigen, but the two kinds of antibodies were quite different chemically, since their complement of amino acids, and probably also the sequence of these, had no overall agreement. Nevertheless, both had the same complement of antibodies, the same antibody characteristics. Obviously it is possible for a tertiary structure (the complementary attachment of the antibody) to be produced by different primary structures.

In the reverse sense, it is possible for an 'ordinary' gamma globulin – from rabbit serum – in a test tube to fold itself up into a specific antibody, specific against human serum-protein, with a dinitrophenyl radical (DNP – its chemical formula is unnecessary here) as additional side chain. The molecule of gamma globulin needs only to be temporarily folded a little (not completely), and to allow the unfolding to take place in the presence of DNP protein. Certainly, the greater part unfolds itself back to the original form, but 0·5 per cent has become specific for DNP protein. This may seem to be very little, but it is actually more than was to be expected, since it is certain that only a few antigens of the partly unfolded gamma globulins could reach the right place at the right time.

An interference by the antigens, or the information carried by the antigens, in the formation of the tertiary structure can thus by no means be excluded, although exact details of how this happens are not yet known. No less a person than Professor Linus Pauling, twice a Nobel prize winner, has worked out a scheme to show how the specific splitting off of the two attachments on the antigen – these attach to the surface of ribosomes – can be represented.

The four necessary polypeptide chains, formed on the ribosome (Fig. 169), are at first

168. Pauling's theory of antibody formation. A ribosome (olive green, with its m-RNA blue) synthesizes a polypeptide chain (orange), which then folds itself up onto a determinant group (green) of the antigen (yellow). Having done this, the polypeptide chain becomes an antibody with two specific attachments (see Fig. 169).

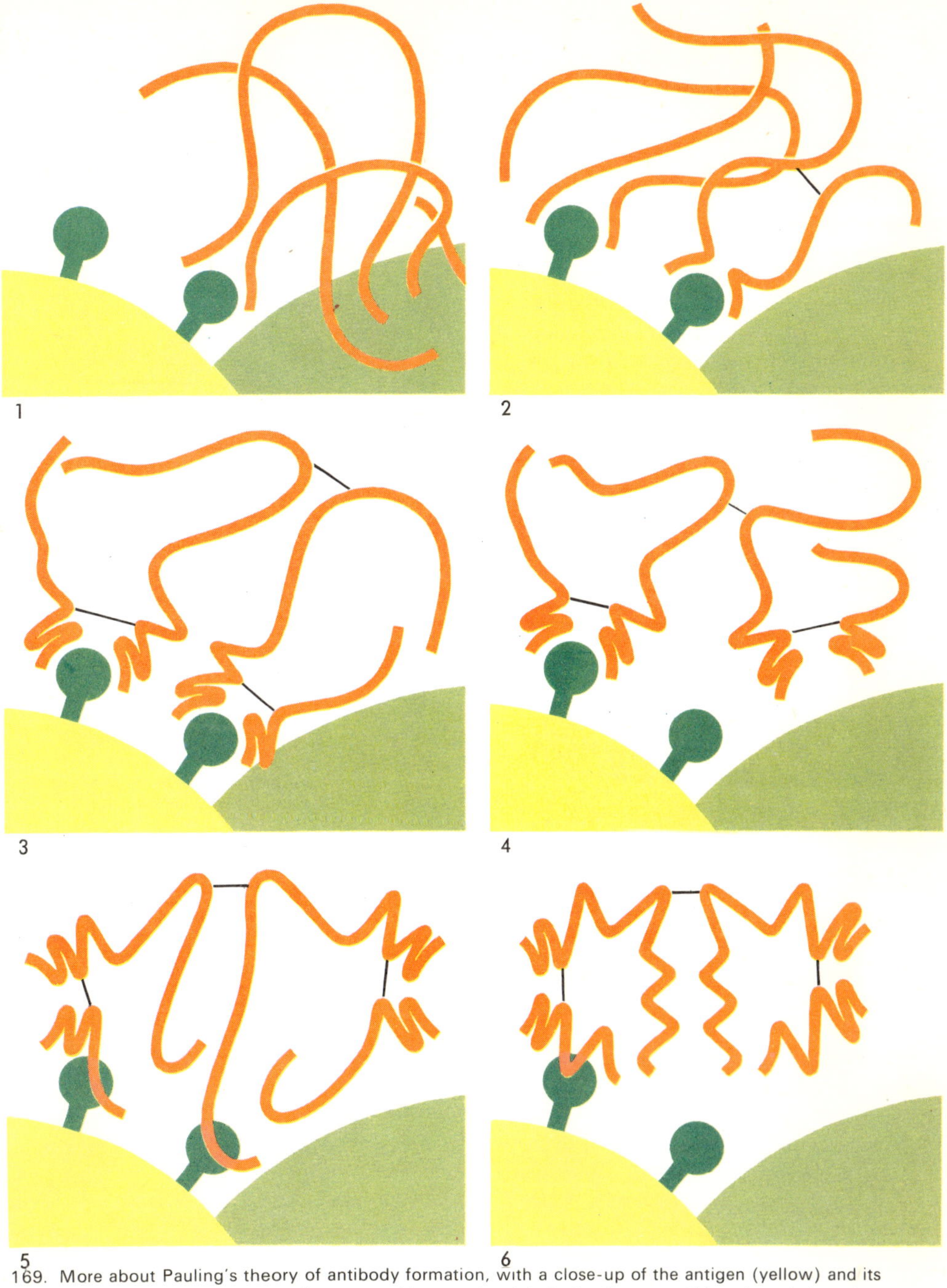

169. More about Pauling's theory of antibody formation, with a close-up of the antigen (yellow) and its determinant groups (green). The olive green object on the right is the ribosome. This enlarged view shows that the antibody is composed of four parts, which are gradually folded onto the determinant groups in a specific way. (It is not strictly necessary to have *two* identical determinant groups, as shown in this diagram.)

not yet folded, but probably they already have a helical structure (1). The free chains come alongside the determinant groups of the antigens, and fold themselves complementary to the attachments, while the middle section at first remains unfolded (2, 3). Then first one chain and then the other detaches itself from the antigen (4, 5): at first this seems unlikely, for antibody attachments and antigen determinant groups should immediately combine to form a precipitate; all the same, the codon of the m-RNA which exactly matches the codogen of the DNA *must* be detached if the m-RNA is to migrate to the ribosome! (Cf. p. 51.)

Finally, the middle sections of the polypeptide chains also become folded, this time obviously including their own primary structure (6), and then the bivalent and four-part molecule of antibody is fully developed.

Thus the hypothesis of folding does not run one into insoluble difficulties if certain additional assumptions are permitted, but it must be admitted there is as little proof of it as there is of any of the dogma mentioned above.

Second theory : plasma cells are clairvoyant. The 'instructive' theory demanded some way of getting round the dogma that DNA information alone was in control – at any rate, to the extent that the antigen might convey new information for the folding of the globulins. The 'clone-selection' theory adopted as its basis the following extreme postulate: *all* information for *all* antibodies, against *every conceivable* antigen, must be present in the organism from the beginning. The immunity system operates over a *very large number* of cells, each of which understands how to make one kind antibody. Each is therefore 'potent,' but until the infection has taken place it is inactive. Its 'hereditary' antibody structure must lie, not in the interior of the cell, but on its surface, and if an antigen now enters into the bloodstream, it must actually come up against this surface structure. This, however, would be the moment when the macrophages were released, which consume the antigens, and pick out the right cell by matching its antibody structure (hence the 'clone-*selection*' theory). Their arrival gives the first indication to the immunologically competent cells that a cell division is to be set in progress, and so there comes into being – through mitosis – a clone (see p. 78), a small group of related individual cells (hence the '*clone*-selection' theory). These now in their turn mature into plasma cells, manufacture antibodies, and give these

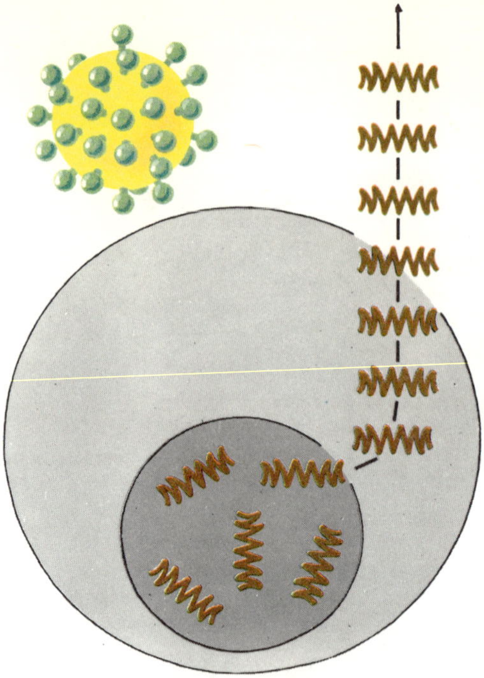

170. Diagram to illustrate the 'clone-selection'-hypothesis. The leucocytes with antigens locate only the appropriate plasma cells, and stimulate them to produce the antibodies (brown symbols), while the antigens themselves (yellow and green, above) remain outside the plasma cells.

up into the serum – and all without the antigen itself needing to penetrate one of these cells.

This theory has many advantages. It does not, for example, challenge the 'dogma' of DNA dominance: the entire sequence of amino acids in the antibody, as well as the spatial folding of the tertiary structure, and even the union of two light and two heavy chains to the complex of 160,000 (the quaternary structure), is controlled *genetically*, while the antigen functions merely as the releasing agent.

This can be expressed in a slightly more modern terminology, such as we learned in connection with the regulation of (enzyme) protein synthesis on p. 205. In all the cells that are capable of producing antibodies the genetic information is blocked before infection occurs, inhibited through a *repressor*. The antigen cancels this inhibition ('derepresses' it), and consequently operates in the same way as the induction of the enzyme synthesis, as an inductor in fact – working here on the antibody system and inactivating the repressor.

A second advantage is that the theory explains in a very plausible way how the rapid multiplication of the plasmablasts, which is always observed, actually comes about: by clone formation. Admittedly it still remains obscure how the antigen, or the antigen-containing phagocyte, is able to stimulate a plasmablast into dividing.

There is something else, too, that needs to be clarified. Since they were derived from each other by mitosis, all cells of a diploid organism contain the same genetic information, and thus they are not by any means immunologically 'competent.' Where do the mesenchyme cells, or plasmablasts (which are innumerable, and from time to time conditioned to different antibodies), get their additional information? It must be assumed that before birth (we shall come back to that later), cells with new characteristics arose through mutation, and remain preserved. In this way there came into existence an extensive population of potent (competent) cells, from which the antigens then picked out the matching individuals. This is perfectly possible, but any kind of evidence for it is still lacking.

The clone theory has its difficulties too. One objection has already been raised. It is really quite unlikely that an organism should hold in readiness many thousands, or even hundreds of thousands, of different cells against antigens that are only likely to come fortuitously (we have seen that just about any chemical substance *might* become an antigen). This becomes even more improbable if we take into consideration such 'unnatural' antigens as the azoprotein mentioned on p. 258. Such azo-compounds are completely synthetic, and were first put together by a chemist.

The same is true of dinitrophenyl groups, of azophenylarsenate groups, and of quaternary ammonium bases – to mention just a few substances which are notable both for their complicated structure and their nomenclature. They are all capable of stimulating the production of antibodies. Is it really reasonable to suppose that living organisms clairvoyantly anticipated which new compounds the chemists of the nineteenth and twentieth centuries would offer to them, and so were in a position to have the appropriate anti-cells in readiness, having carried out additional mutations for this purpose?

If the second theory holds good, then we can only hope that the ability of the plasma cells to foresee the future extends to all the substances that will be brought to earth by space travel from strange heavenly bodies ('extra-territorial' antigens!) – otherwise epidemics will arise against which we shall be helpless.

A further argument against this theory is based upon the results of new research. If the cells that are potent to form antibodies arose at one time by mutation, then on grounds of probability it must be reckoned that one cell will mutate only towards the formation of one kind of antibody, and that it will not mutate equally towards the production of several. In the same way a clone of plasma cells will also be able to synthesize *only one kind of antibody* – i.e., it will be *unipotent*. There are cases, however, in which a cell can demonstrably produce two different antibodies. Indeed, when an experiment was made with four different antigens, it appeared that the cells of one clone could actually evoke four different antibodies! This certainly reduces the plausibility of the clone-selection theory considerably, and the author of the theory has actually admitted this.

The two theories are completely opposed, and appear to be incompatible with each other. Either:

a. The entire information for the antibodies is already present and complete before the infection, and therefore has a phylogenetic origin, and afterwards needs only to be sought out ('selective' theory), or

b. The information in its entirety originates from the antigen, which tells the plasma cells what they have to do ('instructive' theory, 'template' hypothesis).

Perhaps this presentation of the two theories as being mutually exclusive is an exaggeration? During a colloquium on the chemistry of immunology of the Gesellschaft für physiologische Chemie, Professor Eigen suggested that the difference between selective and instructive theories might possibly be a question of relative level. At the lowest, molecular, level, undoubtedly there is only selectivity, i.e., making a choice from among things that are already present. Moreover, the development of a polypeptide chain from activated amino acids is based in the last resort upon a selection mechanism, since the molecules of t-RNA that are loaded with the 'correct' amino acid are picked out. But the building up of certain 'precoded' sequences of amino acids into higher units can proceed in an instructive way throughout, i.e., according to the instructions of the antigen (probably including the uniting of light and heavy chains into the complete antibody?). Then we have before us something like a 'semi-instructive' mechanism.

But we must break off at this point, and postpone to the future any further clarification of this topic; no doubt it will not be left long in its present state. Without declaring for one theory or the other – for which we should have to produce some justification – we have gained a new insight into several points of molecular biology and cell physiology; we still do not understand completely the wherefore of active immunization; but we know more than we did at one time. We have observed how a weak infection, a mere inoculation, causes the body to produce a few antibodies. This production comes to a stop if all the antigens are mopped up, or precipitated, but immunologically competent cells retain the power of making antibodies of this sort; if a new infection occurs they 'remind themselves,' and immediately begin a new production of antibodies. For this reason, they are called memory cells; nowadays there is much discussion about whether these have any relation to the 'memory molecules' of the previous chapter.

Nevertheless, this reaction between antigen and antibodies, which as a rule is carried on in the serum, is at best only one aspect of the biology of immunity, and probably not the most important. In the eyes of many research workers the job of rendering foreign substances harmless to the body is only the rough hackwork; what really appeals to the scientist is the other challenge, to find out how biological individuality is maintained. What do we mean by that?

6.08 Self or Non-self?

Biological Individuality, and Ailments Caused by Auto-immunity

We will approach the problem by asking a question, which is so obvious that we might simply have overlooked it up to now. When the body is reacting immunologically, how does it actually know how to recognize an intruder; how does it distinguish between 'self' and 'non-self'?

That the organism can make this distinction has been accepted without question for a long time, as a necessary ability. Since *every* chemical substance, if it gets the chance, is able to act as an antigen, or at least as a determinant group, this would include all the body's own constituents. If the body could not distinguish these, then it would produce antibodies against its own tissues, and end up by poisoning itself.

Naturally, attempts have already been made to find such 'auto-antibodies' (Greek *autos*, self), and to stimulate their production, but they were not successful. Paul Ehrlich, one of the old masters of immuno-chemistry experimented with goats, immunizing them against the red blood corpuscles of other goats. This worked well. The blood corpuscles used in the immunization acted as antigens, and antibodies against them were produced, but the goats never reacted thus against their own blood corpuscles. An unexpected result of this work was the discovery that goats, like men, have blood groups (but we shall not go into that here), but no auto-antibodies were demonstrated. Instead, Ehrlich invented for this process the much-quoted expression *horror autoxicus*, or fear of self-poisoning, but this is no explanation, only a postulate, or at best, a dogma.

Today we know that nevertheless an auto-immune reaction does exist, and even a whole series of them, and that these – according to expectation – may lead to *auto-immunity sickness*. Indeed it is much more correct to say that there exists a whole range of ailments which can be regarded as arising from auto-immunity. To mention just a few:

1. Multiple sclerosis, widely known and feared; this comes into being through auto-immunity reactions – or these participate in causing it – against substances from the central nervous system.
2. Various heart ailments; these develop when the body produces antibodies against its own heart musculature.
3. Myasthenia gravis, or muscular dystrophy, caused by antibodies formed against the protein of the skeletal muscles.
4. Sympathetic ophthalmia; this will be discussed separately very shortly.

Let us consider the last two. In myasthenia gravis, or muscular dystrophy, a particularly insidious disease, the auto-antibodies can generally be made visible. Serum with antibodies is extracted from a patient suffering from the disease, and injected into several rabbits. As might be expected, the injected antibodies act as antigens to the rabbits, which produce their own antibodies against them. Furthermore, these 'anti-antibodies' can be extracted from the rabbits' serum, purified, and 'marked,' not, this time with a radioactive substance, but with a fluorescent dye – that is, with a dye which when illuminated with ultra-violet light, invisible to the human eye, gives out visible light which can be seen against a dark background. Such a dye is acridine-orange.

Fluorescent anti-antibodies prepared in this way are now allowed to act on a diseased muscle, which has its own tissue-bound antibodies, and only to these can the fluorescent anti-antibodies attach themselves. All the anti-antibodies that are not attached are washed out of the preparation, and under ultra-violet light the areas of muscle fluoresce where the (auto) antibodies are located.

Cases of sympathetic ophthalmia are particularly tragic. 'Ophthalmia' as used in its normal sense refers to an infection of the eye, but 'sympathetic' should be understood not in its everyday meaning, but in its original, literal interpretation of *syn*, together, and *pathein*, to suffer. If part of *one eye*, such as the lens, or the iris, is injured or diseased, eye proteins are released into the bloodstream, where they are not recognized as being 'native,' and so the immunity mechanism goes into action and produces antibodies – and these antibodies spread to the lens and iris of the *other*, intact eye, and destroy it bit by bit.

Here, just as in muscular dystrophy and the other auto-immunity reactions listed above, the distinction between 'self' and 'not-self' has not operated; the body's own constituents have been regarded as extraneous matter, and treated accordingly. How does this come about?

First of all, here is a remarkable thing. If some of the body's own material is treated as 'foreign,' then it must first have found its way into the blood system. It is 'new' to the blood, so to speak, and this means that it ranks as 'foreign' to the blood system, and it is *from this source* that the antibodies arise. We shall return to this extremely important point presently.

A further piece of evidence that helps to clarify this phenomenon arises from systematic experiments which were made in an attempt to find out which of the body's own constituents can provoke the formation of antibodies (a process also called auto-sensibilization). The vigour with which a bodily substance can act as an antigen increases with the distance by which it is physically separated from the immunity apparatus of the body; the immunity apparatus is not restricted to the serum, or even to the blood system, but includes all the organs and tissues that were listed on page 253, from the thymus down to the lymph vesicles and the leucocytes. It is this 'immunologically competent' apparatus that decides what is 'self' and what is 'not-self.'

Now we understand why Ehrlich's attempts to find auto-antibodies which were directed against the animal's own blood cells were bound to be fruitless: the red blood cells are present in the blood system from the beginning, which means not merely from birth onwards, but even during prenatal life in the uterus. They have already been there 'from time immemorial,' and for this reason are not treated as strangers.

Further experiments give general support to this view. If the immunity apparatus (though *not* the blood system!) of a rabbit is slightly impaired by surgically removing the thymus and vermiform appendix at the time of birth, then often auto-antibodies are produced against the body's own red blood corpuscles, while in other cases the formation of antibodies does not take place. In cases of myasthenia gravis, or muscular dystrophy, which is caused by auto-antibodies, we often find tumours in the thymus: loss of ability by the body to recognize its own tissues follows as a normal consequence of destruction of the immunity apparatus.

All this still does not tell us even now *how* the distinction between 'self' and 'not-self' is drawn, but we are beginning to understand reasons why the immunity apparatus sometimes fails to distinguish these, and indeed is bound to fail. We need to study more cases, and to get new angles on the subject. First, though, to finish off the subject of diseases of auto-immunity. The facts that we have learned so far make it possible to indicate the way in which these diseases – often so dangerous, because nothing can be done about them – can be tackled. Either we can try to restore the immunity apparatus to normal, or we can try to prevent a consequential loss of ability to distinguish the body's own tissues, by inhibiting the production of the auto-antibodies. Cortisone, in spite of its many questionable side effects, has given very valuable service in this respect.

6.09 When 'Self' has no Meaning

To the Newly Born There Are No Antigens

Having accepted the fact that a body can treat its own substance as foreign matter, and attack it with auto-antibodies, we must now mention the opposite aspect of immunity, so-called immunity paralysis. This is the name given to the phenomenon in which a very large quantity of an antigen can cut off the entire production of antibodies, *even against this very antigen.*

What goes on is still completely unknown, but immunity paralysis on the one hand and the formation of auto-antibodies on the other show very clearly that the boundary between 'self' and 'not self' cannot always be drawn very clearly. The great flexibility of this boundary first became obvious about ten years ago, when the first discoveries were made concerning immunity tolerance, and it was discovered that self-consciousness, in the immunological sense, is never inborn, but is always acquired. Immediately after birth – and of course earlier, before birth – the organism is unable to distinguish between what is 'self' and what is 'not self'; everything that appears in its bloodstream is looked upon as being part of itself. All the substances, all thc chemical compounds, that in later life can act as antigens, at this early stage are accepted and tolerated. They are therefore called *tolerogens.*

The fetus is immunity tolerant while it is still within the mother's body, in the uterus. An experimental animal at this stage can be injected with tetanus antigen, or with rabbit protein, and then it will remain immunity tolerant throughout its life against these substances, and later infections with tetanus toxin or rabbit protein provoke no reaction.

What at first glance appears to be a remarkable breach of the immunity principle (which is aimed at the preservation of biological individuality) turns out to be basically an extremely sensible adjustment. We only have to bear in mind that the fertilized ovum in the uterus contains both a maternal and a paternal genome, and these direct the production of both maternal and paternal protein. If the object of sexuality, namely the recombination of hereditary elements, is not to be jeopardized, then the two parental contributions to the embryo, however different they may be immunologically, must agree with each other; they must be mutually tolerant.

Quite obviously, in the embryo the production plans for antibodies must be in abeyance from the beginning onwards – maybe this comes about in much the same way as the total repression of the antibody gene in the plasma cells, that is required by the clone-selection theory.

Now the intra-uterine embryo at first is linked with the blood system of the mother, by way of the umbilicus and the placenta, and by this route 'paternal' proteins might pass from the embryo into the adult, intolerant tissues of the mother, and there provoke antibodies. This is prevented by a barrier, which allows all the nutritive substances to pass, but not antibodies. There are, nevertheless, a few exceptions. Soon after birth the Rhesus blood-group factor (Rh factor) can pass from the fetus into the bloodstream of the mother, and in the opposite direction the barrier is open to the syphilis antigen. In addition, this barrier itself is without any immunologically competent cells, and can be thought of as a sort of 'demilitarized zone,' which prevents any exchange of antigens, and thereby any unnecessary formation of antibodies.

The change-over from the immunity tolerance of the fetus to the fully developed immunity reactions of the small child is certainly not a sudden effect, but it corresponds approximately with the time of birth. There are even a number of indications that this change forms part of the onset of parturition. As soon as the fetus begins to abandon its immunity tolerance, to relax the ban on the production of antibodies, and to put its personal immunity mechanism into operation, as soon as the first traces of antibodies appear, then the maternal organism recognizes the fetus as a stranger, and expels it. In addition the immunity barrier is broken down, of course, just as we saw earlier with the Rh factor.

One further qualification must be made. From what has been said it might be assumed that the immunity mechanism first comes into existence at or after the time of birth, but this seems not to be the case. Immunity tolerance may be under the control of the lymphatic system. Certainly all the substances belonging to the body, all 'autoantigens' which come into contact soon enough with immunologically competent tissue, are recognized in later life as being part of the animal's 'self' and are therefore tolerated. However, not all substances, by a long way, that are foreign to the organism, and which are injected into the uterus, will lead to immunological tolerance; on the contrary, their number is comparatively limited. Which will be tolerated, and which not, is obviously controlled by the immunity mechanism from a very early stage of the organism.

And now for a further characteristic of the immunity mechanism, which came to notice in certain tropical fishes. One of these, *Cynolebias adolfii*, a South American fish is comparatively indifferent to the temperature of the surrounding water, and can live as well at 30°C as at 15°C. At 30°C it shows normal immunity reactions, but at the lower temperature (15°C) immunity tolerance is produced; the immunity mechanism is blocked!

At first sight, this immunity tolerance seemed interesting but without much significance, but it soon had sensational implications for the problem of the transplantation of tissues, which up to that time had been very much a matter of luck.

6.10 Transplantation of Tissues and Organs

How Dangerous Immunity Reactions Can Be Avoided

At the present time, with its high accident rate, it often happens that tissues or organs are seriously damaged. Burns may destroy large areas of skin, and internal injuries may involve, say, a crushed kidney that no longer functions. The need then arises to replace the missing skin, to implant a new kidney, and

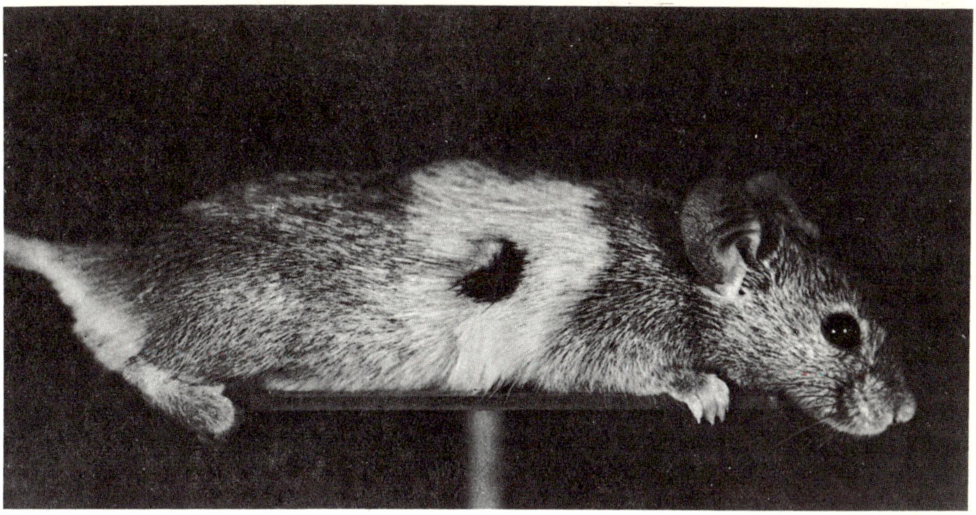

171. A double transplantation of tissue. An immunity-tolerant grey mouse has been implanted with a piece of skin from a white-haired mouse, and inside this graft is yet another graft, of black-haired skin. Both grafts were tolerated.

so on. It has been known for a long time that such operations are normally successful only if the grafted piece of skin or new organ comes from the damaged organism itself, and is thus immunologically identical. These operations almost always succeed between identical twins, which have the same genetic constitution, but brothers and sisters are already sufficiently different for the experiment to fail. The graft may even take comparatively easily at first, but after a few weeks, and often after only a few days, it is rejected.

The donated tissue releases foreign substances into the bloodstream of the recipient organism, where they are recognized as strangers and treated as antigens; an antibody reaction is provoked, and destroys not only the antigens themselves, but also, bit by bit, the implanted cells from which the antigens arise.

This is not the case if use is made of the period of immunity tolerance. Fish frozen while they are still immunity tolerant will accept grafts from others of their kind without difficulty, whereas their companions reared under warm conditions reject every graft. The mammalian fetus and the newly born behave in a similar way, and no difficulty arises if they are implanted with a graft from a fully grown organism. The graft persists throughout the life of the host, i.e., of the recipient individual, which even tolerates the antigen of the graft tissue, accepts it as being 'self,' and so does not develop any antibodies against it. This is assuming, of course, that the graft and the host do not belong to widely different groups of animals.

The experiment can be made even more elegant by injecting into the fetus a skin homogenate (see p. 44) of the animal that is to be regarded as being the donor of the graft. Much later on, when the host animal has fully developed its immunity apparatus, a transplanting of a piece of skin can be carried out, and will heal well because following the injection, the antigens of the skin graft behave as tolerogens (see above).

Birds make equally good subjects for experiments. Hens' eggs are taken – say from one of the white breeds; the eggs are opened at the place where the air bubble lies, and an injection is made into one of the soft envelopes. If the injection has been an homogenate from a black race of poultry, then either the cocks or the hens of the white host can be implanted with a skin graft from one of the same black donors, and it will take successfully, and grow an outcrop of black feathers.

Immunity tolerance acquired in this way is not confined to skin tissues. The graft of skin will heal just as well if the homogenate consists not of skin, but of liver or kidney cells, as long as the tissue of the graft comes from the same individual as the homogenate. The reverse process can be carried out, of course, using an injection of skin cells to make possible a subsequent graft of kidney tissue.

In principle, therefore, a newborn baby can be, so to speak, prepared for any transplantation of tissue that may later become necessary for it. Its parents only need to find for it a donor who (1) has some cells ready for injection, and (2) will, if necessary, later be ready to hand over a kidney, or an eye, or a heart . . .!

The long-term effects of a transplantation are denied to older patients who have not been made immunity tolerant by having just been born, and then it is necessary to use some artificial means to put out of action part or the whole of the immunity apparatus. This is possible with the help of X-rays, or – for bones – by freezing at very low temperatures, with the result that the immunologically competent cells are killed. If, now, bone marrow or cells from the spleen are injected, these are able to detach the dead cells, and to take their place. Nevertheless, one basic condition must be fulfilled; the cells to be injected must come from a *fetus*, since only these are immunity tolerant towards the host.

The result is in jeopardy if the injection is of immunologically competent cells from *adult* donors. If the relationship between donor and recipient is not close enough, then the host reacts – often weeks or months later – with immunity defenses. A complex of morbid symptoms appears ('secondary syndrome'), and eventually the recipient dies.

This way of preparing a protection for a subsequent transplantation is obviously too tricky for clinical application, but nevertheless it can be helpful if, for one reason or another, there is reason to fear that too much use of X-rays might be fatal. Provided that the X-rays are sufficient to destroy the immunity apparatus, they may then be replaced to some extent by an injection of fetal cells.

All the same, even with thorough preparation, a transplantation is not always successful. In certain circumstances, in particular if the graft contains immunologically competent cells, these may rebel against the host, in the sense that they produce antibodies against the host's antigens. By doing this the graft has assumed the dominant role, and continues to produce more and more antibodies against the tissues of its host. The result is that limitation of growth, anaemia, shortage of lymphocytes, and eventually wasting away of the whole immunity apparatus lead in the end to death. The German term for this affliction is *Zwergkrankheit*, and the English term is 'runt disease.'

These evil consequences can be avoided only if the implanted tissue can be guaranteed free from immunologically competent cells, and at present this can be attained only in exceptional circumstances. One such case may be mentioned, where bones are concerned. By freezing pieces of bone at very low temperatures, practically all the living cells can be killed, and definitely all those that are immunologically competent. All that is left is an 'empty' framework of bone. If this fragment is now implanted into a living bone, it is accepted willingly, and at once assimilated by the cells of the host. No immunity reaction takes place, and the graft heals very quickly. It is already the practice of many accident clinics to keep bone fragments of varying shapes and sizes (mostly chicken bones) ready to be used for grafts at any time.

Can Tolerance Be Overcome by Medical Means?

Of primarily theoretical interest is the question of whether an immunity tolerance that is experimentally induced can be either reversed or abolished. This is possible in practice; it is only necessary to take a tolerogen (such as a quite ordinary serum protein) and couple to it a new determinant group, as we have seen on p. 248. One of the twenty amino acids, tyrosine, is very effective for this purpose, and in the form of a tyrosil residue can be bonded to the amino groups of the tolerogen. The 'tyrosilized sero-protein,' also called a tolerogen conjugate, is then no longer a tolerogen, but has become an antigen, and antibodies are immediately produced against it.

This was to be expected, of course, particularly from what we know of determinant groups. On the other hand, it is astonishing to find that antibodies were formed not only against the tolerogen conjugate, but also *against the tolerogen itself,* the original carrier substance, the sero-protein. Obviously this may have the most important clinical consequences; we need only think of the variety of medicines that are poured into our blood. What happens if one of these is able to attach itself as a side chain to the natural tolerogens? The body's own sero-protein, for example, may be changed from a tolerogen to an antigen, and antibodies against it will immediately appear.

Ought we not in future to test every substance used in medicine before allowing it to be used, in case there is any chance of its acting as a destroyer of tolerance? In fact, up to the present such an effect has been very seldom observed, and for the following reason: if the tolerance-breaking substance is administered *together* with human sero-protein, tolerance is *not* broken. The additional tolerogens give an effective protection, though it is still not known how this comes about; probably the tolerance is preserved by immunologically competent *cells* and not by the serum system which has been responsible for 'doing the dirty work.' However, since in practice the serum proteins in the blood are always present in excess anyway, the danger of a breach of tolerance is remote. Of course, it would seem rather different if, for some reason or other, we actually wanted to break down tolerance.

Now we can turn back, with a rather fuller knowledge, and look once again at the subject of birth. As we had established, the changeover from the immunity tolerance of the fetus to the complete functioning of the immunity apparatus in the child or the adolescent takes place not suddenly, but gradually. To simplify things a little, if the mass or concentration of antibodies in the blood is taken as a measure of activity of the immunity apparatus, something like the following sequence is observed: immediately after birth antibodies are virtually absent; the protective reaction first comes into existence in about the second week of life, and then very slowly. The production of antibodies is still weak and insufficient, and it is not until the second year that the quantity of antibodies is 'adequate.' The greatest activity in forming antibodies is to be found up to the tenth year – and this is the time of life when, as a rule, children catch the so-called childish ailments, measles, scarlet fever, and so on. The intensity of immunity reactions falls off a little between ten and twenty years of age, remains at about the same level up to the age of forty, and finally declines at an ever increasing rate towards old age.

Two things about this curve of life are interesting, the slow start of antibody formation, and the loss of immunity with old age. Let us follow this sequence chronologically, starting with the earliest antibodies. For a long time it was thought that these were 'inherited,' but this idea cannot be maintained in view of the fact that, by definition, *anti*-bodies can only appear in response to the presence of foreign matter, i.e., of antigens.

6.11 Immunity in Youth

The Germ-free Breeding of Animals Is Expensive but Necessary

In an attempt to understand this in more detail, we may try to breed animals that are 'germ free,' that is, without antigens. This is easier said than done. Certainly the fetus is germ free as long as it remains inside the body of its mother, but the process of birth brings it into contact with a variety of new substances, both from the maternal body and from outside – and every chemical substance is potentially an antigen (see p. 246). Now, it is already possible to arrange that birth shall take place under germ-free conditions, and in a very few places throughout the world expensive apparatus has been developed for this purpose. But this still does not solve the problem, for the following reasons:

1. The newborn animal must be fed, and food materials may contain antigens.
2. An experimental animal comes in contact with its own excrement, which may provide a range of antigens.
3. Every experimental environment contains dust particles, which cannot be avoided by the customary bacteriological 'sterilization.'
4. An experimental animal may swallow its own sloughings of tissue, hairs, etc., whenever it can, and these may act as antigens ('auto-immune reaction').

It follows from all this that an absolutely germ-free animal probably does not exist, and for the time being, we must be satisfied with an animal that is free from any obvious infection. Using experimental material of this kind it has already been possible to demonstrate that the apparently 'spontaneously' generated antibodies that first appear are not inherited, but are acquired at a very early stage as a result of immunity reactions that are going on even then.

Considering what has been expended on elaborate and costly apparatus, these more or less predictable results seem rather meager; yet beyond them opens up a whole field of research that is still virtually unexplored, but which will certainly lead to new and highly significant discoveries. To give just a hint of this, let us consider two experiments with 'germ-free' guinea-pigs.

Vibrio cholerae, the causative agent of cholera (cf. p. 134), affects only man, and not other animals. If, however, guinea-pigs are bred under germ-free conditions (we will not go into how this is done), then after infection with *Vibrio cholerae* they immediately fall ill, with all the typical symptoms of cholera; they die after six to nine days. This happens just the same if one additional kind of bacterium is injected into them at the same time as the *V. cholerae*. The skin of guinea-pigs contains a whole flora of different kinds of bacteria, and against this mixed bag the cholera bacillus cannot prevail.

There is a kind of 'reverse side' to this experiment. *Entamoeba histolytica* is the name for the causatic agent of amoebic dysentery. This time we are dealing, not with a bacterium, but with an amoeba, a unicellular animal without a firm cell wall. It changes its shape, and propels itself along with the help of protoplasmic extrusions, the so-called pseudopodia.

Amoebic dysentery affects man, monkeys, and domestic animals – and also guinea-pigs. *In a germ-free animal, however,* Entamoeba *is completely powerless by itself*. Only if the animal is injected with a harmless bacterium at the same time does it become critically ill.

From this it is easy to see that the expensive method of rearing under germ-free conditions does not justify itself. On the other hand, this method is presumably the only way in which we can possibly elucidate the mysterious transition between the 'resident' bacteria that are tolerated, and the immigrant ones that become antigens. Perhaps this line of study will help to answer the question of how an antigen that can cause death in one strain of animals is not only tolerated but is not even poisonous in the blood of another.

6.12 Immunity and Old Age

Estrangement of Cells, and Parabiosis

Now let us look at the other end of the curve of life, old age. From time immemorial mankind – and this includes scientists too – has been intensely interested in the incomprehensible laws of old age and death. Hypothesis after hypothesis has been formulated, rejected, improved upon, turned into a generalization, cut down again – and we are still left as far as ever from a final understanding of these phenomena. We still do not even know whether old age is an illness, a natural process of wearing out, or a gradual extinction. The immunologists would have been doing themselves an injustice if they, too, had not tried to contribute something to the understanding of the process of aging.

Let us start with the single cell. In the body of a vertebrate the life of a single cell is relatively insignificant; at any rate, it is very much shorter than the life of the whole organism; in many cases it is only a matter of days. Since the immunity apparatus of the skin has to maintain the individuality of the entire organism, and to protect it against 'strange' cells, it can easily be used to separate old and damaged cells. The idea that these may be classified as 'strangers' may seem surprising at first, but it is by no means improbable. We have already established that the determinant groups of an antigen cell, by which this is recognized as an antigen, lie entirely on the outside of the molecule. More about the surface structure will be remembered from the pneumococci and salmonella polysaccharides (shown in Figs. 160, 161), which play a great part in the selection of the corresponding plasmablasts (see p. 260, clone-selection hypothesis). It is now suspected that in old age a normal body cell loses its outermost structures bit by bit, developing, so to speak, 'bare spots,' where the underlying layers are exposed. These layers, however, are 'new' and are treated as 'foreign'; accordingly, they are attacked with antibodies.

Apparently the antibodies are sufficient to

render such cells harmless as antigens, but not actively to dissolve them. To do this requires another factor, called the *complement,* and to fit this into the picture, let us repeat Table 5, with the addition of a fourth line.

Table 6

Reactions between Antigens and Antibodies

1. Free antigen + antibodies (anti-serum) → precipitation
2. Cell-bound antigen + antibodies (anti-serum) → agglutination
3. Antigen bound to blood corpuscles + antibodies (anti-serum) → hemagglutination
4. Cell-bound antigen + antibodies + complement → cytolysis (cell dissolved).

For a long time the concept of complement was regarded with the gravest suspicions: many immunologists denied it any real existence and at most were willing to concede that it might be a substance that participated under certain conditions. It is only recently that it became certain that the complement actually exists as a substance, admittedly in extremely small amounts; that is why it could not be detected with older research techniques.

On the other hand, several complement *fractions* were already known and eventually it even proved possible to make the operation of the complement visible under the electron microscope. In the cell walls of red blood corpuscles, after treatment with complement, small holes were visible; it is fairly certain that these holes were 'shot into' the cell by the complement (in a way similar to that in which bacteriophages introduce their DNA into the body of the bacterium; see p. 110). It is through these holes that the cell contents flow out until the cell is completely dissolved away.

It can no longer be held in any doubt that immunological processes play a decisive part in the expulsion of cells that have become old or otherwise undesirable, but of course this tells us nothing about the causes of aging itself. Very recently quite new and very bold ideas have been developed in this matter, based upon experiments in the so-called *parabiosis*. Two animals, such as guinea-pigs, can live in parabiosis with one another if their blood systems are joined together functionally; they then have a common blood system. Except when the parties joined together are identical twins, the life of such an artificial parabiotic organism is naturally very short, since the two partners are immunologically different individuals, and as a result, develop antibodies against each other. The parabiotic system breaks down altogether; first one partner dies (because its 'colleague' is a little quicker in producing antibodies); then, later, the second individual also dies. The outcome is a perfect example of the rejection of a graft by a non-tolerant host.

If the partners to be joined in this way have been inbred through thirty or forty generations, they resemble each other more closely (and are more immunologically similar), and then a parabiological system will survive longer, eventually almost attaining the normal life span of the species. This system also breaks down in the end, with first one partner dying and then the other – again because the protective reaction can no longer be restrained.

The immunological theory of aging states no more and no less than that every organism is an individual *that is living in parabiosis with itself*. Aging is unavoidable, because the boundary between what is 'self' and what is 'not self' becomes more and more blurred, until finally antibodies against the body's own structures begin to appear. Certainly still deeper analyses are needed before this theory can be regarded as proved, and further research before there can be any idea of making use of this theory in the therapy of gerontology. (In ancient Greece the wise old men were called *geronts*, from *geron*, grey; gerontology is thus the science of aging.)

One thing is correctly brought out by the theory of parabiosis. The immunological situation in which a person or an animal finds itself is a matter of luck, and *indeed individual luck*. What drama lies in the phenomena (though not in the undramatic terms) that we have learned to speak of as immune reactions in the serum, and in the cells, as partial and complete immunity-tolerance, as a breakdown of tolerance, and as immunity paralysis! What a tremendous number of different influences there are to which a single individual may be exposed during its development in its mother's womb, and during its own life and ageing. However much one tries to keep conditions exactly the same, it is probably quite impossible ever to 'condition' two individuals equally.

It is a property of chance to be irrevocable, and chance when related to immunology cannot be otherwise. If cells producing either antibodies, or memory cells, have once arisen, then they remain in existence; they might be destroyed by intensive radiation, but this would be a very risky attempt.

6.13 Is Cancer a Problem in Immunobiology?

The Origin of Life, and Its Development

Having considered the subject of aging and death, we might have brought this chapter on immunology to a close, but the reader has probably read recently about clinical experiments in an attempt to cure cancer by implanting immunologically competent cells. This invites the question of whether cancer is to be understood as an immuno-biological problem and to be treated accordingly.

This is a good question, but to answer it is very difficult – difficult not so much technically as because under the most favourable conditions, only a partial answer can be given. The problem of cancer, which is so tremendously important to all of us, has many sides and many aspects. We all know that thousands of laboratories and clinics all over the world, with an annual budget beyond computing, are working on this problem; that many different approaches are being tried, and other, new ones constantly being searched for; and that innumerable separate pieces of work are being co-ordinated. Yet not even a very large team of scientists can examine the problem from every angle, to say nothing of all the specialized literature – so how can one individual bring it all together into a book?

The terrible thing about cancer, and the thing that is still incomprehensible, is that normal cells, after behaving normally through thousands of physical, chemical, physiological changes, suddenly react differently, and *always in the same manner*. They change their nature and become cancer cells; their metabolism, growth, and multiplication are all freed from restriction, and go out of control. They overrun the normal cells, and so eventually bring about the death of the individual.

So far we do not even know what the 'de-restriction' of the cells entails. This much only seems to be clear: although the de-restriction appears to be completely uniform, and final, it is in fact brought about by *a variety of different factors in the cell*. These may include changes in the genetic information, through mutation (e.g., after exposure to radioactivity), or disturbance of the normal regulation of metabolism outside the cell nucleus, in the cytoplasm, or many other causes. The whole story cannot be told here.

From the standpoint of immuno-biology the question can be simplified, but we must bear in mind that even in its simplified form the problem is still very difficult, scarcely to be answered as yet. For a start, let us leave aside the question of how the cancer cells come into being, and ask instead, How does the organism – that is, its immunity apparatus – react with the cancer cells when they are present? Are they classified as 'self' or as 'non-self'? Are they recognized as a threat?

At present, by the time a doctor diagnoses a case of cancer, the patient is generally already inundated with millions of cancer cells; the time is near when the patient's body will become tolerant of cancer cells (of all kinds, or only some?), which will no longer be recognized as 'foreign.' The appropriate treatment, suggested by several Russian scientists, would then be to try *to break the immunity tolerance* – for example, by transmitting immunologically competent cells from an immune organism.

The problem can also be seen differently. During a case of blood poisoning millions of bacteria or antigens are present simultaneously in the blood, yet this does not make anyone doubt that the immunity apparatus is still present and functional: it is merely overwhelmed. May it not be the same in the case of cancer, and may it not be that we only recognize the disease in those cases where the immunity apparatus, fully functional in itself, has been swamped?

Degenerate cells, cancer cells, tumour cells, and so on, probably always exist in the body, as a result of mutation, defective metabolism, or some other cause, and they may be widely distributed. They often have surface structures different from those of normal cells, and hence they are normally recognized at once, as soon as they appear, as being 'not self,' and are eliminated by the immunity apparatus. Only if they are too numerous, or of too many different kinds, do they penetrate the immunity defenses and overwhelm the body.

It may also be that not every kind of degenerate cell has its outer surface structure changed, producing antigens or determinant groups. All the same, attempts will be made to analyze the surface structure of cancer cells as precisely as possible, in tissue cultures, to investigate their determinant groups and serotypes, to try to understand their capacity for variation, and whether this is brought about genetically or not, to study their effects, and generally to do everything relevant to the immuno-chemistry and immuno-biology of tumour cells.

Little though we know about cancer cells, and vague though the answers are to the questions that we have asked above, it is nevertheless possible to say that immuno-biology is involved, and that this science will be able to make its contribution to the fight against malignant tumours. This will be only one contribution among many, but one which has already been shown to be important and worthwhile.

7. THE PRIMEVAL BROTH AND THE FUTURE OF MAN

And God made the animals upon the Earth, each one after its own kind, the cow after its kind, and various worms on the Earth after their kind. Genesis 1, 25.

The title of the chapter has a scientific, 'materialistic,' practical, and sensible ring. What, then, is the point of this quotation? Is the word of the Bible being exposed to the derision of the scientist? Nothing is further from our intention. We are not in any sense setting the story of the Creation and science (or molecular biology) *in opposition* to each other, but side by side. By doing this we are merely demonstrating the immense length of time during which the human mind and spirit have been concerned with the greatest of all mysteries: the origin of living creatures. *Here* is the unique, unrepeatable act of creation, which brought forth all living things 'after their kind'; *there* is the primitive ocean, in which molecules came together, partly by chance, partly from inherent causes, to form the first primitive cells, cells from which, bit by bit, the highest living things developed. Within this framework there is room enough for every human effort at understanding, as befits the immensity of the problem.

There is no doubt that science and biology lie closer to the mechanistic way of thinking than do faith and poetry – it is no accident that in the preceding chapters we have continually been meeting molecules and mechanisms. But there is also no doubt that this statement is often misunderstood. It is by no means the aim and object of science to prove that every subject and every process is its proper concern, that the world, and therefore the phenomenon of life, is to be reduced to chemical and physical laws exclusively (though many scientists assume this). The object of science is to *investigate to what extent* things and processes follow physical and chemical laws, to test whether these things can be explained by scientific laws without leaving any unexplained residue, and to determine how far the laws as they are already known must be extended or altered – all under the password 'causal relationship,' and the assumption that cause leads to effect.

This does *not* exclude the possibility that there may be other ways of understanding things. The believer, the poet, the prophet may each in his different way discover relationships that are not based on causality, and we must respect them. They cannot in any case be set up in opposition to scientific research, since scientific methods cannot be applied to these other forms of experience.

On the other hand, respect must also be paid to what science has discovered about causal connections, and respect is due when science has succeeded in throwing light upon phenomena that were previously obscure. It is sad but inevitable, however, that as a result the harmonious world-picture of the man of faith will be more and more eroded away. But it is not the scientist's fault that it is not possible to replace the old harmonies with something that will adequately take their place.

7.01 Spontaneous Generation ?

Maggots Arise from Rotting Meat, Mice from Damp Earth

Let us go back again to the 'alternative' explanation, that living things were created out of the night in one, unique, divine act of creation – or else that living creatures came into being, either by chance or inexorably, from non-living, inorganic material, that *they produced themselves*. The idea that these two processes are alternatives is actually a very modern one, unknown in earlier periods, up to the end of the Middle Ages, and even further into modern times. Everyday observation showed that maggots arose out of rotting meat; worms, fleas, and beetles out of straw and dust; and from stagnant hay and water, infusoria and larvae of midges. Such experiences showed – to all appearances – that the original divine creation was supplemented by a continuous 'spontaneous generation' of living creatures from dead materials, so that at any time life was arising spontaneously and anew, by itself. Aristotle (384–322 BC), who was a philosopher and natural scientist, accepted the idea that plants had always existed on earth, but he thought that he had observed the generation of many animals.

Thus it seemed the normal thing for worms, larvae of bees and wasps, mites, glowworms, and other insects to be generated from dew, rotting dung, and mud, from dry wood, from sweat, and from meat. Every possible kind of intestinal worm arose from putrefying parts of the body or from excrement. Midges, flies, moths, butterflies, dungbeetles, fleas, bugs, and lice arose from the mud of springs, rivers, and lakes, from field humus, from moulds and dung, from decaying wood and fruit, from the droppings of animals, and from rubbish of every kind, from the dregs of vinegar and from old wool. Aristotle believed that not only insects and worms, but also other, more highly organized creatures could be spontaneously generated; for example, crabs and various molluscs from foul earth and mud; eels and many other fish from the mud of lakes, from sand, and from decaying water plants; even frogs and under certain circumstances also salamanders could generate themselves from liquid mud. Mice were believed to arise from moist earth. (Quoted from Oparin, *Die Entstehung des Lebens auf der Erde*).

These ideas had the backing of everyday observation, as well as the authority of the Greek teachers, and were not lacking in either support or criticism. The same thing applied to the view that the sun revolved round the earth! At the onset of the Renaissance period, about the year 1500, men became skeptical of appearances, and open to experiment, particularly to experiments that could be repeated and would give the same result each time.

The arguments put forward by Galileo, Copernicus, and Kepler were quickly accepted, and convinced people that the earth revolved round the sun in the same way as the planets. Yet whole generations of medical men and biologists struggled fruitlessly against the belief in spontaneous generation. Louis Pasteur (1822–95), with his irrefutable experiments, succeeded in winning this struggle, but only very slowly.

The reader will perhaps ask with some surprise why we have not glossed over the errors of the past, as we have always done so far in this book. Now it was certainly not a crude taste for the comic that caused us to dwell so long on the idea of spontaneous generation. It was, first, because the disproof of this idea by experiment led to the modern conception that life did come into existence at some time or other, but that since then it has never arisen again; that, on the contrary, a living creature can arise only from another living creature.

Today the question is framed thus, Did life arise out of the night, by a divine creation, or did it arise automatically ('mechanistically'), or by chance, from non-living matter? This, however, presents scientists with the problem of finding out how far it is possible, according to the laws of chemistry and physics, for an inorganic, non-living material to give rise to life, i.e., to a living organism.

7.02 Or Evolution?

First Traces of Unicellular Animals in Graphite and Corycium

This alternative is not as simple as it appears to be, because it immediately calls for a further decision: were all living creatures formed in the immeasurable variety that we see in the plant and animal worlds today, simultaneously, in one operation, or did they arise bit by bit – first quite simple organisms, and then, *by development from these*, ever more complicated ones? In other words, has an evolution of living creatures existed, and does it exist now?

If we think it over, it seems extremely unlikely that so highly differentiated an animal as a horse or a dolphin could arise by *chance* out of disorderly, non-living material. Appearances suggest such animals arose from a much more simply constructed ancestor, and that they have gradually become more highly developed. In this book we have implicitly assumed that the organisms of the present day have been derived from very much simpler ones, eventually indeed from unicellular animals, the most primitive of which did not even possess a real nucleus (see p. 72). Thus we have already anticipated the answer to this question: evolution has indeed taken place.

Since we know very well that appearances can lead to entirely false conclusions, we must surely be a little more cautious, and look around for some outright proof of the progress of evolution. The best thing would obviously be experimental proof, but if the experiments are going to require not merely a few decades, but millions and millions of years, then unfortunately we shall no longer be here to finish them. Then we must leave the experimental field, and take up palaeontology, though even in this field we shall find that a few questions require the use of modern science to answer them.

The organisms of earlier times have not disappeared entirely without trace; many have left traces behind them in the shape of fossils (from the Latin *fossilis*, dug up), which present themselves to us as bones, shells, and even soft parts of animals, turned into stone. Who has not heard the touching, sad story of the fossilization of a female ichthyosaur – a fish lizard from the Jurassic – which had just given birth to an infant when catastrophe overtook them both? Five others remained unborn, and were preserved as fossils inside their mother (see Fig. 172).

Fortunately very many fossils exist, in almost all the older layers of the earth's crust, from the present time (the alluvium being the newest deposit of the Quaternary period) back to the Cambrian, one of the lowest levels of the Paleozoic, which is about 500 million years old.

Paleontologists – who concern themselves with the fossil plants and animals, and arrange into order the different strata – can tell us with a good deal of certainty that the abundance of species and other forms becomes progressively less as we go back to older and older strata. In the alluvium are to be found all the bigger groups of organisms, including man, as far as they had not already become extinct in earlier periods. If we go back to the Jurassic (i.e., about 150 million years) man is absent; only the oldest, most primitive birds are there, and there is no trace of flowering plants. In the Devonian, about 350 million years ago, we look in vain for conifers, insects, or any crawling, terrestrial animals, while, finally, in the lowest of the Cambrian there is nothing but algae, corals and sponges, brachiopods, together with ancient versions of the cuttlefish of the present day.

Things were even more difficult in the period that is still older than the Cambrian, at Precambrian levels (Archaicum, the Urzeit). This level is almost without fossil remains, and the only ones that have actually been identified from there are the 'Algonkian algae,' so named after the most recent of the Precambrian strata. These are single-celled algae of simple construction, though of course they are not so simple that one can proclaim them to be the oldest of living creatures. Rather, they resemble the most primitive of the algae *that are alive today*, and in their turn must have been able to look back to a long line of ancestors. Consequently the Algonkian era must have had its own fauna, and if no trace of this remains, it may be because the creatures of that day were not able to produce a firm shell of carbohydrates; this is consistent with the known fact that the concentration of CO_2 in the atmosphere at that time was very high, suggesting that it was not being used up (see later, p. 289).

Moreover, even if there are no fossils of that era, there is other evidence that organisms were already present in the Algonkian. There is a remarkable structure with the name of corycium, sack like, and carbonaceous, and there is a part of the Precambrian graphite. What have these to do with living organisms?

Besides the 'normal' carbon with atomic weight 12 (written C^{12}) there is another, heavier isotope, C^{13} (as well as the radioactive isotope C^{14}, which is used in modern chemistry, biology, and medicine, and which we saw used as a radioactive marker on p. 60). In the non-living carbon compounds of any period, in carbonates and carbides, the ratio

172. A female ichthyosaurus which was overwhelmed in some catastrophe just as one of its young had been born. Five other young were still unborn, and were fossilized inside the body of their mother.

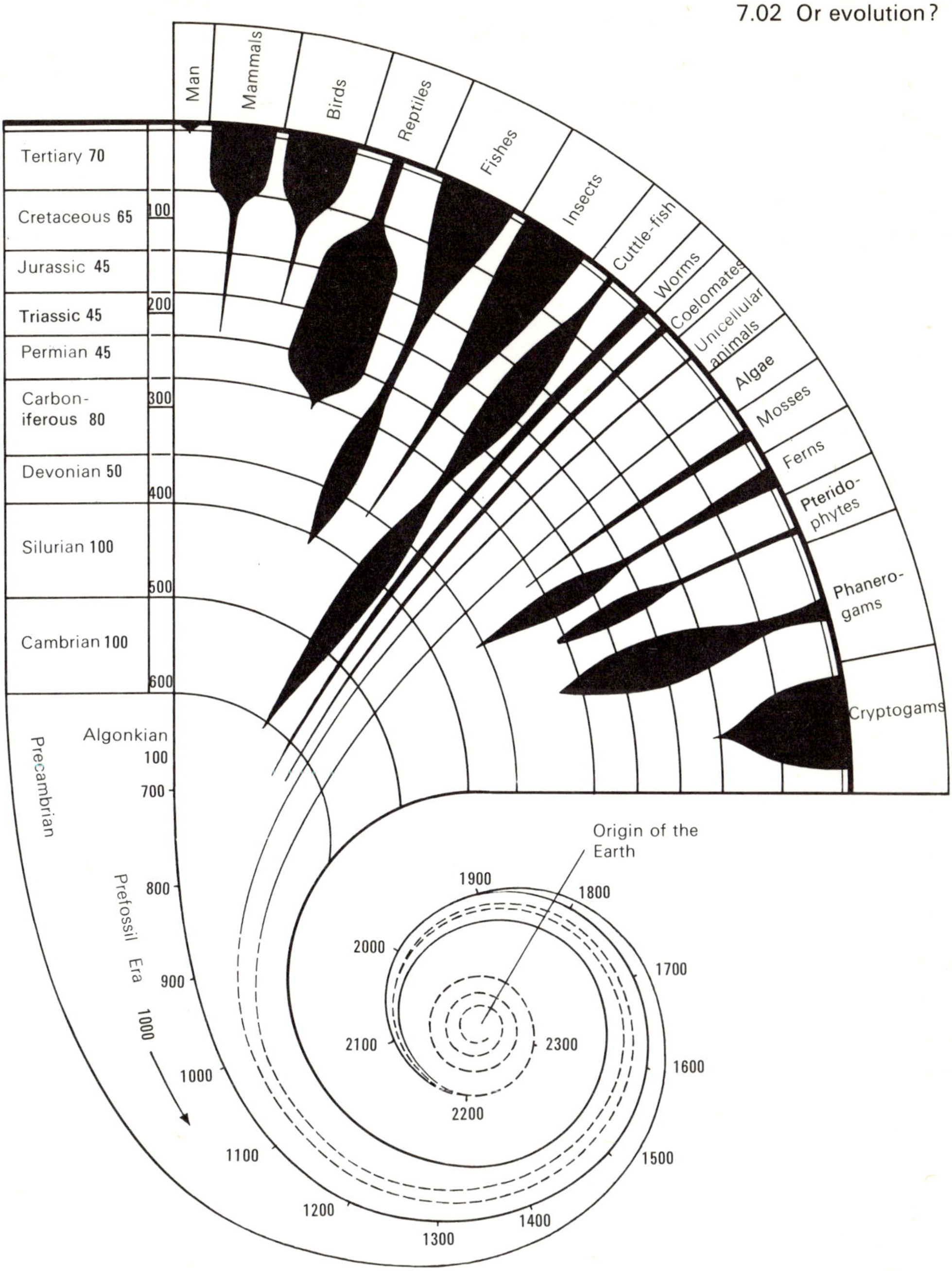

173. A 'time spiral' of the history of the earth, showing when the different groups of living organisms arose.

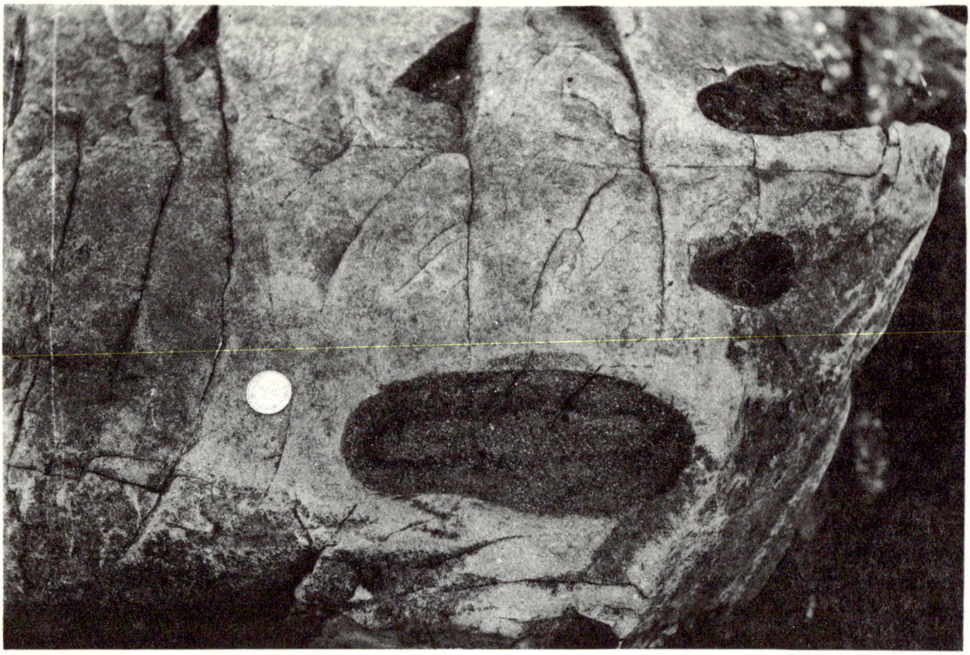

174. A nodule of pure carbon in a rock deposit (corycium or phyllite). It is derived from early organisms swept together from the sea bottom by wave action. The coin on the left shows the approximate scale.

of C^{12} to C^{13} is very constant. In graphite and in corycium it is different; here the ratio is displaced in favour of C^{12}, and we know only one process that favours the lighter isotope, and which preferentially manufactures it. This is the metabolism of living organisms, and today it is considered certain that in the graphite and corycium of the Precambrian we have before us the oldest traces, if not the actual likeness of early life.

Two things follow from this: first, that life has existed on earth for fifteen hundred million years – later on we shall have some ideas about this; second, that the sequence in which the various groups have arisen (see Fig. 173) is more than a mere sequence in time; the ones that come *after* also come *from* the others; the later groups have developed out of the earlier ones. Of course, this does not mean that the birds have descended from the ferns, or the mammals from the cuttlefish; but rather that the gymnosperms (e.g., the conifers) arose from the ferns, the angiosperms (i.e., most flowering plants) from the gymnosperms, the birds from the reptiles. These few examples must be enough. Occasionally we are able to find intermediate forms, which convincingly demonstrate the truth of this phylogenetic development. Although this detail or that may still be debatable, no natural scientist any longer doubts that life began with forms of the very simplest kind, unicellular, or even more primitive. From these arose the evolutionary stems, which with time, went on branching more and more, until they produced the multifarious forms of today.

Now that we have come so far, it is unfortunate that we can no longer avoid using the correct technical terms. As applied to living creatures, the expression 'to develop' came into use in connection with an idea that has long since been discarded; this was the concept of the 'development' of an individual creature from birth (or else from conception) until it was adult as an 'unfolding' of a tiny, miniature man, or homunculus. It was thought that the head of the human sperm contained the whole of the future being, enclosed in an envelope. This led to logical difficulties, because inside the homunculus, inside the sperm head, there ought to be a still smaller man of the next generation, and so on. This need not concern us any further.

In spite of these false assumptions, the term 'development' has persisted, and it is still used today, *but only for the development of an individual* (ontogenesis, from the Greek *on*, a being, and *genesis*, creation). Development along a line of descent, or phylogeny (Greek *phylon*, family or racial stem), which is what we are concerned with – and which was hardly discussed at all during the period of the false theory mentioned above – is better called *evolution* (Latin *evolvere*, to roll out). We cannot avoid making this change of terms (from 'development' to 'evolution') in future, and we shall understand evolution as being more or less irreversible, never repeating itself. And now, let us get back to the subject.

We are a long way from knowing in every detail what has been the course of evolution, but the *principle* is well established: mutation and selection – that is, changes that are inherited, coupled with the relentless selection of the 'better fitted' of the new forms thus produced. This is bound to lead to organisms progressively more complicated and progressively more highly differentiated, as long as the principle is valid in the first place.

We know for instance, that no living creature of the present day arises *de novo*, but we do not understand *why not*. How can we find out what actually happened one and a half billion years ago? Did the same laws of physics and chemistry that operate today operate then, and if so, do we know how they, in their turn, came into existence? If life cannot arise anew today, did it perhaps arise contemporaneously with the laws of physics and chemistry?

The problem is deadly difficult, but not so intractable that every attempt to solve it is doomed in advance. We must appeal for help to every branch of science: not only to paleontologists and phylogeneticists, but to geologists and – strange as it may sound – to astrophysicists, i.e., to every physicist whose work concerns the stars.

7.03 Impressions of the Early Days of the Earth

The Distinction between Organic and Inorganic Matter Is Blurred

Men have a peculiar urge to divide things into categories and set these up in opposition to each other: good and bad, light and dark, body and spirit. Antitheses of this kind remain in use even though everyone is aware that they denote nothing more than opposing *tendencies*, and that a sharp division in any kind of precise terms is quite impossible.

We, too, have an antithesis in our subject, which divides the material – or as the chemists say, the substances – concerned into organic and inorganic compounds; we have met this division before (on p. 17). Organic compounds are those which are based upon the carbon atom, and the rest are inorganic.

This is not entirely correct, since there are carbon compounds which are classified chemically as inorganic: carbon dioxide (CO_2), carbon monoxide (CO), as well as the carbides, compounds of carbon with metals or with boron or silicon (the best-known carbide is calcium carbide, CaC_2). With these exceptions, all carbon compounds are rightly grouped as organic, including such substances as sugars, amino acids, nucleic acids, and urea [$CO(NH_2)_2$], which is excreted by the kidneys.

The distinction between organic and inorganic was not really based upon the presence or absence of carbon, but upon the supposition that only living organisms contained organic compounds, and that organic compounds could be manufactured only by living systems. Hence the name 'organic.' These compounds were not to be found in the inorganic world, the constituents of which were fundamentally different.

This was a neat and tidy distinction – or almost tidy, since CO_2 is unquestionably produced by organisms, as a waste product of respiration, yet it is classified as an inorganic compound. For a long period this distinction between organic and inorganic was accepted as valid, even by chemists, and the boundary between the two was regarded as inviolate.

It was the chemists who first found a link between the two worlds. In 1828 Friedrich Wöhler made urea 'synthetically': he synthesized it from carbon dioxide and ammonia abiotically, i.e., without using an enzyme and without the intervention of any living system (not counting the experimentor himself, of course!). In this case the synthesis was really quite simple: Wöhler only needed to use a sufficiently high temperature and pressure.

The reaction of Wöhler's contemporaries to his discovery was mixed. Some jeered, saying he had only achieved a crude imitation of nature, and a very imperfect one at that, since he needed a temperature of 150°C and 100 atmospheres' pressure to do what a living organism can do at body temperature and in normal surroundings.

Most of the specialists of the time, however, were fully aware of the significance of Wöhler's discovery: what could be done with urea could be done with other organic substances. 'Organic synthesis' developed from this beginning until today thousands of organic compounds, including such complicated ones as vitamins and antibiotics, can be made synthetically.

Of course, only a very few chemists suspected that Wöhler's achievement breached the wall between the organic and the inorganic worlds, that perhaps organic compounds could come into existence from inorganic without needing a chemist to arrange it; in fact, this brought us nearer to an understanding of how life originated on earth.

The new discovery was hardly known outside a group of specialists, or at least outside the world of natural science, and consequently its implications were not appreciated. The dogma of the two worlds, organic and inorganic, persisted. This was one of the biggest handicaps to research into the origin of life, on a scientific basis of causality. It was still thought that the basic materials from which living organisms were constructed, the organic compounds in fact, arose exclusively from living creatures, and could never be formed abiotically. It was thought out of the question that life might at some past time have arisen from the inorganic world by natural processes. *Either* life does not 'come from' anywhere, but is eternal and unending, implying an initial act of creation, *or* else it must always be reinforced by spontaneous generation. This explains the stubbornness with which atheists denied the evidence against spontaneous generation, right down to the time of Pasteur.

Observation supported these views. If we look round the present-day world in search of carbon, the indispensable constituent of all organic compounds, we find it almost exclusively as carbon dioxide or calcium carbonate, that is *in an oxidized form*, and useless as a source of energy. Carbon does exist, of course, in a more or less pure – and combustible – form in lignite, soft coal, anthracite, and graphite; but these result from the activity of living organisms, whether they are the ferns and horsetails of the 'Carboniferous' period, or whether they are the unicellular animals of the Precambrian (see p.278). For the presence in the ground of petroleum, that mixture of carbohydrates with molecules of different lengths, we probably have to thank the activities of bacteria.

How does this fit in with the division of the world into organic and inorganic? Not at all. An error (among others) has crept into our thinking. If today we can find carbon only in an oxidized form (in inorganic compounds), this is not the least evidence that it must have been so in earlier periods. Perhaps things were quite different fifteen hundred million years ago. We ought to look into this quietly and without prejudice, and try to find out.

We can hardly find out anything about this on the earth itself, but what about the other planets, or the sun, from which the earth came, or finally, the fixed stars? Among all these heavenly bodies perhaps one or another has not yet evolved as far as the earth; this may be in the state today that the earth was in all those years ago. In any case, it would be worthwhile to find out how and from what source the carbon in the universe arose; this might give us a more reliable indication of what things were like on earth in Precambrian times than the most ingenious speculation.

Spectroscopy as a Detector of Carbon in the Universe

This is not nearly as difficult as it at first appeared to be. We receive information in the form of light rays from the fixed stars, which shine with their own light like our own sun; all that is necessary is to be able to decode this information, by means of the spectroscope. Naturally we cannot go into all the physical details of spectroscopy in this book, especially since we are interested only in the results, not in the technique itself. The most important results to us are those relating to the presence of carbon, and only these will be mentioned.

Sunlight, which appears yellowish white to us, is, of course, not homogeneous; by using suitable methods, it can be resolved into a *spectrum*, which ranges from red, through yellow, green, and blue, to violet, just as happens in the rainbow, and after this into ultra-violet. These colours are produced by light of different wave lengths, the red being long-wave light, and violet and ultra-violet light of short wave length. Substances in the solid and liquid states, when heated to incandescence, radiate a so-called *continuous spectrum*, with all the colours of the spectrum joined together through all the intermediate hues.

Gases behave differently. The atoms or molecules of substances in a gaseous state emit at any one moment only light of a quite restricted range of wave lengths. Use is made of this in the sodium (vapour) lights which are used to light highways; these contain sodium in gaseous form and emit light which is almost exclusively of an orange-yellow colour. The other colours are missing from sodium light, so that multicoloured objects seen in this light appear to be fawn.

When light of this nature is examined through a spectroscope it appears as a series either of sharply defined bright lines (a line spectrum), if the gas consists of atoms, or of rather blurred bands (a band spectrum), if the gas consists of molecules. The identification of certain lines with certain gases has made spectroscopy a valuable tool.

But this is not all. If an incandescent body, emitting a continuous spectrum, is enclosed in a gaseous envelope (i.e., if we give it an 'atmosphere'), then part of the radiation from the glowing body will be absorbed by the gases of the envelope, so that the resultant spectrum will contain certain *dark lines*. These will be absorption *lines* or absorption *bands* according to whether the gas is in an atomic or molecular state. Thus spectroscopy (or spectrometry, as soon as precise measurements are made) will give a detailed picture, which will indicate all the different kinds of atoms and molecules that are present in the body that is emitting light.

We know from other measurements that the fixed stars consist of a fluid core surrounded by a gaseous envelope, and so the light that they emit includes both emission bands and absorption bands – just what we need in order to obtain as much information as possible about the distribution of carbon throughout the universe.

The fixed star that is best known to us is, understandably, the sun, the nearest, and the most important to our lives. As might be expected, the sun contains all the atoms that occur on earth – including those of carbon, hydrogen, nitrogen, and oxygen – as well as several more. The carbon in the sun never occurs in atomic form, but always in combination either with itself (dicarbon, C_2), with hydrogen (CH), or with nitrogen (cyanogen, CN). The carbon in the sun is never present in combination with oxygen, as carbon monoxide or carbon dioxide, and this is all the more surprising since the presence of other oxides, such as those of boron (BO) and aluminum (AlO), as well as that of hydrogen (HO) can be established. So it can be regarded as conclusive that in the sun there are no oxides of carbon.

What is true for the sun is also true for other kinds of stars. The sun is a very old star, and so belongs to the so-called G type, that of yellow stars, and the temperature of its atmosphere is 'only' 5,000–7,000°C. Other, younger stars belong to the very hot O type and have temperatures round about 25,000°C, and they are entirely lacking in compounds of carbon, which is present only in atomic form. It would be tempting to follow the formation and evolution of the universe, and so to find out still more about carbon, but we must concentrate on the solar system, of which our earth is a part.

However important it is to know that no oxides of carbon exist in the sun, it tells us very little about the comparison between the earth of today and that of the remote past. So now let us look at two other members of our solar system, the largest and smallest planets, Jupiter and Mercury. Spectroscopic observation of these two planets is substantially more difficult for two reasons: first, because neither of them has any light of its own, but merely reflects, with modifications, the light of the sun, and second, because the outer layers of our atmosphere contain small amounts of ozone (O_3), and this has the fatal peculiarity of absorbing the spectrum from 2,900Å onwards – and this is just the most interesting part.

In spite of these difficulties, we have quite a useful amount of information about our neighbours. Suppose we begin with the smallest, and nearest to the sun, Mercury.

Mercury certainly has an atmosphere, but this is so extraordinarily thin that we can ignore its existence, because its possible effect on our measurements falls within the margin of error of the methods. Things are quite different on Jupiter, the biggest of the planets, which has a diameter eleven times that of the earth, a volume 1,350 times, and a mass 318 times as great. It is surrounded by a mighty atmosphere several thousand kilometers thick. Jupiter can easily be observed with a simple telescope, and then is seen to bear great spots, which change their shape, and occasionally disappear. They may be clouds, but clouds not of water vapour, but of methane (CH_4) and ammonia (NH_3). Thus Jupiter, like the sun, is lacking in oxidation compounds of carbon, but contains instead *reduction* compounds, i.e., those which contain an atom of hydrogen in place of one of oxygen. They are also more strongly reduced than the carbon compounds of the sun, which means that they contain more hydrogen. Where the sun has C_2 and CH, Jupiter has CH_4.

Attempts to find further hydrocarbons, such as ethane (C_2H_6) and acetylene (C_2H_2) in Jupiter have proved fruitless. This is actually not very surprising, since Jupiter is so far from the sun that its surface temperature is very low, in the region of $-135°C$, and so these higher hydrocarbons, if they were present at all, would exist as liquids and not as gases. Some authorities suggest that the famous red spots of Jupiter are nothing more than islands of heavier carbohydrates in an immense ocean of lighter carbohydrates. Line spectra cannot be obtained from such fluids.

What has been said about Jupiter applies also to Saturn, which is only slightly smaller, but even farther from the sun. Saturn is shrouded in an envelope of methane and ammonia vapour, but a part of the ammonia is frozen solid, because the surface temperature is even lower than that of Jupiter. Finally, Uranus and Neptune, both almost one hundred times as big as the earth, show only the spectroscopic bands for methane, and none for ammonia.

This leaves only Venus and Mars, the two planets which most closely resemble the earth. Venus possesses CO_2, and Mars has H_2O, O_2, and some CO_2.

Thus our earth occupies an intermediate position between the big planets, which have methane, and Mercury, without any atmosphere worth mentioning. What can we deduce from this? What has all this to do with the origin of life? To answer these questions, we must once again turn to the astronomers, and ask them something about the evolution of the solar system.

The Evolution of the Sun and Its Planets

Astronomers are in widespread, but not universal, agreement that stars and planets arise from the condensation of gases – the astrophysicists are not all of the same opinion about this. Space is not empty. The fixed stars are its principal content, but material exists between the stars. Part of this is gaseous, such as individual atoms of hydrogen, helium, and neon; part of the material is dust, the smallest possible crystals of ice, solid methane and ammonia, metallic oxides, and metallic carbides. Of course, this material is unbelievably tenuous in space.

Even with the naked eye, irregular, glowing 'clouds' can be seen in certain parts of the heavens. These consist of the same material that is found throughout interstellar space, but in the 'clouds' it has become appreciably denser. The expression 'denser' requires clarification in this context: if these 'dense' clouds could be compressed until they reached the same density as the sun, they would occupy only one thousand millionth of their original diameter.

For various reasons, clouds of this kind can thicken up still more, and become spherical bodies, ending up eventually as stars like our own sun. An enormous condensation of matter takes place, and in addition to tremendous pressure, the system releases thermonuclear reactions, like those in a hydrogen bomb. The mass of matter, the young star, begins to glow. So much for the sun.

In all probability the formation of a star does not always incorporate the surrounding matter completely, but leaves a residue, which arranges itself in the gravitational field of the young star into a flat disk or a spindle. This residual matter is not continuous, but splits up into a number of aggregations, larger and smaller, which are the precursors of the planets, and may be called protoplanets.

The protoplanets in their turn become denser according to the same principles as did the sun itself, though of course their masses are very small in comparison; at the present time the mass of the sun is 333,434 times that of the earth. Hence the planets do not achieve thermonuclear reaction like the sun, but only a moderate degree of heating up, to between 1,000 and 3,000°C. This rise of temperature probably does not continue for very long, since the planets soon cool down in the icy cold of interstellar space. Nevertheless, while it lasts, the high temperature is above the melting point of metals, and this results in the heavy nucleus for the planet. On top of this nucleus 'float' the lighter substances, while the lightest materials of all form a gaseous envelope. Since this 'uratmosphere' consisted of material from interstellar space, it was predominantly methane and ammonia. Cyanogen, hydrogen, and water should be there, but possibly not oxygen or carbon dioxide.

Because of its composition, the uratmosphere was a reducing agent (hydrogen in excess), not an oxidizing agent (oxygen in excess) – *and that was true for all the sun's planets*. We shall come back to this highly important discovery later, but we must first pursue the tale of the planets a little further, because their fates have been very different.

The cooling that succeeded the period of condensation led to the formation of a firm, solid crust, and perhaps to primitive oceans, on the planets, though at first there would be too little water for these. The uratmosphere was naturally the first part of the planet to show the effects of the cooling off, which brought those planets farthest from the sun down into the neighbourhood of absolute zero (−273°C). The nearer a planet's orbit lay to the sun, the stronger the radiation that it received, and hence the higher the temperature that it reached on the sunny side.

Today we are fairly sure that two thousand million years ago, i.e., 'shortly' before the origin of life on earth, the surface temperatures of the earth and the other planets were not very different from what they are today: Jupiter, −135°C; earth +20°C; Venus, +60–90°C; and Mercury up to +400°C on its sunny side.

In consequence of these enormous differences in surface temperature, the uratmosphere developed very differently on the various planets. It is well known that every constituent of a gas is in rapid motion, and the higher the temperature, the more rapid the motion. If the temperature is high, the velocity of the atoms of gas may result in their escaping from the force of gravity, and being lost into space. To take an example: on earth the velocity of a molecule of hydrogen (H_2) at 0°C is about 1·84 kilometers per second, and it is many times greater at temperatures of 20° or 100° or even 400°C – at any rate, greater than the so-called 'critical velocity' of 11·2 kilometers per second. It is known from rocket technique that this is the velocity that a body must reach in order to escape from the earth's gravitational field.

It follows that the planets nearer to the sun, and warm to hot in the sun's rays, will lose more gas (and more kinds of gases) from their atmosphere than the cold planets farther away. Temperature is not the only factor, however, that determines the composition of the atmosphere. Exactly opposed to the temperature effect is that of gravity, which is operative on every fragment of material. The greater the mass of a planet, the more firmly does it attract every molecule of gas in its vicinity.

Gravitational attraction, however, is always a matter of two partners, which attract each other, so that the gravitational attraction between a planet and its gaseous envelope involves not only the mass of the planet but also the mass of the particular particle of gas. The heavier the particle of gas the more strongly it is attracted to the planet and the less the likelihood that it will escape into space.

Let us consider the planets Jupiter and Saturn from this point of view. Both planets have a mass that is great compared with the mass of the earth. Their surface temperature is very low, and the velocity of their gas molecules is correspondingly low. Both factors operate in the same sense, and tend to hold the atmosphere together. It can therefore be inferred that the present-day atmosphere of these two planets is almost exactly what it was in their earliest history. Spectroscopic examination confirms the presence there of methane and ammonia.

Conditions on Mercury are exactly the reverse: a tiny mass, and therefore minimum power of attraction, coupled with high temperature, which means high velocity of the gas molecules. So, little by little, the entire uratmosphere has been lost, and today the planet Mercury circles the sun as a naked surface of rock fragments, without any appreciable atmosphere.

Finally, the earth. Once again, this stands midway between the heavy, cold Jupiter and the light, hot Mercury, and the effect of this on its retention of the uratmosphere is as follows. The lightest gases, such as hydrogen, helium, and indeed neon, have been partly or completely lost. Helium, for instance, which is one of the 'inert gases' and does not combine with anything, has been lost without trace, and any helium found on earth today is a secondary product of radioactivity. Most of the free hydrogen in the earth's atmosphere was also lost at an earlier period, when the earth was less dense and its temperature was higher; however, enough hydrogen was present in the heavier compounds (such as H_2O, NH_3, and CH_4) to make the atmosphere still have some of the properties of a reducing agent.

Free oxygen is heavy enough, anyway, to be retained, yet for long periods it was absent. This is because it is so readily taken up by metals, and other hydrogen-replacing substances, and used to produce metallic oxides and the like. CO_2 was also completely absent in earlier periods. As quickly as it was produced – we shall come to that later – it reacted with the silicates that were present in vast quantities, forming carbonates, and silicon dioxide (SiO_2) in the form of sand or clay.

So the sun and the heavier planets retain their uratmosphere, but on earth, even two thousand million years ago, the atmosphere had become distinctly poorer in hydrogen and helium. The amount of nitrogen had been reduced, and water was present partly as liquid water in the primeval sea and partly as atmospheric water vapour. In spite of this, the atmosphere as a whole was a reducing agent, having hydrogen in excess, even if only in combined form as methane (CH_4). Exactly corresponding to this reducing atmosphere was the fact that the carbon was also present in reduced form, in the *lithosphere* (Greek *lithos*, rock).

From Uratmosphere to Primeval Broth

We have given a very full description of the origin and modification of the uratmosphere, and the reader has followed this patiently (we hope!), even if he has been puzzled about why we went so far into astronomical, geophysical, and geochemical matters – whatever he may have expected in the book about biology, it was not this. Why have we spent so much time on the uratmosphere, and given so much attention to the fact that the uratmosphere was a *reducing* agent, and held its carbon, the basic substance of all organic compounds, in a reduced form? And since, in any event, life must have arisen in water and not in the air, why have we bothered about methane, cyanogen, and ammonia in the atmosphere?

What is at least a provisional answer can be given quickly. The precipitation from the uratmosphere literally 'fell into' the primitive ocean, which was already present in the middle period of the earth's history, and which actually contained water and not, as on Jupiter, liquid hydrocarbons. Of course, it was not pure water, because it dissolved out a variety of substances from the earth's crust. But the primitive ocean did not derive its dissolved substances only from below; it also derived them from above, from the uratmosphere. Rain, and thunderstorms, may well have been even more frequent and violent than they are today. The rain water brought down some of the constituents of the atmosphere into the ocean: ammonia as ammonium hydroxide, NH_4OH, and cyanogen as hydrogen cyanide, HCN (commonly known as prussic acid). Both substances are exceptionally active as reducing agents, and this is why we have been so interested in the fact that the uratmosphere could be chemically classified as reducing rather than oxidizing. Carbon dioxide and nitrogen are very inactive as oxidizing agents, and little can be expected from them, whereas compounds containing hydrogen readily give up their hydrogen to other substances – thereby reducing these – while at the same time themselves becoming oxidized.

So chemical reactions of various kinds already took place in the primeval sea. Of course, these were still *inorganic* reactions, and for a long time no trace of life can be seen. Quite the opposite, indeed: the presence of even minute traces of hydrogen cyanide would have made the primeval ocean exceptionally poisonous to any living creature of the present day, and so this was a hostile environment to life.

This is only the beginning, however. Our earthly atmosphere today is an oxidizing agent, since it is made up of oxygen, nitrogen, carbon dioxide, and water. Sometime and somehow a change-over must have taken place from a reducing atmosphere to an oxidizing one. We must give a little thought to this.

The original sense of the term 'oxidizing' is the taking up of oxygen. Where does this oxygen come from? Almost exclusively from water, from the water vapour in the earth's atmosphere. But the water in the ground is also hydrogen that has been completely oxidized. We all know from our school chemistry that a gaseous mixture of hydrogen and oxygen reacts with exceptional violence, exploding to form water, and releasing a large quantity of energy as heat. If the reverse process is to be carried out, splitting water into hydrogen and oxygen, then the same amount of energy must be put back into the system. Where is this energy to come from?

The source of energy is sunlight, and in particular the short-wave, ultra-violet component. At the present time only a small part of this reaches the earth's surface, because most of it is absorbed, either by the high-level layer of ozone or by dust in the lower layers. This is fortunate for us, because the little that does penetrate can easily burn the human skin. If we were to receive the full strength of the ultra-violet rays, we should have to take precautions against it, or succumb.

Two thousand million years ago the atmosphere contained only methane and water; there was not yet any ozone in it. The molecules of methane and water equally absorbed the energy-rich ultra-violet, with the result that the molecules of water split into hydrogen and oxygen, while the methane split into hydrogen and a hydrocarbon containing less hydrogen, such as CH_3, CH_2, or the radical $-CH$. The result of such reactions in the atmosphere was that the hydrogen produced, being light, was mostly lost into space, but the hydrocarbons, and in particular the oxygen, remained behind. So the atmosphere became *progressively poorer in hydrogen and richer in oxygen*.

The minerals of the lithosphere, insofar as they were not already completely oxidized, readily took up oxygen, but one day they were saturated with oxygen, and then it was the turn of the constituents of the atmosphere. NH_3 was oxidized to molecular nitrogen (N_2) and then to nitrogen dioxide (NO_2); any sulphuretted hydrogen (H_2S) was oxidized to sulphur dioxide (SO_2); and methane (CH_4) became carbon dioxide (CO_2). All of these reactions were possible only after free oxygen became available in the atmosphere.

These things did not happen in rapid succession. It has been calculated that the transformation of the reducing atmosphere into the oxidizing atmosphere of today occupied between one thousand and fifteen hundred *million* years, and that the process came to an end eight hundred million years ago. *If life ever originated from inorganic matter on our earth, it must have been within this last period.*

Moreover, oxidation did not progress in such a way that every molecule of CH_4 immediately became one of CO_2. Oxidation took place (and still takes place today) in many intermediate steps, passing through alcohols such as methyl alcohol, aldehydes such as formaldehyde (H CHO) and formic acid (HCOOH), and others. These compounds did not remain exclusively in the atmosphere, but were carried by the never-ending rain down into the ocean, where at first they were protected against any further oxidation. On the other hand, in the ocean these compounds reacted with ammonia and hydrogen cyanide, oxidizing these bit by bit, and as all these compounds were chemically active ones, a lively reaction was soon in progress. Some of the products of this reaction were amino acids and organic bases of the adenine type.

In this way the primeval ocean, which at first was nothing more than a solution of minerals, gradually became a primeval broth, in the sense that it contained a large number of organic substances which provided abundant nourishment for the first, primitive living creatures. (The prussic acid must meanwhile have become oxidized and rendered non-poisonous to life.) Living creatures arose in this nourishing broth without any outside assistance – *they had a completely inorganic origin.*

Introducing a Time Scale. Man Appeared Forty Seconds before Midnight

The reader is probably feeling a little impatient with this progress from astrophysics to primeval broth – a slovenly term, but one that is current among scientists – even though it has given him something new to think about. In the beginning he found the story interesting, but theoretical and speculative. At the time it never seemed to become real – how could the geochemists know so precisely what had actually happened? And in the second place he remembered the first chapter of this book, where it is stated that organic compounds – in contrast to inorganic substances – react together exceptionally slowly, which is why they have made for themselves the accelerators that we know as enzymes. We are now saying that at one time the same organic compounds were vigorous in reaction. How does this hang together?

Let us take the second objection, about the speed of reaction of organic compounds. On page 16 it was stated that if we bring together glycerine and a fatty acid, although these are the ingredients for the formation of fats, yet after several days we shall still not find any fat present. This is just as true now as it was then, but we have to remember that when we are talking about the uratmosphere and the primeval broth we are working to quite a different time scale from experiments in a test tube. For the transition from uratmosphere to the oxidizing atmosphere of today, and for the formation of the primeval broth, the geochemists have postulated a thousand to fifteen hundred million years. How can one compare that with a few days?

Since a thousand million is a number that no one can really visualize, let us try to make it a little more significant. Taking the whole period from the formation of the earth's crust and of the uratmosphere (two thousand million years), let these be compressed into the twenty-four hours of one day. Each hour of this day is then roughly equivalent to eighty million years, one minute is about 1·2 million years, and one second equals 20,000 years.

To put it differently, midnight, zero hour, we start with the transformation of the uratmosphere and the formation of the primeval broth. The Precambrian period would then begin at about midday. The change from Algonkian to Cambrian would take place at about 18:00 hrs. And when would man appear for the first time? – forty seconds before midnight!

The whole of the 'historical period,' back to the Sumerian culture of 6000 BC would compass half a second, and the lifetime of a single individual would flash by in one two-hundred-and-fiftieth of a second.

If we put it in this way, it is clear that the alcohols, aldehydes, and organic acids in the atmosphere had plenty of time to react with each other even without having the right enzymes present. Looking at it from the opposite standpoint, the origin of life took so long because the synthesis of its basic materials proceeded so slowly.

Now to other questions. Is there actual proof that the story of the earth happened as we have described, or must we dismiss the whole thing as fantasy? If we want to be honest, then we must answer no to both halves of the question. Every historical science constructs pictures of the past, but no historian deceives himself into thinking that the course of events was really so simple.

Instead, he goes back to 'sources,' to 'findings,' follows indications, and so on (the archaeologist, for example, finds evidence in contemporary writings, buildings, implements, coins, and so on), then tries to fit all these together into a picture that is as free from contradiction as possible.

Anyone who undertakes the task of telling the story of the living organisms through two billion years cannot make use of the testimony of contemporaries, but he can find other forms of evidence: bones and fossils, the stratification of rocks in which the smallest trace of any constituent can be exactly determined. By these means the scientist may arrive at a precise knowledge of chemical and physical conditions, and so decide what would or would not be possible. All these support the view that our reconstruction of the story of life is not an untenable speculation, but leads to more or less the correct conclusions. Of course, these speculations are not exactly proved, or even provable. The final certainty can come only from experiment, and this is out of the question – or is it?

Experimental Evidence that the Primeval Broth Was Nourishing

A group of American scientists, led by Stanley Miller, were not impressed by the reiteration of the view that experiment was impossible in the realm of history. These scientists said to themselves: Let us assume that the uratmosphere actually did consist of methane, ammonia, and water vapour, as well as some hydrogen. What is there to prevent us from taking a volume of gas of similar composition, enclosing it in a glass container, and then supplying energy to this system, either by means of ultra-violet rays, or – even simpler – by a spark discharge? The system then ought to manufacture organic compounds, such as amino acids.

The idea of using an electric discharge was not so farfetched as it may seem. The geologists and geophysicists already maintained that over the primeval sea it not only rained but also thundered, probably more frequently than it does nowadays – and during the passage of a million years quite a large quantity of reaction products may accumulate!

Miller and his team set up an extremely simple piece of apparatus (fig. 175). A sample of artificial uratmosphere, consisting of methane, ammonia, and hydrogen, was enclosed in a glass sphere, into which two electrodes

175. Miller's apparatus for producing organic substances from methane, ammonia, and water.

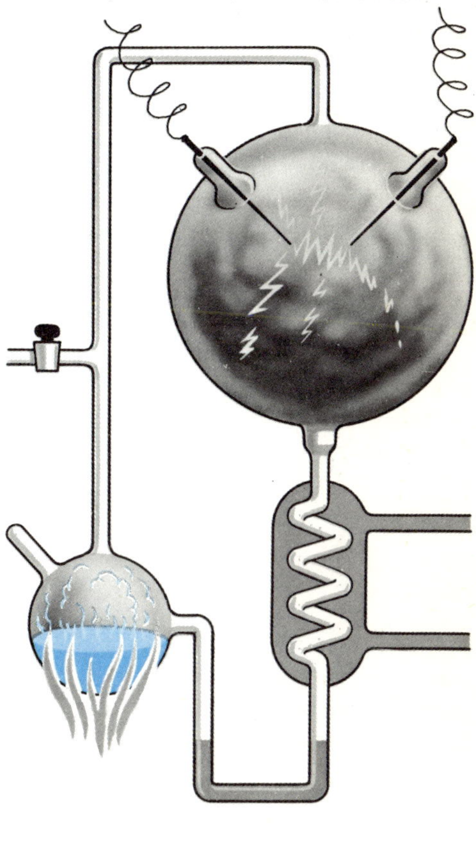

projected. A steady stream of sparks between these electrodes represented thunder and lightning. Water was heated in a separate flask, and its vapour passed into the reaction chamber. The reaction products passed down through a condenser, into a U-tube, where they accumulated as a liquid. Water vapour from the farther surface of the U-tube passed back into the water flask, and the cycle began again.

After the experiment had continued without interruption for a whole week, the 'primeval broth' in the U-tube was analyzed, and at the first attempt five amino acids were found: glycine, alanine, β-alanine, aspartic acid, and α amino-butyric acid. In addition, there was a yellow substance with a relatively high molecular weight.

Encouraged by this very interesting result, the team repeated the experiment many times, often under different experimental conditions. The first step was to make certain that the amino acids were not being produced by bacteria accidentally included. The apparatus was sterilized for eighteen hours at 130°C, so that it was certainly free from bacteria, yet the experiment still produced the same compounds. Then it was demonstrated that it was, in fact, the electrical discharges that produced the organic compounds: without sparks there were no amino acids.

Finally, instead of leaving the analysis until the end of the experiment, tests were made at short intervals. These intermediate analyses showed clearly that the amount of ammonia started to fall immediately the experiment began, and continued to fall at a uniform rate; on the other hand, the synthesis of amino acids made a delayed start, and only reached a constant rate after about five days.

The most important finding from these further experiments was that amino-acid production was not limited to the original five, but that the apparatus could produce twice as many. In addition to the amino acids were to be found urea and methyl-urea, and finally nitrogen-free organic acids, among them acetic acid, lactic acid, and succinic acid – acids well known to us as intermediate products in the metabolism of carbohydrates in living cells. A total of about two dozen organic compounds of low molecular weight were experimentally produced from this artificial uratmosphere, and that completely *abiotically*, i.e., without the intervention of any living organism.

This happened in 1953, and at first the report of Miller and his colleagues was received with hesitation and skepticism. It was soon established, however, that the experiment really did mark a turning point in the search for the origin of life. Keosian spoke the truth when he wrote (in 1964), 'The aura of metaphysics which surrounded this problem was pushed aside, and discussion and experiment were possible for the first time.'

And experiments were carried out. All over the world, scientists reported that they could confirm Miller's results, even using different methods. Instead of the lightning flash, some of these used heat, visible light, ultra-violet light, X-rays, ionizing radiation, even ultrasonic vibrations. In place of methane as a source of carbon, some used formaldehyde, carbon monoxide, and even carbon dioxide; in place of ammonia, nitrogen. Some tried to see if the synthesis of organic compounds could take place *in a watery medium*.

All these variations gave the same result, and produced amino acids, nitrogen-free organic acids, and even sugar. A particularly comprehensive result was obtained from an experiment in which fluid hydrogen cyanide (prussic acid) and an aqueous solution of ammonia were heated together at 90°C for eighteen hours; no fewer than seventy-five amino acids and similar compounds were obtained, as well as other compounds that were not amino acids. Over and above these were a number of *polymers,* consisting of a number of small molecules piled upon each other.

One condition must always be fulfilled; the reaction mixture must be a reducing mixture.

The last experiment mentioned above may not be exactly comparable with conditions in the primeval sea, but what takes place in eighteen hours at 90°C may perhaps be achieved at 20 to 30°C if several million years are available!

At first the reports made no mention of purines or pyrimidines, the series of organic bases (see p. 35) which supplies the ingredients from which amino acids are constructed. By 1961 adenine had been found; in 1963 came guanine and adenosine, and – hardly credible – adenosine mono-, di- and tri-phosphate, AMP, ADP, ATP, the last being the universal source of energy for the cell (see p. 57). *All this took place under conditions similar to those that we have postulated for the primeval broth:* e.g. the formation of adenine from methane, ammonia, and water under a powerful electrical discharge.

Thus there is no doubt any longer that the primeval broth originally contained nothing living, but it did have a miscellany of organic compounds, almost all of those that today can be found in a living organism. Fifteen hundred million years ago the world was not divided into organic and inorganic sectors; the organic compounds of the time arose abiotically as a result of 'normal' chemical reactions between inorganic compounds. It might indeed be said that the organic compounds were a development from the inorganic in response to changed conditions, or to a new environment.

This experimental work has solved one problem about the origin of life on earth. We now know that the primeval broth was indeed a nutritive solution, and contained the first food for the first organisms, sufficient both in quantity and variety. The next problem that immediately arises is this: how did the organisms themselves, the first cells, come into being? To anticipate the answer to this one, we know almost nothing, at any rate less about this question than about how the organic compounds arose. In contrast to our earlier certainty, we now have to depend almost entirely on speculations, which are hardly susceptible of proof, at least not at the present time, and not experimentally. Nevertheless, these speculations are interesting enough to be worth going into a little further.

All of these speculations arise from the assumption (or fact?) that the organic compounds that were moving around in the primeval broth did not just confront each other inactively, but reacted together according to their respective chemical capacities, so that the primeval broth was a hotbed of chemical reactions. Perhaps these terms are an exaggeration when applied to substances that were present in concentrations as low as 1–10 per cent, but once again we have to remember that almost unlimited time was available.

Do such chemical reactions deserve the name of metabolism?

7.04 Early Metabolism

Were the First Cells 'Coacervates'?

Metabolism is to be understood as *physiological* chemistry in the classical sense, and does not include every chemical change, but only those changes that are associated with living cells. It was in this classical sense that we used, in the first chapter of this book, the headline 'Life Manifests Itself as Metabolism.' Yet in that very chapter we learned that metabolism as a whole is constituted from a large number of individual, well-understood chemical reactions, each of which can take place, on its own, in the test tube, and removed from any living system.

Objection may be made at this point that an essential feature of metabolism is to co-ordinate the individual reactions, and that this is possible only in a living cell. This can be disputed. The synthesis of protein is undoubtedly a co-ordinated system made up of many individual reactions; on page 56 we set out a list of constituents and their functions. Many of these may still not be known very accurately at present. Protein synthesis, and the mechanism of coding of information, are to be regarded as the most essential part of metabolism. If, however, the right components are brought together under the right conditions, then metabolism *can* take place outside the cell, in the test tube. Hence the co-ordination that is a mark of metabolism is not exclusively displayed inside the cell.

How can we draw a line between 'simple' chemical reactions and 'co-ordinated' metabolism? We can hardly blame the younger generation of scientists if, in face of such a dubious definition, they ignore traditional niceties and unconcernedly talk about metabolism in the primeval broth.

This metabolism was, of course, simple, not to say primitive, compared with that of a living organism, but basically it followed the same laws: energy-rich compounds released energy, and so became poorer in energy content; if several reactions were linked together, then the energy released in one reaction was used in another reaction to enable an energy-demanding synthesis to take place. In such a metabolizing medium it was 'only' necessary for each little focus of metabolic activity to become divided from its neighbours by a membrane, and lo! we had a primitive cell!

That little 'only,' however, represents such a decisive step, and one that is even less understood than the rest of the metabolism, that the quotation marks can only be regarded as an ironic gesture. Certainly we have already noted (p. 155) that mono- and bimolecular films can arise spontaneously out of lipoid material, and it is at least plausible that such films might have formed round adjoining spherical regions of the primeval broth, giving *droplets*. Such a space was at first surrounded, but not completely enclosed, since the lipoid film was permeable to a certain extent, and so fresh nutritive material could pass in when it was needed.

Loose, fortuitous formations of this kind are certainly very unstable, and they decline and disintegrate if they cannot continue their metabolism. The probability of their surviving for a long time is not very great. They may increase, however, if the number of reactions between droplets is 'sufficiently' high, and if the individual reactions are 'correctly' coupled together. Under these conditions metabolism can be maintained, and with it, the 'system' of a droplet surrounded by a lipoid membrane.

Even at this level, where we cannot yet speak of living cells, but only of tangles of reacting molecules, it is already possible for selection and evolution to begin to operate. We shall have more to say about this later.

Other scientists have designed different models, and one of these is the much-discussed *coacervates*.

In Latin *acervus* means a heap or pile, and *coacervare* means to pile up. In the present context this word refers to a particular situation in a solution. The solution contains large numbers of hydrophile molecules, each of which therefore surrounds itself with a water envelope, just like the one we saw in the protein molecule with hydrophile groups attached (Fig. 10). Normally this aqueous envelope has no sharp boundary; those molecules of water that are attached immediately to the surface of the macromolecule are held on with great force, and so are in a firm arrangement. The farther outwards we go, the looser becomes the attachment, and the less definite the arrangement, until finally the aqueous envelope of the molecule grades away imperceptibly into the 'free' water surrounding it.

Under certain conditions, for example, after certain salts have been added, the electric charge of a molecule, or of a cluster of molecules, is altered, and the strength of the attachments also alters. Then the aqueous envelope becomes smaller, but at the same time more definite in outline. Throughout the solution, therefore, a segregation takes place: a denser phase, with less water (which is called a coacervate) separates itself off from the continuous fluid phase. The latter still contains a few dissolved molecules, and therefore is not quite reduced to pure water. The coacervate does not form a sediment, but remains in the form of innumerable droplets of the smallest possible size suspended in the liquid phase.

These droplets are still much too big to be thought of as containing only a few molecules, or bundles of molecules; on the contrary, they contain a large number of molecules, each originally still with its sharply defined aqueous envelope, and these are *'heaped together' (coacervate), and surrounded by a common aqueous envelope and a common boundary against the surrounding liquid* (see Fig. 176).

Such simple coacervates can be produced comparatively easily, but when produced they are as short lived as the single droplets with their lipoid membranes. The smallest change in water content, in electric charge, or in temperature, and so on usually brings about

176. A coacervate is formed when individual molecules surround themselves with a sharply defined envelope of water, and then join together into bigger complexes. (After Bungenberg de Jong.)

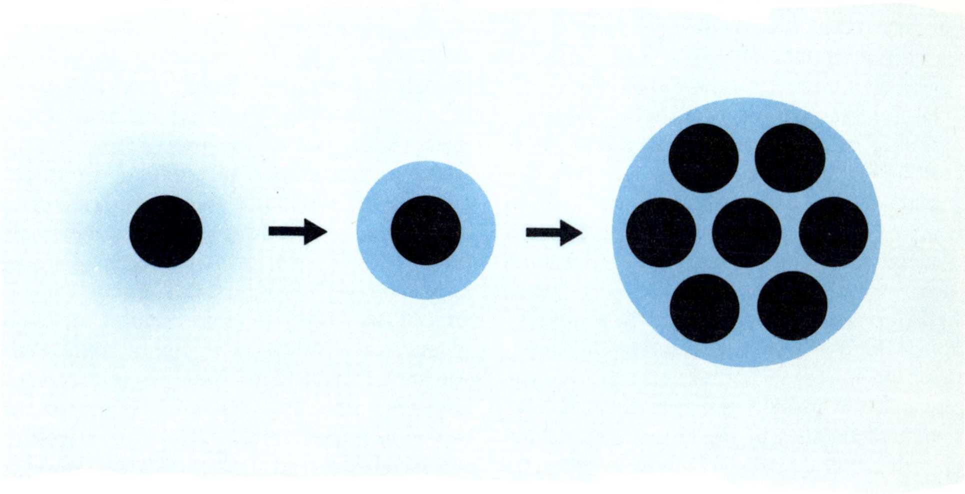

the dissolution of the coacervate, and the segregation of constituents is reversed.

The so-called 'complex coacervates' have a somewhat better chance. These formations arise between molecular aggregates of opposite charge, which are attracted together, but only as close together as the aqueous envelopes allow. An equilibrium is set up between the forces of attraction and repulsion, and this makes the system more stable than a simple coacervate.

The physical chemists have made complex coacervates from different partners, and studied their reactions. They frequently found that something like a second small coacervation had come into being in the interior of the first. This contained vacuoles (see p. 178), and might be coloured with the same pigments that are found in the vacuoles of living cells. There is a striking resemblance; is it possible that the first living cell was a complex coacervate?

It was not as simple as all that. Of course, the idea occurred of using the convenient phenomenon of coacervation to separate off a portion of the primeval broth, but we should not deceive ourselves; the product is still very different from a living cell.

Firstly, we need comparatively big molecules in comparatively large quantities, peptides, or even proteins, and how these arose in the primeval broth is still obscure.

Secondly, a coacervate is formed from many molecules *of the same kind;* a complex coacervate from only two groups of molecules *each of one kind.* Coacervates are, so to speak, statistical aggregates, in which the molecules of one kind are mutually interchangeable; this contradicts our impression that even the molecules of protein and nucleic acid have the characteristics of individuals – we need only think of enzymes in this connection.

Thirdly, from what the chemists and physicists know about the origin of coacervates, it is extremely unlikely that in such droplets, apart from the actual formation of the coacervate, there will be any substantial amounts of substance of low molecular weight enclosed. Yet this must take place if the coacervate is to be anything more than a bundle of large, inactive molecules – if it is to be an actively metabolizing, co-ordinated portion of the primeval broth, or to put it less facetiously, a metabolically active region of the primeval ocean.

It is advisable, therefore, not to overrate the idea of coacervation. It has by no means brought forth the first organism; on the contrary, it is always possible that coacervation brings about a subdivision *inside an already existing cell,* forming 'compartments' (cf. p. 188).

7.05 From Metabolism to Propagation – or the Reverse?

Growth and the Primitive Genes

Thus we still do not know how the first discrete metabolic region came into existence, and we have a choice between several models or mechanisms. But even if we had some degree of certainty, we should still be nowhere near calling these droplets living things, whatever appearance and function we might attribute to them. The overwhelming majority of them are quite certainly unstable, temporary formations, which quickly dissolve away again. The more stable among them – we can hardly yet refer to them as the 'survivors' – have at least a chance of going on to further improvement. Since their envelopes have only a limited permeability, substances can accumulate in the interior that otherwise would be scattered throughout the primeval broth. Increased concentration within the droplets may bring about a general increase in the speed of reactions (which is undesirable), but it may also lead to the production of longer molecules. The most important thing is that with higher concentrations the different processes can be better linked together, better co-ordinated, and proceed more smoothly. The total effect has the aspect of a kind of higher development, or evolution, and in fact the biochemists speak of 'chemical evolution' – to the annoyance of many biologists, who object to such a concept of an 'organism.'

We shall not take sides in this dispute; it is enough for our purposes to note that this is a possible way in which droplets might be improved, have an effective future, and be acted upon by selection. Nevertheless, however effectively the metabolism of the droplets may have become co-ordinated and smooth running, it is still not that of a *living system*. Remember a heading we had in the first chapter of this book: Metabolism as a Manifestation of Life. This expresses, though in a very condensed form, precisely the problem that is before us in this matter of the droplets. Metabolism is *one* aspect of life, even a very important one, since no life is possible without metabolism; but the reverse statement is not true, as we have just seen. Metabolism *is* possible without life.

But life consists of more than just metabolism. Even in its most primitive form, a living organism must grow and multiply. How about that?

If the products of metabolism are simply accumulated within a droplet, the droplet certainly gets bigger, but this does not constitute growth in the true sense. We can only speak of growth if all the reactions are so co-ordinated and harmonized that there is a multiplication not only of the products of metabolism but also of all the vital constituents of the system – for example, of all the enzymes that may have been present at this early stage. However, since 'chemical selection' would no doubt ensure that the system with the best co-ordination would survive, it is not too difficult to imagine that a genuine form of growth would eventually evolve. But what about reproduction, procreation, multiplication?

The longer a droplet goes on growing, the greater is the probability that it will split up. For purely mechanical reasons alone it is likely to fall apart into two or more fragments, which will have the same make-up as the original droplet, and so in their turn can carry on metabolism, and can grow. All this goes on, however, at the expense of the surrounding medium, the primeval broth, which becomes poorer in organic compounds; hence 'living conditions' for the droplets become 'harder.' Rapidly growing droplets are more effective in competition for the nutritive substances than those that grow slowly. Droplets would be more successful, and be able to 'divide' more frequently, if they themselves could synthesize one or more components of the primeval broth, and thus become independent of outside conditions.

The fragmentation still depends upon some chance mechanical disturbance in the vicinity of the droplet, and so is not a true procreation in the sense of an orderly reproduction of identical offspring. True procreation at the present time involves replication of the genes, and handling them on to daughter individuals. Since today only the nucleic acids function as carriers of information of the genes, and handing them on to guarantee that the effective components in chemical reaction are going to lie next to each other, it follows that nucleic acids must have come into action in the droplets at some time or other.

Nucleotides were already present, if not actually abundant, in the primeval broth, and the first primitive genes may have arisen from them. But where – in the broth itself, or within the droplets? Our ignorance on this point is underlined by the fact that there are different and contradictory hypotheses about it.

One hypothesis is based upon the undisputed fact that all the earliest peptides and proteins arose without the help of nucleic acids. DNA (or RNA) must have arisen at a later period, to free the synthesis of peptides and proteins from having to depend on chance, and to bring order into it. DNA itself thus appears in the role of an improved regulatory mechanism, which gradually evolved *inside* the droplet (or inside the *cell*, as the case may be).

Another hypothesis sees the gene (or *a* gene of some kind), rather than droplet metabolism, as the starting point of life. Such a gene must have possessed three characteristics right from the beginning:

1. The ability to reproduce itself.
2. The ability to mutate.
3. Heterocatalytic possibilities.

We already know what reproduction and mutation are, but what is *heterocatalysis?* We know already, from enzymes, the meaning of a catalyst: something which sets chemical reactions in motion, and determines their direction and speed, but without itself being changed – or more precisely, without undergoing any obvious change. The phenomenon is called *catalysis*. Horowitz (1959) defined heterocatalysis as being the ability to influence the surroundings catalytically in such a way that they provide everything necessary for the maintenance of the system. One thing, at least, follows from this definition: one gene, or one molecule of DNA floating round by itself in an empty environment, without relations with anything else, is of no interest to us. It is completely ineffective. It can at any rate mutate, but it cannot duplicate itself, and it cannot carry out any heterocatalysis. So it cannot function without a matching environment. But can it, perhaps, operate without a droplet, just in direct relationship with the primeval medium?

Self-duplication in the Test Tube

First, let us consider the phenomenon of self-duplication, with which we are already familiar because this is what dividing cells do. All the same, it seems very unlikely that a gene, isolated from any cell, would be able to duplicate itself in a test tube. Because it was so unlikely, for a long time it was thought to be impossible, but exactly a year before this chapter was being written, in September 1965, a very remarkable paper appeared in the *Proceedings* of the US National Academy of Sciences. The authors were Professor Spiegelman (Urbana, Illinois) and four of his colleagues, and the paper was entitled 'Synthesis of a Self-reproducing and Infectious Nucleic Acid with a Purified Enzyme.' This title really says all there is to say, but for our present purposes perhaps it would be helpful to expand it a little.

The nucleic acid in question is RNA, and in particular, the RNA of a bacteriophage – the phage Qβ, which is able to infect an Hfr strain of *Escherichia Coli* Q_{13}. The details are not important, but if need be, the reader can turn back to the discussion of bacteriophages and Hfr strains on page 130. This RNA can be isolated and purified by using appropriate methods, which are far from simple, and then we have in the test tube pure RNA, or more precisely, a complete phage genome (not just individual genes!).

If this is brought into contact with host cells, namely, *E. coli* Q_{13} Hfr cells, these become infected. On the 'plates' (see p. 115) appear the plaques, or disks, that we have met before – but things are not quite as simple as that, because first the stable cell-membranes of the cells that are to be infected must be weakened; otherwise the RNA would not be able to penetrate into the cell. So the complete phage has enzymes which punch holes in this membrane, as described on page 111.

The purified enzyme is a so-called RNA-dependent-RNA-polymerase. The term *polymerase* means that the enzyme builds up long chains of RNA out of individual nucleotides, such as the phosphates of adenosine, guanosine, cytidine, and uracil, by polymerizing them (cf. p. 61). On the other hand *RNA-dependent* means that this polymerase can operate only in the presence of RNA, and that it needs this RNA as a matrix, or template, from which it can be stamped out. Furthermore, the polymerase can detect whether this is the 'correct' RNA, and remain inactive if it is supplied with the wrong RNA. The polymerase must thus be derived from *coli* cells which have been infected with the phage Qβ, and the induction of the synthesis of the polymerase is one of the consequences of infection.

If everything agrees, then all that we still require is the four bases, adenosine, cytidine, guanosine, and uracil, in the form of triphosphates – ATP, CTP, GTP, and UTP – with the addition (once again) of magnesium ions, and then the polymerase synthesizes new RNA *following the template* of the phage-RNA. All this happens very quickly, and the synthesis is complete in forty minutes.

Because the polymerase brings about a multiplication of the phage-RNA it is known for short as a 'replicase.' The title of the paper published by Spiegelman and his colleagues uses the term 'purified enzyme,' and the problem of purification is the root of the experimental difficulties. It is a most laborious process to obtain the enzyme free from ribonuclease, which has the tiresome habit of turning up as an impurity in almost every enzyme extract. The slightest trace of ribonuclease, of course, will immediately break down the phage RNA, or at least the newly formed RNA.

Thus the polymerase is able to reproduce phage RNA in the test tube, and to reproduce it so exactly that the newly formed RNA is just as infectious as the original template RNA extracted from *coli* cells or bacteriophages. Of course, the objection has to be met that possibly the new infection is being brought about merely by the 'old' material.

Using the centrifuge – as described on page 68 for the separation of 'light' from 'heavy' DNA – to separate the new and the old RNA, one operation did not give a sufficient degree of purity, so the following experiment was started. A total of sixteen samples were centrifuged; each of them comprised all the material that was necessary for a complete synthesis of RNA. Only the template-RNA of the bacteriophage was missing, and this was introduced only into samples No. 0 (zero) and No. 1. Sample zero was *immediately* tested for infection, and so provided a 'null point' for the experiment. Sample No. 1 was tested after forty minutes, when, as we have seen, the synthesis of new RNA would be complete. Shortly before this a small amount (50 cu. mm. out of 250–300 cu mm.) had been taken out of sample No. 1 and added to sample No. 2. Forty minutes was allowed for the reaction to proceed in sample No. 2, and then a small quantity of this was transferred to sample No. 3, and so on from one sample to the next, up to sample No. 15. In the final sample, as a result of the progressive attenuation, there could not be, at most, more than a single strand of RNA present: theoretically it should be one sixth of a strand.

Since, however, the number of new infections shown by the plate test remained high *and unchanged,* this means that the newly formed RNA must be infectious, and just as infectious as the original template-RNA. Thus it is actually possible to cause a phage genome to multiply, and replicate itself exactly, in the test tube, and *cell free*.

The whole scientific world paid the highest tribute to the work of Spiegelman and his colleagues. The successful synthesis of a biologically active RNA by enzyme action is in itself a magnificent achievement, but an equally important result is that from now on, the phenomenon of reduplication itself, in all its stages, can be studied with much greater precision than was previously possible. Details that have remained obscure until now may soon be cleared up.

'Artificial Reproduction' After All?

The daily press did not fail to seize on this 'sensation' and to spread it around, and occasionally to print headlines exaggerating it into 'artificial reproduction,' 'test-tube babies,' and so on. We can pass lightly over these, but since the announcement of 'test-tube babies' has been made every two or three years, with great regularity, for many decades – just as the Loch Ness monster emerges again each year – perhaps we ought to devote a short section of this book to considering what artificial replication implies.

In Spiegelman's experiment all the substances used were 'natural.' The RNA, the polymerase, and the four nucleotides were none of them the products of laboratory synthesis, but were entirely derived from living cells of the bacteriophage Qβ. It was only the experimental conditions that were 'artificial,' in the sense that they were an imitation of natural conditions. To that extent, the 'test-tube environment' is very precarious, because if the concentrations are not exactly matched, or if impurities – e.g., ribonuclease – are present, then it does not work any more. It is surprising enough that the system will work at all outside the cell, and that replication of the genome can go on with the greatest precision in a test tube.

It is even more surprising, however, that this labile system, which has to be operated with such great care, not only operates in the cell, but does so *with much greater certainty* than it does outside. In the cell there are thousands of metabolic products, and thousands of enzymes (including ribonuclease!), with ribosomes, m-RNA, and ions of all kinds, as well as many other substances. Many of these, by themselves, would be sufficient to cause the 'test-tube system' to break down, whereas in the cell they are obviously not harmful, nor does the replication process interfere with any of the other processes going on at the same time.

This is a question of *organization,* of course, but organization in this context means more than just a trifling arrangement in space and time or the carrying out of a short-term plan. The term implies *harmony* between all the partners and all the phenomena, and organisation on a scale far beyond our present powers of conception. This organization is self-contained, and yet it maintains contact and reciprocal action with its environment, ready to react, to transmit, to incorporate. Finally, it is the result of a process of selection and evolution, which has been in operation for millions of years, and which has left traces behind it.

It is not very likely that in the near future a computer will be built that can enable us to make a model of all this. Even if it were built, we should have far too few data to feed into it, probably not even one hundredth of one per cent of what would be necessary. Only when we could feed the computer with one-hundred-per-cent data would it be able to tell us how a cell equipped with this 'knowledge' would function (ought one to say 'live'?) – and we should need to know all this before we could build such a cell.

This makes the prospect sound gloomy and outside practical possibility. Somewhat more definite, and more relevant to the problem of artificial life is the following: instead of the phage-RNA we could obviously use an artificial genome, a genome made to our own design. In the test tube the experiment would fail, because the system there is adapted to using Qβ-RNA, and not any other kind. If we take RNA from the phage MS-2 – which is equally capable of infecting *Escherichia coli* Q_{13} Hfr – in place of Qβ-RNA, then the polymerase refuses to act. The living cell is certainly more flexible than this. If we supply the living cell with artificial RNA, it either refuses to accept it (case 1), or the cell takes it up, but recognizes it as 'foreign,' and immediately demolishes it with ribonuclease (case 2). A possible case 3 would then be as follows: the foreign RNA would be taken up, but would *not* be recognized as 'foreign.' In that event, it would destroy the harmony of the cell, and the cell would die, exactly as if it had fallen a victim to a lytic phage.

Only in one particular situation could a positive result be expected, and that would be if the foreign genome only departed from the normal to a trifling extent, and so could still be accepted as 'normal.' This would give rise to the possibility of experimentally adjusting the 'genome compatibility,' and the relationship between the genome and the other constituents of the cell.

The reader will have to decide for himself how small is the probability of ever being able to produce 'artificial life' or 'synthetic cells.' The author will now bring to an end his comments on life in the test tube, with a return to the question, Which came first, metabolism or the gene?

If complete genomes can now be replicated in the test tube, then it should certainly be possible to do this for a single gene. And in the primeval broth, too? This is not yet so sure. The experiments to which we should be committed have shown themselves to be very sensitive to the participation of unsuitable reagents. Obviously the multiplication of genes goes on in the living cell with very much greater certainty than it does in an experimental sample; and the experimental sample is more like a living cell than is the primeval broth, since it is an enclosed space, not a region without boundaries, as is the broth. In the latter not only the products of a reaction but the participants themselves, the enzymes, template-RNA, UTP, and so on, would quickly be lost.

If this is true of reproduction, it is also true of heterocatalysis, the third criterion of the 'ur-gene,' or original, primitive gene (see p. 298). What is the use of a macromolecule from the most perfect heterocatalysis if the product that has been extracted from the surrounding liquid under direction is immediately scattered again? How can it remain in the vicinity of the all-powerful gene if the gene cannot create its own environment and screen this off as quickly as possible from the outside world? A gene that is wandering around by itself in the primeval broth, is in fact powerless and meaningless. Hence the defender of the gene hypothesis is confronted with the same difficulty that we met several pages earlier, that we must find some way in which a 'portion' of the primeval broth could be isolated, either by a lipoid membrane, or by coacervation.

A third hypothesis runs as follows. Polymerization in the primeval broth gives rise to macromolecules of DNA, many of which have a 'meaningless' sequence of bases. Nevertheless, a few may act as primitive genes, and at first they need carry only a little information. Life might have arisen if a gene of this kind became enclosed in a droplet that was reasonably 'matched,' so that the two could work together, rather like a primitive nucleus in a primitive cytoplasm. The two might then gradually adapt themselves to each other, to form a co-ordinated, harmonious system, able to carry on metabolism, growth, and reproduction.

No one at present is able to say which of these three hypotheses is correct, or whether all three are wrong, and a fourth must be found. And so, perforce, the question of whether reproduction is older or more recent than metabolism must remain unanswered.

7.06 Order out of Disorder

Is It Too Improbable that the 'Right' Macromolecules Should Find Each Other?

The origin of the first gene is lost in a fog of vagueness and uncertainty, and we have dodged the most serious problem by saying simply, that 'Polymerization in the primeval broth gave rise to macromolecules of DNA. Many of these had a meaningless sequence of bases; but a few may have acted as primitive genes.' These two sentences seem so obvious that one is apt to read them over without thinking any more about it. Actually they get round an awkward fact which otherwise might reduce every speculation about the origin of life on earth to an absurdity. Remember our 'exercise in permutation and combination' on page 26? How did it go? We learned that from one protein with 150 amino-acid groups attached, about 10^{130} combinations were possible. This time let us make another calculation, on a somewhat larger scale, thus:

The earth may be about 3,000 million years old. A protein molecule arises every second, and later on – depending on what is its half-life – it is demolished again. Thus over the earth as a whole, *at any one time,* there can be only about 10^{60} different protein molecules though we have seen that it is theoretically possible to construct 10^{130} different kinds, i.e. 10^{70} times as many. The probability that the 'right' ones will happen to be among those 10^{60} is extra-ordinarily small. However, the chance is even slimmer than that. These millions and millions of molecules would be spread over the entire earth in a layer one meter thick, and in every cubic centimeter of this layer there would be 10^{19} (ten million million million) macromolecules. The probability of the right ones coming together in such a layer is effectively nil. How could the right molecules come together if one of them was in Iceland and the other in Australia? In the face of such masses of molecules no imaginable selection mechanism would work.

What applies to the proteins applies equally well to the nucleic acids, and the reader will surely agree with the author that the two sentences above are an irresponsible oversimplification. Natural selection of random polymerization throws no light on the origin of the gene.

Juggling with Balls in the Balance

Shall we then give in, bring our investigations to an end, and attribute everything to a

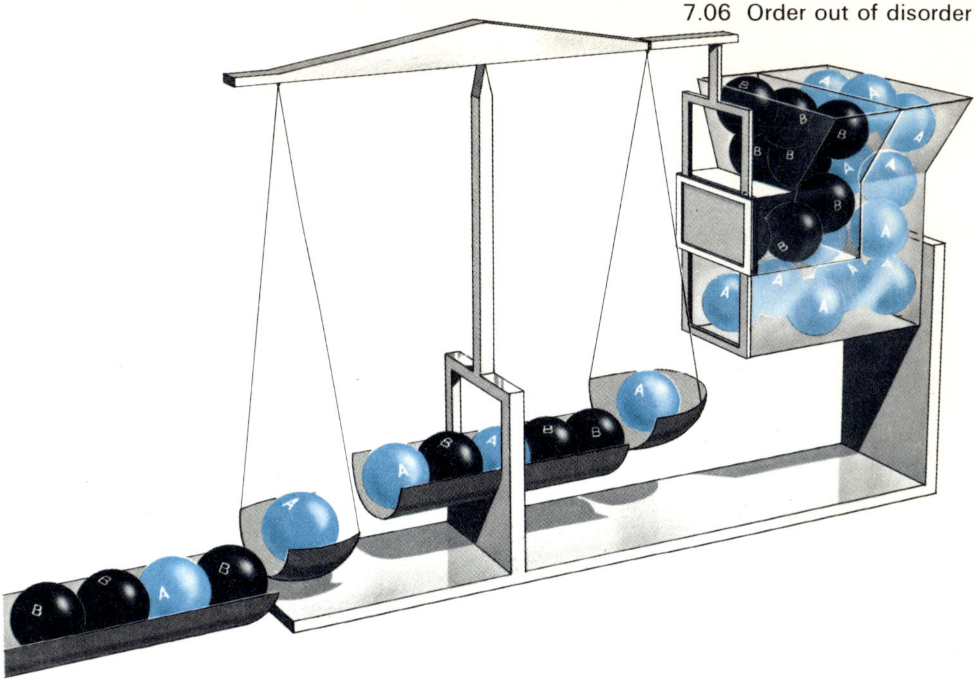

177. This balance produces an orderly sequence of heavy (blue) and light (black) balls, as described in the text.

single act of creation? This is a matter of conscience, of faith, or of temperament, but it is not a scientific answer. The scientist demands, and searches for, causal phenomena, i.e., for cause and effect. Since he cannot very well experiment with selection that has taken place long ago, he chooses instead to research into polymerization. If polymerization by pure chance is not good enough, then perhaps there is some mechanism that shortens the odds. Perhaps the sequence of bases is not random, or not all sequences have the same probability, or some sequences are favoured over others. Perhaps there is some device by which order arises out of disorder.

This sounds even less likely than the affair with 10^{130} protein molecules, yet mathematicians and calculation experts have shown that such a thing is possible; co-operating with biologists and biophysicists, they have thought up a computer that 'produces order.' We will try to follow their train of thought – and to do so without involving ourselves in mathematics, or with electronic brains.

Let us begin by describing a mechanical model, a relatively simple balance, which has two scale pans at different heights (see Fig. 177). The base of the balance also supports an arrangement of two storage boxes, Box A and Box B, which in the diagram contain blue and black balls respectively. The movement of the beam of the balance opens Box A or Box B, but only one at a time. Finally, there is a trough along which the balls run from one pan to the other, and this trough is just long enough to hold *five* balls. The crux of the matter is that the blue balls in Box A are heavier than the black balls in Box B.

1 … 29
B B A B A A B A B B A A A B B A B B B B A B B A B A B B A

30 … 58
B B A A B A A B A A A B B B A A A A B A B B B B B A A B A

59 … 87
B A B B B A A B B A B A A A B A A B B B B A A A B A B A A

88 … 116
A A B B A A A A A B A A A A A A B B B B B B B A B A B A B

117 … 127 128 … 145
A A B B A A B B B A B B B A B A A B A B B A A A B B A B B

146 … 174
B B A B B A B A B B A B B A A B A A B A A A B B B A A A A

175 … 203
B A B B B B B A A B A B A B B B A A B B A B A A A B A A B

204 … 232
B B B A A A B A B A A A A B B A A A A A B A A A A A A B B

233 … 254 255 … 261
B B B B B A B A B A B A A B B A A B B B A B B B A B A . .

178. Sequence of light and heavy balls produced by the balance shown in Figure 177. The grey shading shows up the fact that this seemingly random pattern repeats itself every 127 terms.

The balance works as follows. In the situation shown in Figure 177 the balance has a blue ball in each pan, and so is in equilibrium. The construction of the balance is such that under these circumstances Box A is open, and so another blue ball runs down into the right-hand pan, dislodging the ball already there. This pushes the five balls along the trough, and the blue ball on the left falls into the other pan. The balance remains in equilibrium this time, but when the next blue ball rolls out, and the balls move along the trough again, this time a *black* ball falls into the left-hand pan. Since the black balls are lighter than the blue balls, this allows the left-hand pan to rise, and the door moves, closing Box A and opening Box B, thereby releasing a *black* ball next time.

If the model is allowed to continue in action, and if the two boxes are kept full of balls of the right kind, a long row of balls will accumulate on the left, with a variegated arrangement of blue and black balls, which seems without any regular plan, quite random. Figure 178 shows the sequence that will arise in this way – it looks rather boring.

Now the sequence *is* boring when it merely concerns balls, but it becomes quite different if instead of balls, we take A and B *amino acids* and build these up into a macromolecule – i.e., polymerize them into a protein or a nucleic acid.

As soon as we talk about proteins, etc., we are back on familiar ground, and we can see the relevance to the problem of the primitive gene. If we consider the sequence in detail, we begin to see that it is not quite random. *The series repeats itself every 127 terms* (as is shown by the shaded area in Fig. 178), and the mathematicians can prove to us that this must be so.

At any one time the model balance involves

seven balls, one in each pan, and five in the trough between; hence the system must have a 'maximal period' of $2^7 - 1 = 127$. The model balance is a kind of computer, and has therefore given a notable advance over the total randomness of chance polymerization, yet actually the computer is not responsible for the degree of order that has appeared, and this is independent of the opening sequence. The order is merely the result of the laws of combination.

The computer in this case, like all computers, does no more than carry out the certain operations. In this case, the mechanism:

1. *Chooses* between A and B – i.e., decides which ball to release.
2. *Compares* the new ball with one already present – by balancing them against each other.
3. *Adds together* – a new ball, and according to laws which are simply related to the nature of the two preceding balls.
4. *Has an output* – of a ball which is added to a row – determined by the same laws as in No. 3.
5. *Keeps a running total.*

The result of this series of operations is a degree of orderliness that depends only on the nature of the two elements, A and B, and on whether it is possible to compare them and decide between them.

Can Molecules Count?

The question, Can molecules compute? was asked by Professor Pattee in an enquiry into the origin of sequences in macromolecules. The reader may ask the same question, since it is in the highest degree improbable that in the primeval broth, with its random, statistical splitting up of molecules, there was a computer that could produce molecules of protein or

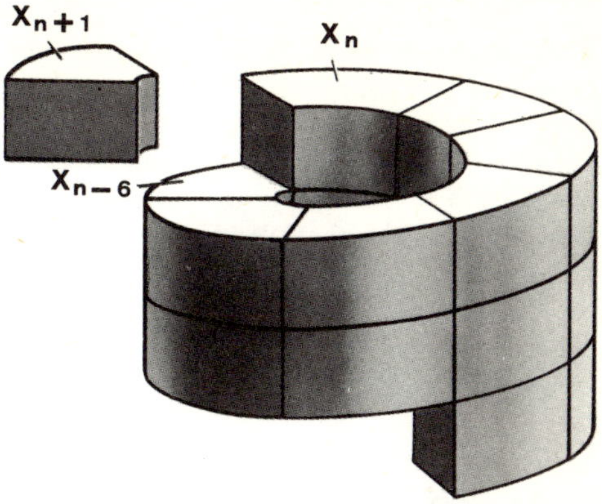

179. How a 'calculating spiral' can produce an orderly sequence of units, provided that each new unit (X_{n+1}) is determined by the one that precedes it (X_n) and the one in contact with it below (X_{n-6}).

of DNA with orderly sequences in them. Even if there had been, we should then have to point out that the computer itself is a highly organized system, and so it is not surprising that it could induce order in other things. Our original questions would still remain unanswered, since no one would know where this computer itself came from: did it drop from Heaven? Are there, then, macromolecules which, by their own efforts, can carry out the five steps listed above?

The best molecules to study are those that we know can operate as a spiral, such as the α-helix of proteins, or the double helix of DNA. If such a molecule does not originate in the extended form and then become coiled, but *grows as a spiral from the beginning,* then the following interesting situation arises:

To preserve the analogy with the balance model, we will assume that the macromolecule consists of only two different kinds of unit, A and B, and let there be seven such units in each turn of the spiral. Figure 179 shows such a spiral, with two coils in existence; and already after the first turn the place where a new unit will be formed can be seen as a kind of niche.

This niche is bounded *on the one side* by the preceding unit (X_n), onto which the new one will be joined, and *on the other side* by the unit numbered X_{n-6} which lies immediately underneath it (i.e., six units backwards). It is easily possible that these two adjoining units, X_n and X_{n-6}, in some way or other will exert an influence on which kind of unit is chosen to fill the space X_{n+1}. They may chose between A and B in the sense that A may 'match' and B may not.

We may formulate the following rule: if unit X_n is of the same kind as unit X_{n-6}, then unit A will be taken on in the vacant place; if unit X_n is not of the same kind as unit X_{n-6}, then unit B will be taken on.

Such a rule is quite feasible, and indeed reasonable, as we shall see later. For our purposes it has the advantage that it corresponds exactly with the rule which operates in the balance-computer in Figure 177, where if the balls in the two pans are similar ($X_n = X_{n-6}$), one kind of ball is released, and if the balls in the pans are different, the other kind of ball is released. Then the sequence of units in the macromolecules will be that of the balls in Figure 178. For the sake of completeness, we may add at this point that, beyond the periodicity of 127 units, there is a third-order periodicity of $7 \times 127 = 889$ units, and this leads on to a number of further deductions.

Thus molecules can, in fact, count, and even calculate. If the units, because of the number and arrangement of their individual atoms, have a quite distinctive spatial structure, and a quite distinctive set of attachment bonds (one day the chemists will tell us exactly what these are), then they will not merely grow to macromolecules, but they will also keep count of themselves; they will be able to 'calculate.'

Of course, a macromolecule that has had its arrangement dictated in this way is no more than a reflection of some kind of chemical order that was already present in the component units. Once again we see how molecular biology helps to clarify any kind of biological problem.

The reader will see the importance of such an ability to create order out of disorder. He

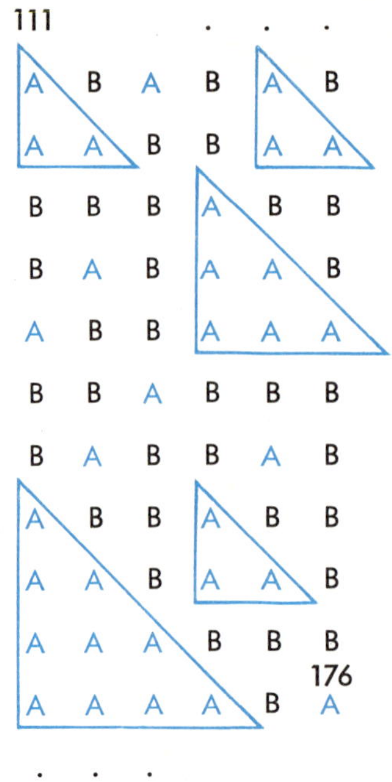

180. Terms 111–176 of Figure 178, rearranged into a table of six columns, produce a pattern of A's arranged in right-angled triangles.

may prudently ask, however, whether in fact any such orderly sequences exist in any of our macromolecules today.

There is one bit of evidence for this. The sequence of amino acids is known very precisely in the virus of tobacco mosaic. We can find the exact probability that the

181. A similar pattern to that of Figure 180 can be produced from the sequence of bases in the tobacco mosaic virus.

125 ASP	ILEU	ASPN	LEU
ILEU	VAL	GLU	LEU
ILEU	ARG	GLY	THR
GLY	SER	TYR	ASPN
ARG	SER	SER	PHE
GLU	SER	SER	SER
GLY	LEU	VAL	TRY
THR	SER	GLY	PRO
ALA	158 THR		

distribution of the amino acids found in tobacco mosaic virus is not random – but only as long as it is read off from one particular end; if it is read from the other end, a completely random distribution emerges! This result conveys very little to anyone who is not interested in mathematical methods, but it can be made a little clearer in another way.

Let us write down once again part of the sequence of A's and B's we arrived at from the model with the balance, and from the 'calculating molecule,' but in a slightly altered form. We will write the terms down in six columns, beginning with number 111 – i.e., the first line of the new series will consist of the last six characters on the fourth line of Figure 177. We then get the sequence shown in Figure 180, and within this pattern it is possible to draw a number of right-angled triangles each of which encloses *nothing but A units*. This is no accident, but is a consequence of the peculiarities of this particular model.

Now let us take a sequence of groups from the tobacco mosaic virus – which, we remember, has been accurately mapped – and write them down in a similar fashion, except that we will put them into only *four* columns because the helix of this molecule has only 3·7 amino-acid groups per turn. It is completely unexpected to find that in this pattern we can immediately draw a triangle which encloses *only serine* residues (Fig. 181). As far as it goes, this seems to be a parallel of the model we used above.

Of course, this evidence is nowhere near enough for us to conclude definitely that sequences of amino acids and bases are not random, or to attribute them to some pre-existing, orderly system, but this is one of the indications that we were hoping for. And we can say with some justification that we had not tested anywhere near the total of 10^{130} possible variants in the sequence, while the number of macromolecules actually present was far fewer than this, and we can claim that we were dealing, not with a random polymerization, but with one that shows more or less evidence of orderliness.

Thus order really can be produced out of

disorder without the intervention of a living organism. Even so, we must remember that this order did not come out of the blue, but was already implicit within the molecules themselves, in such sub-units as amino acids and mononucleotides.

7.07 Drawing a Picture of the First Living Things

From Fermentation to Photosynthesis

In this way we can get round at least the most serious difficulties that stood in the way of our attempts to account for the first gene, and we can form some conception of the primitive gene, and of the 'metabolically active droplets.' On the other hand, we are still in the dark about the mode of origin of the first *cells,* and it should be stated clearly that we know next to nothing about them. What we do have is the chance both to think around this subject and to perform experiments, and it is a good deal better to do this than to speculate wildly as in the past.

We have gone as far as to give the first living things a name: *eobionts,* from the Greek *eos,* dawn, or 'very early.' In order that the reader may form at least some conception of an eobiont, let us try to describe one, even if we do know nothing about its origin.

The specialists' journals are always bringing out more accounts of how the electron microscope has been used to demonstrate the remains of the most primitive organisms in Precambrian rocks. Mostly, these take the form of spherical or rod-like fragments, which are a long way from resembling the traces of bacteria or carbonaceous fragments that are known from more recent deposits. In general, the view that these are the remains of eobionts has not stood up to criticism from specialists, because we have been too often deceived by completely inorganic formations which have an amazing resemblance to cell structures. Only a few months ago pictures appeared in an American scientific periodical of organic traces obtained from a deposit that was three thousand million years old. We shall wait and see what the specialists say about those before we commit ourselves to describing what the eobionts *looked like.*

We can be a little less non-commital about the biochemical performance of eobionts. One thing about them is even *certain!* They were not self-sufficient in carbon, and in order to form their organic carbon compounds, they did not make use of CO_2; instead, they drew nourishment from more or less reduced carbon compounds just like those that were present in the primeval broth. They were thus heterotrophic (feeding upon a variety of things), and they could not perform photosynthesis under any circumstances.

This is reassuring in one respect. If the living process had been based on photosynthesis from the beginning, then we should have been hard put to it to explain how the quite exceptionally complicated mechanism of photosynthesis, with all its component parts, could have sprung into being. On the other hand, a metabolism that made use of a variety of materials fits in very well with what we know to have been present in the primeval broth.

This also fits in very well with the fact that most living organisms of the present day are heterotrophic (most bacteria, fungi, animals, and man). Even modern green plants themselves have chlorophyll in only a tiny minority of their cells; in the interior of the parts of the plant that are above ground (leaf stalks, branches, stems), as well as in all the underground parts, there is no chlorophyll, no photosynthesis, and no self-sufficiency in carbon – only a consumption of organic compounds that are drawn in from outside.

Another aspect is worth noting. We again fall back upon familiar topics, this time respiration and fermentation. Both start with sugar and lead along parallel lines, to pyroracemic acid. At this point the alcoholic fermentation diverges, splitting off carbon dioxide from the pyroracemic acid, and the acetaldehyde that arises is *hydrogenated* to ethyl alcohol. 'Hydrogenated' means having hydrogen attached, hydrogen that has been split off from some intermediate product, and the acetaldehyde could be said to serve as a hydrogen acceptor. Consequently the fermentation goes on without any oxygen, anaerobically, just as it would have in the primeval atmosphere.

It is a different matter where *respiration* is concerned. The pyroracemic acid is completely broken down by dehydrogenation (loss of hydrogen) and decarboxylation (loss of CO_2), and the hydrogen produced combines with the oxygen of the air to form water. Putting it in other words, in *respiration* the part of hydrogen acceptor is played not by acetaldehyde, but by oxygen. So, of necessity, respiration is aerobic.

When expressed in this way it is clear that fermentation does not really 'diverge' from respiration, but works hand in glove with it and can confidently be looked upon as the oldest and most primitive method of breaking down carbohydrates. On page 166 we noted that fermentation, or anaerobic glycolysis, took place in the almost structureless cytoplasm, whereas the associated phenomenon of respiration went on in the highly organized mitochondria.

At first the eobionts would not be troubled very much by the fact that the output of ATP by fermentation is very small (see p. 165), because they could easily obtain fermentable substances from the primeval broth. As an immediate consequence of the activity of the eobionts, the primeval broth became steadily thinner and the atmosphere progressively oxidized (because of the withdrawal of hydrogen), and so materials were no longer so lavishly available. The process of respiration had the advantage of using the oxygen of the atmosphere as hydrogen acceptor, as well as that of making a great deal more ATP.

Quite certainly the eobionts did not go to work in such a crude and above all so eminently practical and purposive a way, but selection and evolution would compel them in this direction.

What about photosynthesis? We have always thought of this process as being tremendously important, and as the basis of all present-day life, and so we must give it a few lines. Undoubtedly photosynthesis arose *after* CO_2 had begun to be formed as part of the process of oxidation and reduction of the atmosphere; the process did not come into being all at once, but by instalments.

The reduction of CO_2 to the level of a carbohydrate ($-CH_2O$) requires, first, hydrogen, and second, energy in the form of ATP. There was no shortage in the primeval broth of compounds containing hydrogen, and so hydrogen was easy to come by. ATP, on the other hand, can be obtained only inefficiently from the breakdown of carbohydrates. From this breakdown the cell obtains ATP, which it uses in the reduction of CO_2 – that is, for the construction of more carbohydrate, which will be broken down again to obtain more ATP. It is like a cat biting its own tail!

Hence ATP must be produced by some other means as well. At the remote period with which we are concerned it may have been obtained by a primitive form of photosynthesis. Chlorophyll was present, and was able to absorb energy in the form of light, using it (by very complex methods of which we still do not know all the steps) to convert ADP into ATP.

In the course of millions and thousands of millions of years the primeval broth became steadily weaker, eventually consisting of nothing but water and inorganic salts, and naturally this was the end of 'cheap' hydrogen. However, since hydrogen in some form or other is essential for the reduction of CO_2, living organisms were forced to turn to the one source of hydrogen that was readily available – that is, to *water*; but to use hydrogen from this source needs energy (see p. 289).

Those who stress the importance of photosynthesis will at once think of light as this source of energy, to split water into hydrogen plus oxygen, and so to make hydrogen available again. This is the 'second' light reaction (and since the 'first' light reaction goes on producing ATP as before, the necessary energy for the reduction of CO_2 is assured). Four of the OH radicals that result combine together to form two molecules of hydrogen peroxide, H_2O_2, each of which then splits into a molecule of water and one of oxygen – which is thus photosynthetic oxygen.

The two light reactions were linked together in the most ingenious ways, which we will refrain from describing. As it is, we have described the two reactions too crudely, and oversimplified them to the point of error. To describe the complete photosynthesis in all its details (as far as these are known) would take up far too much space in this book. Somewhere or other we have to draw a line between textbook and a book for general information. Since ours is the latter, we can omit the details, other than those which indicate the general direction in which evolution has taken place.

So we have had cells with fermentation, then with respiration, then with the 'first' light reaction, and finally, with the 'second' light reaction. It is still debatable whether this is the order in which these processes actually came into being, or whether, say, the 'first' light reaction came before respiration. The important thing is that a process of *evolution* can be detected, and that this time it is an evolution of *organisms*, and not a chemical evolution of molecules and macromolecules. This new evolution leads from the fermentation of the eobionts right down to the living cells of today.

This chapter really comes to an end at this point. What follows – the progress of evolution from the original single-celled (or *acellular*) organisms to plants, animals, man – is already known to us in principle. It is Darwin's natural selection operating upon mutations. To go into the full details of this would require not just another chapter, but an entire book, with a totally different aim, and asking quite different questions.

So let us recapitulate. Biology cannot *prove* that life came into being without the help of a Creator: yet enough evidence – including experimental evidence – has been accumulated to show that the origin of living organisms on earth can be explained by 'natural' causes, using only the same physical and chemical laws that hold good in the inanimate world. Both chemical and biological evolution are involved.

This is rather a weak conclusion to leave as the climax of this chapter, and of the whole book, and perhaps we ought not to evade considering the implications of it. We might look forwards instead of backwards, and as counterpart of the primeval broth, ask ourselves, What happens next? And since man's interest has always centered upon man himself, this question comes down to asking, What have biologists to say about the future evolution of man?

7.08 Where Is Evolution Leading?

Pros and Cons of 'Progress'
Controlled Selection

It is indisputable that biology can do a lot to help human beings, medically and in their daily lives. Inoculation, immunization, correction of a disturbed metabolism or of a hereditary defect, making available new sources of food for human consumption, pest control – these are only a few of the benefits of modern biology. Can we expect similar benefits to accrue from the study of evolution?

The reader may remember, if only vaguely, alarming headlines in the papers a short time ago about the 'population explosion' that is expected all over the world. The most startling headlines were taken verbatim from remarks of scientists, medical men, and molecular biologists, all of the highest standing, and among them many winners of Nobel prizes, who had been attending a symposium in London. Examples were to wonder aloud if man had the 'right' to have children, and to suggest that one million tons of men, in England alone, were 'surplus.' If the fruit of modern biology on the tree of evolution is like this, should one not have serious doubts about it?

We cannot and will not evade such questions, and therefore this final chapter shall be concluded with a section about man in the future, as many biologists see him. Though 'research' in the scientific sense is hardly possible, we shall at least give some thought to asking what influence the new discoveries in molecular biology may have on the future of the human species.

Let us consider evolution from this point of view, and from today onwards. It looks as if it will be to some extent homogeneous, but not absolutely along a straight line, because otherwise we should also expect to be able to trace one straight line, without gaps or branching, back from man (considering man as being the most highly evolved organism) right to the eobionts. Instead of this, what we actually find in the past is a 'dendrogram,' or tree-like arrangement, leading to a great many species of animals and plants.

Nor is evolution entirely homogeneous. We have already heard of a chemical abiotic evolution, which itself is only a later phase of a cosmic evolution. The first stage after chemical evolution is organic, or biological, evolution, which leads from the eobionts to fungi and trees, to jellyfish and vertebrates.

Many scientists are of the opinion that this biological evolution has been supplemented, or even replaced by a *third phase*, which they have named *psycho-social evolution* (Greek *psyche*, the spirit). By this they mean an evolution of human society in all its aspects, 'from machines and works of art to religions and sciences' (Sir Julian Huxley). Human societies, and social groups, are based upon characteristics of man which were newly acquired by him, and which he shares with no other animals: spirit and intellect, thought and speech, tradition and education – or to put it prosaically, he has a much bigger surface area to his brain. But let us leave these debatable matters alone and go on to consider what are the *consequences* of these things for the future of evolution.

The motive force of biological evolution was natural selection, or rather, the pressure of natural selection. In the psycho-social phase this is supplemented by a *psycho-social pressure*. Social ambitions, wishes, and hopes, intellectual and technical abilities, sense of tradition and hopes of progress, even love and hate, may all be stronger than the simple pressure of natural selection, and may be determinant factors in evolution.

The line, or more accurately the direction,

of evolution may be altered as a result. Whereas in the biological phase of evolution the overwhelming aim of life was to survive, to continue to exist (hence Darwin's 'struggle for existence'), today man's urge is towards *fulfilment*, towards *self-involvement*, and in this pursuit he makes special use of his spiritual and intellectual powers, and of his social abilities. This means no less than that 'blind' biological is replaced by conscious, self-accountable selection, selection that is both organized and carried out by man himself.

Now, every form of evolution, whether chemical, biological, or social, is ruthless in its own way, as well as being irreversible, and permitting no return to a previous state: evolution is *total*. Only those who make themselves as 'fit' as possible will survive; hence man must try to use all his potentialities – economics, science, culture, education, and even sport, games, and recreation – in the cause of his own evolution.

Biology has a particularly important part to play in all this. In the first place, biology, like any other science, and indeed intellectual, spiritual, and social activities generally, must contribute in its own way to the improvement of human affairs. It does this by safeguarding and improving food supplies, making agriculture more scientific, trying to cultivate algae in practical quantities, elimination of important pests among plants and animals, and so on; and in co-operation with medical men, by eliminating the agents of various diseases, perfecting and standardizing methods of medical treatment and operation, making artificial organs for the human body, and in many other ways.

This, however, is merely the conventional aspect of biology, in which it has already given results, and in which it has been so fruitful that the name 'biological engineering' has been invented. The contribution that we look for from evolution is much more significant. As evolution learns how to explain and then to *control* all aspects of life, it must exercise its new power. Evolution will establish a new line of selection for man, the 'perfect social being,' and must see to it that this happens as directly and as quickly as possible, and not just after ten thousand generations or so. Thus it seems to be the objective of molecular biology and molecular genetics to speed up evolution by genetic control and genetic engineering. At least this is what we were told at the London symposium!

Stating a programme in general and noncommittal terms like this sounds comparatively harmless, and has its positive side, but the details, admittedly, are disturbing, calculated to cause alarm and shock. Some of these we cannot ignore, since we have already mentioned them at the beginning of this book. But before touching on these, let us begin with a few prosaic reflections, just to ease the shock.

It must be obvious to everyone that we are no longer dealing just with the 'good old' biological selection. Man has already interfered with nature, and by deflecting it more and more from its natural course, he has upset the much-vaunted 'balance of nature.' The air above our industrial areas, the dirty, smelly streams running through our landscapes, the sprayed fruit, the food that is *not* 'untouched by human hands,' and many other things are eloquent witness of this human interference, and make everybody aware of what previously was known only to well-informed people.

This destruction of the natural order of things did not begin only in the twentieth century, but goes back much further than that. This process probably began when man, after having been a hunter and a collector of food for hundreds of thousands of years, living on berries, roots, fungi, on the 'fruits of the sea' and the 'spoils of the chase,' finally went over to producing things for himself: first food, then consumer goods. The industrial revolution was responsible for the most unpleasant and oppressive features that are still with us today.

There is no question of man having tried to alter natural selection deliberately, in order to improve his lot; in early times he neither knew of nor wished for anything better. The reasons for man's restlessness are much simpler – simple egoism, and looking for an immediate advantage for himself. That will concern us later.

If biological evolution *is* being given a twist by man, then this is certainly not on a world-wide scale, nor equally in all social groups, but the effect is essentially confined to industrial countries. This is shown by the way in which the so-called developing countries are trying so hard to turn themselves into industrial states. It is not the material means to do so that they lack, but the phycho-social assumptions, and these are such that the process of emulating industrial countries will be long and hard. From this it can be deduced that 'productive man,' without himself realizing it, has already changed the course of selection and evolution, and begun to adapt himself in a different direction.

But in what direction? Will this lead to a desirable and self-sufficient direction of life and evolution from now on? Considering how this diversion of evolution came about – as we have seen, from pure self-interest – it is very doubtful if it will, but it is still much too soon to say. Finally, what are two hundred years of industrial development in the face of fifteen hundred million years of biological evolution? On the clock system that we described on page 290 this would be a fraction of a second out of twenty-four hours. If a slight deflection comes at the end of a long straight highway, who can tell, immediately, where it is leading to? The third stage of evolution, which starts with such a 'diversion,' might of course continue to be a psycho-social evolution based on self-interest. We will call this Hypothesis I (or the London Hypothesis).

However, things might go quite differently. Man, or one of his social groups, might quite simply eliminate natural selection, step by step, but very systematically, according to a plan. He has no new aim or goal of selection in mind: instead he puts in place of natural selection the goal of a progressive universal culture, with the help of which he hopes gradually to supersede natural selection. Of course this is to overlook the fact that this will alter the whole course of evolution in directions that will by no means be desirable but undesirable and unwanted. This would be Hypothesis II, which sounds as harmless as Hypothesis I, but involves equally unpleasant details. So now is the time for us to consider the details of both hypotheses.

7.09 Hypothesis I: Self-sufficiency

The Government Decides Who Shall Have Children

The growth of scientific knowledge and technical skill is in contrast to the very much slower progress in the social and cultural field. If we apply the term 'social' not to 'social welfare,' the 'welfare state,' and 'achievements in the social field,' but to the spiritual, moral, ethical, and cultural impulses of every social group, from single families to entire nations, then what has been achieved so far throughout human history is simply depressing. Great nations have collapsed, whole civilizations been wiped out, vast landscapes laid waste, whole peoples uprooted. Conflict, revolution, war, famine, and destitution have been commonplace. All these, and much more besides, have been brought about by men who tried to direct the course of evolution consciously and relying on their own efforts.

Does this not all lead to a lowering of standards, to a deterioration, and even to worldwide collapse? Many biologists are of this opinion. Sir Julian Huxley, former president of UNESCO has said, 'It is clear that the general quality of people throughout the world is not very high, that it is beginning to fall, and that it can, and must, be improved.' If things are to be altered for the better, then men who are socially and culturally fit are needed as quickly as possible. But which qualities should such men actually possess, and how can they be obtained?

One thing is clear from what has gone before in this book. The molecular geneticist can define man genetically, in terms of a double helix of DNA, with five billion pairs of nucleotides, but he is only too well aware that there is no *social gene* among them, and so he has no hope of being able to change man for the better by chemical or physical means. So we have to fall back on the method of *eugenics*, well known, and with something of an evil reputation. The idea is to improve the genetic stock or hereditary material by breeding, to put it bluntly. There is widespread agreement in principle, but what should breeding *aim for*, and how can this aim be attained?

If we aim to breed the 'perfect social being,' at the highest cultural level, then we must have some means of picking out those individuals who should propogate; we need some kind of test of social and cultural potentialities.

The Intelligence Quotient

Any discussion of this subject soon introduces the *intelligence quotient*. This is a number, worked out from a carefully planned questionnaire and psychological tests, which puts a quantitative value on a person's 'intelligence.' The average value of a whole sample population is set at 100. If the I.Q.'s of different social groups within one population are compared, it cannot be denied that the higher the social level, the higher the I.Q. Particularly high values – 150 and above – are found among scientists, artists, statesmen, and leaders of industry; especially low values are found in the lowest social levels, among layabouts and criminals. So what could be simpler than to select people for their I.Q.'s?

No one would be presumptuous enough to take the I.Q. as the *sole* test for possession of the 'social gene,' but in spite of the shortcomings of any single test, or test procedure, it can be used temporarily as a substitute for a whole range of further tests that have still to be worked out. It must be borne in mind, for instance, that intelligence and mental instability are by no means mutually exclusive. What good would it be to select intelligent psychopaths, or habitual criminals with high I.Q.'s. *Any and every* mistake in choice of tests could have fateful results. So much – or rather, so little – for the aims of human selection.

How could such an objective be attained, once it was decided upon? It could be achieved without compulsion only if people were well enough informed, and had a sufficient sense of responsibility, to marry only such a partner as could guarantee offspring with a still higher I.Q. than their parents. This would be a very long process, occupying hundreds or thousands of generations, so long perhaps that man would have exterminated himself in the meantime. Moreover, it is questionable whether any decisive change in traditional attitudes could be achieved by exhortation. Finally, it is just the people with the lowest I.Q.'s who are the least enlightened; they would go on as before, while the people with higher I.Q.'s would have fewer children – as in fact happens already.

People oppose any restraint, whether mild or harsh, and do their best to weaken its effect. Since selection according to I.Q. cannot be done without some form of restraint, supporters of this method of eugenics believe that people should be forcibly restrained in their own interests, in order to promote social and cultural improvement as quickly as possible.

The most radical method that was put forward at the London symposium was as follows. The government (!) should mix a suitable contraceptive chemical with our food, so that in the ordinary way every conception would be prevented from the start. An antidote to this substance would be held in store, and released only to carefully selected couples. This would not be such an extreme suggestion if we accepted the idea that the right to have children was not a private decision.

Certainly a 'child license' of this kind would considerably shorten the time needed to raise the average I.Q. from 100 to 110, or from 160 to 200, and perhaps this could be achieved in a century. But has such a plan any chance of being generally applied? Not to mention the frightening aspects of such an idea, it presupposes a world-wide supervision, without any exceptions, and this makes it hopeless from the start. In every country, there are, and will be, a certain number of people, usually with low I.Q.'s, who are very crafty and cunning in avoiding every form of restraint, and these would reproduce as vigorously as before.

A further consideration is this. Since the third phase of evolution is confined to the technical and industrial countries, it may be expected that the people of the rest of the world would see no reason why they should submit to this sort of restraint, and would not do so. Even if the idea did catch on, one country would have to make a start with birth control, and carry it through on a large scale. Then, after twenty-five or thirty years, when this country had begun to produce Nobel prize winners, other countries would rush to adopt the idea! Now the reader will doubt – and the author does too – whether commercial and military interests would be particularly impressed by such a system. Any idea of population control by 'baby pill' is a fantasy.

So far we have considered only a general supervision or restriction of conception, but the second part of the 'child-license' programme seems even more mischievous. Who will operate this licensing system – the government of each country, or some international authority? And by what standards shall the 'desirables' be chosen – by regional or international standard of I.Q.? The one is as dubious as the other. To apply an I.Q. regionally would classify people into first- or second-class categories; to use an international standard would condemn whole peoples as outsiders because they would not produce anybody with a sufficiently high I.Q. – and this just at the time when we are trying to protect such Stone Age relicts as the Australian aborigines from extinction.

Obviously any committee that had to take decisions about 'child licenses' would be controlled by Parliament, but even in the best democracy – to say nothing of pseudo-democracies and dictatorships – it is difficult for the government to withstand pressure from groups with a vested interest.

Now, in the last resort these are matters, not of biology but of administration, and could presumably be solved, if it ever became necessary. And it might actually happen, one day, that all countries might be obliged to work out a common policy.

There is another objection that is much more serious. The desire for sexual fulfillment and the desire for children are two quite different things, and as the foregoing shows, man can now separate them, satisfying the first and denying the second. Both, however, are legitimate desires, because they have a biological basis. In women the desire for children is often stronger than any other; it is a mistake to think that this desire is 'educated' into them by giving them dolls to play with; on the contrary, little girls have an instinct to play with dolls from a very early age. As far as we can tell, this instinct has a genetic basis, and so far we have not found any means – nor have we the least intention – of suppressing it. The maternal instinct will always exist, but in the future, with the world population at the predicted high level, it must remain unfulfilled for the great majority of women. How does this tie up with the stated aims of modern society, namely fulfilment and self-involvement? Probably women would rebel against this situation – at any rate, those who could afford it would provide themselves with black-market children, the offspring of liaisons not blessed by the authorities.

AID – Artificial Insemination

All the same, it would not be fair to dismiss all plans for the improvement of the human race just because they would be inconvenient, or because they sound frivolous to us. They were certainly not devised in a spirit of frivolity, but out of deep concern, and a sincere sense of responsibility for the future of mankind. Possibly these ideas were couched in such frivolous terms in order to startle those in responsible positions out of their indifference.

A quotation from the London symposium may be mentioned in this context. The speaker said: 'I should be very much happier if I could be certain that our governments were aware of even a part of what goes on in biology'; this shows very clearly what the situation is. The idea is only very slowly gaining ground that while the existing frameworks of law and political economy are still as necessary as they ever were, yet they are not adequate by a long way, to govern a modern state. Without scientific, and particularly biological knowledge, governments repeatedly arrive at grotesquely wrong decisions, as has been seen often enough in recent years.

When we realize that the scheme for 'child licenses' was thought up in all sincerity, then we are all the more disposed to question other plans which have even wider and more profound aims, and which want to bring about improvements in the human condition even more quickly and certainly. AID is one of these ideas. The abbreviation, which not quite unintentionally forms the English word *aid,* stands for **A**rtificial **I**nsemination with semen derived from a **D**onor.

The question of artificial insemination has been very widely discussed as a problem in morality, law, and hygiene, and it is not our intention in this book to take sides in this dispute. We are concerned essentially with the biological aspect of AID.

AID is based upon the fact that human sperm can be kept for long periods by suitable treatment. It must be frozen immediately in liquid nitrogen; then when it is needed it is quickly thawed out, and introduced into the uterus along with a ripe ovum, so that fertilization can take place at once. This procedure has been known for a long time, and has been practiced, for example, in cases where a natural insemination was undesirable or impossible. As a rule, the name and character of the donor is a closely guarded secret.

Biologists remain non-commital about using artificial insemination for deliberate breeding purposes, but for the adherents of the Third Evolution the idea is fascinating. One cubic millimeter of human sperm contains millions and millions of spermatozoa, enough for hundreds or even thousands of artificial inseminations. Hence we 'need' no more than a few dozen of the 'best' men (those with the highest I.Q.'s, who are also culturally and socially of the highest category). Suppose that all natural conception were to be prevented by legal restraint and chemical action, and only artificial insemination were permitted. The sperm from the few top men would be enough to provide an entire nation with paternal genomes for the highest intelligence. It is under these circumstances that it could be said, as we have quoted before, that 'in England alone there would be about one million tons of surplus men.' They would be superfluous only in a genetic sense, of course; neither the Minister of Labour nor the Minister of Defence would agree to have his manpower so drastically pruned!

Future practice will not be quite so ruthless as the earliest schemes suggested. A start has been made in selecting donors, and hence types of sperm, and the wishes of the parents will be respected. Just the same, the AID system remains a trifle macabre, especially when we think that sperm may be stored in sperm banks long after the donor has died. Such objections should not be dismissed simply as provincial narrow-mindedness.

Human Beings Made in the Laboratory?

Anyone who can stomach the above will not deny the justification of another method, which would allow logical breeding to take place. In AID only the sperm is manipulated, and fertilization and the development of the fetus take place in the uterus. Recently an attempt was made to carry out AID outside the uterus, and with such success that fertilization and fetal development were carried out in the test tube, *in vitro – extra-uterine development*. The press briefly reported the results obtained by an Italian scientist. Have we, then, finally come to 'laboratory-bred man,' the first truly artificial human being?

Yes and no. The only parts of this process that were 'artificial' were:

1. The ways and means by which the sperm and the ovum were brought together.
2. The surroundings in which growth took place.

The sperm and the ovum themselves were completely natural products from a living person; the genomes and the other constituents of the cells were completely unaltered, and certainly were not synthesized from chemicals. So there is really no question here of 'synthetic man.'

In conclusion, let us consider again our three systems: 'anti-baby powder' and child licenses, AID, and extra-uterine fertilization and development. In sequence these three show a progressive improvement in ability to arrive at the desired qualities in human beings more quickly and more surely, but on the other hand, they also show a progressive *depersonalization*. In the first method the parents would still participate as individuals, as personalities. Under AID only the mother participates personally; the father is no more than a disembodied I.Q., and an anonymous donor of sperm. Finally, in laboratory culture even the mother would be of no importance as a person, and all that would be wanted from her would be fertile ova containing a genome that could guarantee a high I.Q. She might perhaps be needed as a nurse for the test-tube baby! I wonder if our mothers would ever be content with that.

The reader may not be unduly prudish, and may be willing to let everyone please himself, but still he will wonder whether the (somewhat questionable) aim of all these methods justifies so much compulsion and such a degree of depersonalizing. Is it actually necessary to pursue such investigations? Not that they should be forbidden, but could not the people and the money devoted to them be better employed elsewhere?

Of course, this is a question that men have been putting to themselves from time immemorial, and which they have always answered differently, each according to his own religious, moral, ethical, or practical standpoint. The contribution that the natural scientist can make to this dispute is to point out that the progress of the human race from collecting food to producing it, the development of the primeval mob into the modern industrial state, the development of the scientist from Aristotle to Einstein, involves more than simply becoming more liberal, more complex, and better. It is an evolution, and thus an irreversible process, and this is true whether it is 'still' a biological evolution, or whether it has 'already' become a psycho-social evolution. Man cannot prohibit evolution, which he himself wants, and in which he himself participates; in the Middle Ages doctors were forbidden to use cadavers for improving their anatomical knowledge, but they did so just the same.

If we look back at the Third Evolution, as it appears to be according to Hypothesis I, we shall find with surprise that even if it becomes a deliberately directed evolution, it will still be conducted according to Mendel's Laws. We may call it a biotechnical evolution, but even so, molecular genetics and molecular biology will play nowhere near as big a part in it as has been attributed to them. This is no accident – the explanation has already been anticipated when we quoted earlier from Professor Spiegelman's work: 'It will be a long time yet before we are in a position to make artificial phage-genome, which matches, and which will be "accepted" by, another organism.' This is all the more true for human genomes, which are incomparably bigger than phage genomes, and which, in addition, would need to be enriched with 'social genes.' We can confidently say that modern biology has contributed little to deliberate, self-centred evolution (in the sense of Hypothesis I) and will have nothing to contribute in the immediate future. Is this equally true of Hypothesis II?

7.10 Hypothesis II: Selection by Elimination

The Artificial Environment

As we saw on page 313, it is still too early to say in what direction we are being led by the 'kink' in evolution that many people think they have detected; we cannot even be sure that it leads in the direction of a psycho-social evolution. It is at least equally probable that man has simply been able to abolish natural selection, or that he is on the point of being able to do so, and that in this way he will be able to interfere with the course of (human) evolution, if not to stop it altogether. This time let us look at the details of this hypothesis – which at least sound harmless enough.

We must not forget that man has controlled his environment from those very early times when he first began to keep herds and to grow crops. His attitude, justified in those simple times, was: What is useful to me is good; what harms me is bad. Good things included useful plants, and domestic animals; bad ones included his enemies and competitors, weeds, and anything that devoured his crops, particularly insects. These had to be kept in check, and man wanted to eradicate them if possible. The good plants and animals, however, had not only to be protected but also to be improved, and improved not for their own sake, but for the benefit of man: 'Green herb for the service of man.'

Man continued to follow this line of evolution, and by now has almost gone as far as he can. He has increased production to an unparalleled extent, and his plant and animal products are almost unnaturally uniform. (Soon we shall be able to breed pigs without ears – which are superfluous and undesirable anyway – and so fit in a few more ribs.) All this would never have progressed so far without genetics, and above all, without chemistry. Our agriculture would lag far, far behind its present state without artificial insemination, without vitamin, hormone, and defoliating sprays, without the innumerable means of pest control and weed killing, without artificial fertilizers, all applied according to timetable and weather calendar. Nor can agriculture get along without all these methods, if it wants to retain and improve its standard.

With these methods, man has created a genuinely artificial environment for the living things that are useful to him (the reader may perhaps have visited a chicken farm, and will have realized that 'egg *factory*' is a fair description of it), and in this environment natural selection is more or less completely eliminated. Any selection that does take place is by human hand, and then it is exactly in accordance with human plans for safeguarding and increasing production. Not one of our cultivated plants or domestic animals could survive for long under natural conditions (as far as these exist today); natural environmental conditions would eliminate them immediately.

182. A modern chicken farm provides the birds with a completely artificial environment.

Hence it is essential that the artificial conditions that man has created should be maintained and perfected.

This has one consequence, which was at first treated as an unimportant side effect, but is now taken very seriously. The reader may have heard of the great areas of the Middle West of America which have suffered soil erosion as a result of monoculture. Miles and miles of pure wheat or pure corn were grown. The harvesting of these crops robbed the soil of its humus, and allowed it to dry out; storms blew away the topsoil, and what remained was waste land, useless to man. A start has been made, laboriously and at great cost, to rebuild the original cultivated landscape – with trees and hedges to give protection against the wind, with birds (as destroyers of insects), with ponds and lakes to give moisture to the air, and so on – but how much time will it take to put back what man has destroyed? In her book *Silent Spring*, Rachel Carson described the unforeseen consequence of pest control with chemicals. Many other examples could be quoted, but our business is not with the ills of civilization, but with selection and evolution.

Things do not seem to be quite so bad in industrial production as they are in agriculture. All the same, many industrial plants are made tolerable only by complete air conditioning, giving constant temperature and humidity and illumination. This does not seriously harm anyone, though it does mean that those who work in such places will find one day that they have forgotten how to adapt themselves to sudden changes of weather. And it is another matter if we look for the side effects of industrialization. The pollution of streams with industrial sewage and detergents (anyone who travels on the Rhine will be surprised to hear that right down to modern times the members of the city council of Cologne were forbidden to make their servants eat salmon more than twice a week – so abundant were the salmon in the Rhine!); gloomy skies over industrial areas; the danger to road users through excessive exhaust fumes; the trouble that has to be taken by synthetics factories to dispose of their waste products, which cannot be burned like paper or wood – all these must be admitted, and taken into account here.

But what has all this to do with selection and evolution? Nothing, possibly, but the whole thing cannot just be ignored. The pressure of competition, which operates on every producer, whether in industry or in agriculture, is a kind of selection pressure. A successful production manager in either industry or agriculture must devote himself to increasing production by a ruthless use of every possible means, chemical, technical, and political, without paying any attention to side effects unless these are actually illegal. And experience shows that laws to control these need a lot of time and a lot of patience before they can be made effective.

Have we not here the germ of a kind of evolution, which may not be a conscious one, but which one day may have evil consequences? We will postpone the answer to that one for a while.

Automation and Standardization of Existence

Man has left it rather late before starting to devise an artificially controlled environment, not merely for the sake of production, but also for himself to live in agreeably. This is all the more reason for making up for lost time. This can easily be demonstrated by considering air conditioning – using this as one example out of many, just as in the present chapter we used I.Q. to demonstrate psycho-social problems. Air conditioning is applied not only to working and living spaces, but to shops, places of amusement, and hospitals, and even vehicles, whether private cars or in public transport. With the addition of escalators and travellators, roofed over and heated, the day is not far distant when a man will spend his whole life under full air conditioning. Only in his free time will he occasionally sample the 'free air.'

The immediate convenience of air conditioning is that the environment remains always the same, that man no longer needs to sweat or to freeze, and that he need not continually adapt himself. The intention, at least, is to make man independent of changes of weather, with the air conditioning system working automatically, without human intervention. In this respect it is like a number of other examples of hasty automation: escalators and lifts, washing machines and dishwashing machines, automatic cookers, automatic circuits and switch gear, automatic carburation in cars (perhaps eventually we shall have no need to steer, because the car will have an automatic pilot, which will weave in and out of the traffic for us). A long list of such automatic devices could be drawn up.

If we are to believe the engineers, a completely automated environment, needing no supervision, is a thing of the immediate future. Man will be relieved of many of his worries, and many responsibilities will be taken away from him. Every morning he will start by setting his private computer in accord with his wishes for the day. The private computer will at once get in touch with a central direction, and supply its owner with a complete programme for the day, telling him when he has to go out, and where, what pill he must take at what time, how his mealtimes shall be arranged, and so on. His life will be easy and comfortable, free from having either to give or to make decisions.

Not only that – it will be *standardized.* A man's life will be adjusted to the social group and country in which he lives, so precisely adjusted, indeed, that friction will be avoided. People will even learn to submit themselves to automatic restraint. Man will discard and forget many skills and abilities from lack of use, because selection pressure will have moved away from these.

All this will certainly not be without its effect on man's evolution, but we cannot say what that effect will be, because we are just at the beginning of things. Only time will show the result, and it may be both unexpected and unpleasant. But it cannot be avoided. An evolution can express itself in automation and standardization just as well as it can in the perfection of medical science. There is just as much chance of forbidding the development of these things as there is of forbidding medical men to do post-mortems. Perfection is very expensive, however, and it may be that sooner or later we may not be able to afford it. One day the great big National Computer may write: 'In one year's time the expenditure on the automation and standardization of daily life will become so great that it cannot be financed any more. From then on these processes will have to be restricted to three-quarters of the population.' Will there not then be a new kind of struggle to be one of those who 'belong,' which will undeniably be a new kind of selection pressure; and will it not hit hardest those who have forgotten so much that their best skills have withered away?

Now, this is pure speculation, which the reader may accept or not, according to his taste and temperament. This much, however, is certain: automation is not only expensive, but it is very dependent upon the reliable functioning of all the technical arrangements, and upon a reliable supply of energy from all sources, so that no shortfall of any kind shall cause a gap in the output.

Some idea of how far this dependence has gone already can be gained from the catastrophic effects of the interruption of electricity supplies in New York in 1965. In fifty years time we may have perfected automation and standardization to the point where such a breakdown may last for weeks! How quickly easygoing and complaisant men might find themselves back at making fire without electric gas-lighters, without matches – back to flint and steel! Will anyone remember how to do this?

7.11 Back Again to the Artificial Environment

Medicine and Hereditary Biological Deterioration

Professor Huxley, whom we have already met as the defender of self-sufficient evolution, is probably the author of a bon mot which has been ascribed to a number of different doctors: 'Medicine has advanced so far that today there are no more healthy people left!' Is not this another way of saying the same thing as Hypothesis II, namely that man has made an artificial environment for himself? Are the doctors particularly to blame for having set aside natural selection, interfered with evolution, and thereby contributed to the decline of quality and hereditary excellence?

The sentence quoted above, like most witticisms, contains both an exaggeration and a grain of (bitter) truth. It is an exaggeration, certainly, to say that no healthy people exist any more. Of course, there are always countless numbers of people who enjoy the best of health, and who up to now have needed neither a doctor nor his medicine. And of course, they are not healthy *just because* they have never been to a doctor!

On the other hand, it is true that today there are more sick people than there ever were. There are different reasons for this, and medical reasons are the least important of these. The general susceptibility to illness is probably higher than it was, perhaps as a result of the lowering of quality we mentioned on page 313. Another reason is that many people feel themselves outpaced by modern civilization, and so they fall ill. Nor should we forget the chemical treatment of all our foodstuffs. Many pesticides are harmful, and traces of them sometimes remain even after the directions for use have been scrupulously observed. They may accumulate little by little in the human body, and produce slow poisoning. And finally, the increased radiation that results from atom bomb tests is not without its consequences. The number of mutations is increased, with the likelihood that babies will be born with 'fall-out symptoms'; most of these, unfortunately, relate to mysterious illnesses which doctors will not be able to understand and treat without long and painstaking study.

Modern medicine is not answerable for all of this, and in other ways it has contributed to raising the general level of health. How many of the hundreds of thousands of traffic injuries would have become traffic deaths if it were not for the many excellently organized emergency hospitals? These injured have to thank modern medicine for their survival, but not all of them are *sound* afterwards. We must therefore hope that this is a temporary phenomenon, and that further progress will reduce the number of these pitiful cases. Unfortunately their number will never fall to zero, because the final remedy lies not in medicine but in *prevention*. A much more direct influence on human affairs is exercised by medicines, produced by the pharmaceutical industry, and prescribed by the doctors. By this we do not mean the misuse of pills, since the doctor, unfortunately cannot control that. What we do mean is all those other remedies which after brief tests appear to have the most beneficial effects, but which have side effects that are not discovered until later – perhaps too late.

One such substance, which gained a sad notoriety, was a widely known and well-tried sedative, used against sleeplessness. Expectant mothers who had taken this drug during a particular period of their pregnancy gave birth to babies with the gravest deformities. Recently it has been disputed whether the drug itself was responsible for the deformities, which – it was suggested – might have had quite other causes, and might be much more common than was realized. Normally such children would not be delivered alive, but would be aborted at an earlier stage. The effect of the drug was to prevent this, and so to allow these unfortunate babies to be born alive. It is to be hoped that we shall soon know the answer.

In justice, the blame here should be given not to medical science, but to insufficient, or too hasty, testing. Greatly improved methods of testing would, we hope, make drugs less dangerous, though whether any chemical can be completely harmless is a matter that we would rather not go into.

The cases described so far are not particularly convincing if we are still thinking of the quotation that began this chapter. If we want to be charitable we can say that they are a temporary phase, and that when we have progressed a little further such patients will recover, or even not fall ill in the first place.

The real kernel of the quotation above lies elsewhere. Today doctors heal millions of people who in earlier times would have died of their ailments. In the industrial countries, at least, the most serious epidemic diseases, from plague to diphtheria, have been eradicated, while others can be brought under control; some diseases, however, cannot be controlled – syphilis, for example, is still evasive. The same is true of some of the less serious ailments, though this will be received with astonishment when we think of the common cold or of ringworm and other fungi in the skin, which are no laughing matter. Mycosis – infection with a fungus – occasionally spreads rapidly in the skin, perhaps because it has suffered under the use of modern soaps and detergents and is no longer able to resist the fungus.

The achievements of surgery are equally impressive. Operations are carried out today that would have been impossible only a short while ago. Artificial kidneys and artificial hearts are made out of synthetic materials, because the patient does not form antibodies against these substances. Every big hospital will soon have at its disposal a well-filled store of 'spare parts.'

Illnesses, and injuries do not always heal completely, but life can often be maintained, and the general condition of the patient be improved. This does nothing, of course, to improve the total number of 'healthy men,' but what does that matter when set against the really remarkable performance? Nobody would want to dispense with the services of medical science for this reason – if only because any one of us might need them one day! If there are a few people, deeply religious on the one hand, or fanatics of every hue on the other, who would reject medical help, they cannot stop medical progress. The development of medicine can no more be turned back than can biological evolution.

Imperfect Repair of Hereditary Defects

Nevertheless, there is a hidden danger here for mankind. To put it in one sentence:

medical science not only repairs injuries and heals infections, but it also treats *hereditary defects*, and by doing so, it *defeats the operation of biological selection*. This sounds a terse and slightly menacing statement, which had better be given some basis as well as some explanation by a few examples.

1. A woman who, for genetic reasons, is unable to carry her babies for the full term, and so regularly has a miscarriage at four months, would previously have had to remain childless. Today such premature babies can be placed in an incubator and carefully reared to the point at which they would normally be born; there is no particular problem about doing this. Such children are often somewhat unstable in later life – but they do have a later life, thanks to medical skill.

2. Phenylketonuria is a defect of metabolism, in which the occurrence of a particular gene prevents transformation of phenylalanine into tyrosine – both amino acids important in living systems, which we have already discussed (on p. 60) in connection with artificial synthesis of protein. Instead of this, phenylketone is excreted in the urine. This would not be so bad in itself, but about 63 per cent of newly born babies suffering from phenylketonuria are feeble minded, if not imbecile. Their I.Q. lies below 20, and only about 2·5 per cent of them have an I.Q. above 60. If this defect is recognized in time, it can be rectified by careful treatment, in particular, by putting the patient on a diet that is weak in phenylalanine. The infant improves, and becomes 'normal,' and later on, can even dispense with the diet.

3. In hereditary diabetes the body's own production of the hormone insulin, which regulates the metabolism of carbohydrates in the human body, is interrupted or stopped entirely. Even with insulin injections, this cannot be entirely restored, but the patient usually gets back more or less to his usual physical level.

Many other ills can be 'repaired,' though not always completely. However, it is only the *symptoms* that are rectified; the genetic defect that caused them is left untouched, and if sufferers from such illnesses have children, they will pass on their genetic defect to them.

To avoid any possible misunderstanding, let it be stated clearly that, fortunately, not all premature births, nor all cases of feeble-mindedness or physical deformity, are genetically induced. Very often the cause is a simple disturbance of the development of an otherwise normal child, and in these cases there is no reason why the brothers and sisters should not develop normally.

However, insofar as defects *are* of genetic origin, they will be passed on to any offspring, whereas previously, before medical skill had advanced so far, such defective individuals would have been eliminated by selection. Thus, instead of such defects fading out, they now spread through the population.

Sometimes the victims remain ill throughout their lives, and are never fully active again. They would perish if they were not cared for by our social services. If necessary, the cost of doing this is borne by the community (the 'social group'), and so this is another way in which biological selection is set aside.

In conclusion, an example may be mentioned that is not quite so obvious as those mentioned above, but which is very widespread, and which is equally a frustration of natural selection. This is the unrestrained use of antibiotics, the modern remedy for every ill, which kill bacteria and viruses, or at least prevent their multiplication. In our study of resistance factors (p. 132) we met penicillin, streptomycin, sulphonamides, etc. Nowadays a variety of infections are summarily lumped together, and attacked with antibiotics in an almost perfunctory way. 'Give him a dose of penicillin,' the doctor says of his patient, with the implication that the patient will be kept 'under penicillin' for as long as may be necessary.

The bacterium or the virus, however, is only one partner in the infection, the other being the human body itself. It would be better to treat the infection by strengthening the immuno-biological defenses of the body, a method that is more difficult, of course, and which formerly was almost all that could be done. Nowadays we have moved away from that, and too easily make use of standardized chemical methods. The results were – and still are today, to a large extent – so overwhelming that for a time it was thought that now we should be able to master every kind of infection.

All the same, this is very doubtful, biologically speaking. At any one time there are always some people who have strong powers of resistance (because their body readily produces antibodies) and others with very little or no resistance, with all the intermediates between these extremes. Formerly a person either survived an infection (and thereby acquired an even stronger immunity against a later infection), or the infection killed him; in that case he died because he had too little resistance. That is how selection worked – simply, and inexorably.

This contributed to biological selection by ensuring that the individuals with the strongest immunity were the ones who survived. This was selective breeding (though not conscious selection) for the retention of the immunity reaction (see p. 239). Nowadays patients are treated with antibiotics regardless of whether they have any immunity of their own or not. The selection mechanism is put out of action, since the individuals who are 'immuno-biologically weak' are no longer eliminated, but survive just the same as their stronger brethren. At the same time, those individuals who have a strong immunity reaction get no opportunity to exercise it. As a result, the proportion of the population that has no natural immunity grows steadily bigger.

Now, fortunately, there are still many doctors to whom the patient is more important than the opportunity to use a quick, standardized treatment. And even if the patients – as often happens – actually demand antibiotics, they are given a careful test, to make sure of not using a sledge hammer to crack a nut. But doctors are also overworked, and often quick results are really needed. The doctors therefore cannot prevent natural immunity from being slowly eroded away.

Professor Koprowski Tells His Grandson about Antibiotics

Antibiotic treatment, moreover, has another effect, for which we only need to recall the chapter about genetics, and especially the section about resistance and the RTF. The bacteria and viruses do not remain idle when they are swamped with antibiotics, but mutate to resistant forms (in which they get great help from the RTF), and these must then be attacked with newer antibiotics (because the natural immunity of the patient has been weakened). It is a race between the speed of mutation and the arrival of new antibiotics, and it is an even match, which either might win. We have not, even yet, gained the mastery of infectious diseases in general, as we had hoped to do in the first flush of success.

Professor Koprowski (Philadelphia), one of the participants in that London symposium that we have so often talked about, concluded his paper on the future of infectious diseases and cancer with a 'Letter to My Grandson, in Case He Thinks of Becoming a Doctor.' This was a cheerful and witty letter, but with a deep understanding, and some resignation. Some extracts from this letter contain everything that we need to say in conclusion.

'1. Use a computer to diagnose ailments that you would easily have recognized anyway. If you have to deal with an infection that is spread by some agent that is not mentioned in the handbook of infectious diseases, you will have to use so much time and energy to feed the data into the computer that you might just as well use the good old do-it-yourself medicine. . . .

'6. Don't have anything to do with any scheme for totally eradicating infectious diseases. You will only drive them underground, and start a never-ending guerilla war with them.

'7. If a universally effective antibiotic is eventually discovered, you must immediately start an association to prevent its being used. You will then be able to keep it under control, as we should have done with the atom bomb (and not have allowed it to control us). You must make use of every possible national or international deterrent to stop the antibiotic from falling into the hands of stupid people, which will probably mean the majority of people, even in your time.'

7.12 The Outlook

In conclusion, let us try to show that this book has achieved something, and see if we can answer the question, What will be the future of man?

The evolution of organisms, and in particular human evolution, is still often a long way from moving into its third phase, of deliberate and self-reliant progress, in accordance with Hypothesis I. Plans exist for doing this, but the prospect of actually realizing them is slender. It may need a world catastrophe to force us into this. Up to now man has confined himself to outwitting natural selection, and putting obstacles in the way of 'positive' evolution, but he has hardly produced any other principles of selection to replace this (Hypothesis II). This has had tremendous results in the short term, but it is becoming clear that every gain will have to be paid for. Yet who could have known that we should be paying, today and in the future, for past improvements? We are paying today for the pollution of the air and of the water, for the resistance factor RTF, and so on; tomorrow we may be paying for the hereditary mishaps, for the 'genetic deterioration' that we are permitting; and there is no end in sight.

This much is certain. We have interfered with natural selection and evolution, and the consequences are unavoidable, since evolution is irreversible. Even if we want to, we cannot reverse the trend of what we alone have started by interfering with *only a few* of the mechanisms of natural selection. Man as an individual is a fighter, but man as a link in the chain of evolution is much too vague a figure for people to bother about. This conflict between the individual and evolution (and not, as is often said, between the individual and society) has made man what he is now, and what he will be in the future: a tragic figure.

Biology can only find out what laws exist, point out false trails for what they are, and try to make the struggle between the individual and society turn out to man's advantage.

POSTSCRIPT

You have now worked your way right through this book. I am glad if it was something of an effort, because the field of research, and the thought processes of modern biology are of such a novel kind that they cannot be mastered without taking trouble. I thank you all the more for having done so, and I sincerely hope that you have no regrets.

You now know something of the results of research in this field. Not all of it, since a great deal could not be brought into a book like this. Many aspects of the subject that are equally new and equally important had to be left out. Perhaps, for example, you would rather have heard about modern studies of behaviour than about the complicated regulator mechanisms. However, I was forced to make a choice, and to take the chance that you (and my specialist colleagues) might be annoyed because the choice was such a personal one. This book was not meant to be a textbook, setting out to demonstrate the whole of the subject; it was not designed to turn you, the reader, into an expert. Its aim is merely to tell you something about molecular biology, and to arouse your interest in the subject, in the hope that you will maintain this interest in biology in the future.

I think that it will now be easier for you to follow the articles in the press, and to assess them more critically, and that you will have a better understanding of when more care is necessary, and when we are being served up with empty phrases intended to shock. This in itself would be quite a lot, and would go far to repay me for the work that I have put into this book. If at any time you want to penetrate further into biology, if you want to find out more from specialist books, or if you have questions to ask, I am happy to be at your disposal.

Finally, I have one thing to ask. If, as I hope, you are convinced that biology must have a special place in modern society, and in national and international affairs, then please help to bring this about. Talk to your friends and acquaintances, to your Member of Parliament or Congressman, so that the importance of biology, and especially of biologists, will be proclaimed to governments and local authorities, and in every place where responsible decisions have to be made. Not because biologists could cure all our troubles at one blow, but because by making more use of biology we could at least avoid laying up still more trouble for ourselves in the future, and avoid having to go on paying for the mistakes that we have made in the past. What matters is not the next election, but the whole future of us all (if you don't feel embarrassed to talk in such grandiose terms).

Hans Joachim Bogen

INDEX

N.B. Numbers with an asterisk (*) indicate illustration on page.

PICTURE CREDITS

Amelunxen, Dr F. Münster, fig. 20
Anderson, T. E., E. L. Wallmann, & F. Jacob, *Ann. Inst. Pasteur* 93:450 (1957), figs. 61, 71
Arnott, Prof. H. J., Austin, Texas, U.S.A., fig. 105
Bleecken, S., G. Strohbach, & E. Seifert, *Z. allg. Mikrob.* 6:121 (1966), fig. 31
Bonnett, Dr H. T., jnr., Madison, Wisc., U.S.A., figs. 81, 82
Brenner, S, R. W. Horne, and others, *J. molec. Biol.*, fig. 56
Bresch, C., *Klassische und molekuläre Genetik,* (1964), figs. 58, 59, 132, 134, 137, 138, 145
Buvat, Prof. Dr R., Marseilles, France, fig. 102
Cinader, Dr B., Toronto, Canada, fig. 151
Cramer, Prof. Dr F., Göttingen, fig. 33
Fawcett, Dr D. W., Boston, Mass., U.S.A., fig. 90
Freeman, Prof. Dr J. A., New Orleans, Louisiana, U.S.A., fig. 91
Frey-Wyssling, Prof. Dr A., Zürich, Switzerland, figs. 107, 114, 116, 117, 118
Hasek, Prof. Dr M., Prague, Czechoslovakia, fig. 171
Hayes, W., *The Genetics of Bacteria and their Viruses* (1965), figs. 36, 37, 38
Heitz, E., *Ber. deutsch. Bot. Ges.* 54:362 (1936), fig. 106
Ito, Dr S., Boston, Mass., U.S.A., figs. 84, 85
Kleinschmidt, Dr A., New York, N.Y., U.S.A., fig. 62
Kröger, Prof. Dr E., Oldenburg, fig. 68
Laetsch, Dr W. N., Berkeley, Calif., U.S.A., fig. 108
LKB Produkter AB, Stockholm, fig. 79
McConnell, Prof. Dr J. V., Ann Arbor, Mich., U.S.A., figs. 146, 147
Mechelke, Prof. Dr F., Stuttgart-Hohenheim, fig. 136
Menke, Prof. Dr W., Cologne, fig. 83
Museum Hauff, Holzmaden-Teck, fig. 172
Oehlkers, F., *Das Leben der Gewächse* (1956), figs. 34, 41
Porter, Prof. K. R., Cambridge, Mass., U.S.A., fig. 94
Rankama, Prof. K., Helsinki, Finland, fig. 174
Revel, Prof. J. P., Boston, Mass., U.S.A., fig. 95
Ross Group Poultry Division, Grimsby, England, fig. 182
Schacht, W., Munich, Germany, fig. 98
Siemens, Pressebild, fig. 78
Sievers, Dr A., Bonn, figs. 99, 100
Sitte, Prof. Dr P., Freiburg, fig. 86
Tromans, J. and R. W. Horne, *Virology*, fig. 58
Westphal, O., *Neuere Ergebnisse und Probleme der Immunbiologie: Jahrbuch 1965 der Max Planck-Gesellschaft*, figs. 149, 156, 157, 158, 159, 162, 163
Whaley, Prof. Dr W. G., Austin, Texas, U.S.A., figs. 96, 112
Wohlfahrt-Bottermann, Prof. Dr K. E., Bonn, figs. 93, 111
Zeiss, Carl, Oberkochen, Werkfoto, fig. 77